D1528957

CONTENTS

TAKE NOTE!

TO ACCOMPANY

CELL AND MOLECULAR BIOLOGY

Concepts and Experiments

Fourth Edition

Gerald Karp

WILEY

JOHN WILEY & SONS, INC.

Copyright © 2005 John Wiley & Sons, Inc. All rights reserved.

No part of this publication may be reproduced, stored in a retrieval system or transmitted in any form or by any means, electronic, mechanical, photocopying, recording, scanning or otherwise, except as permitted under Sections 107 or 108 of the 1976 United States Copyright Act, without either the prior written permission of the Publisher, or authorization through payment of the appropriate per-copy fee to the Copyright Clearance Center, Inc. 222 Rosewood Drive, Danvers, MA 01923, (978) 750-8400, fax (978) 646-8600. Requests to the Publisher for permission should be addressed to the Permissions Department, John Wiley & Sons, Inc., 111 River Street, Hoboken, NJ 07030, (201) 748-6011, fax (201) 748-6008.

To order books or for customer service, please call 1-800-CALL-WILEY (225-5945).

ISBN 0-471-66909-1

Printed in the United States of America

10 9 8 7 6 5 4 3 2 1

Printed and bound by Courier Westford, Inc.

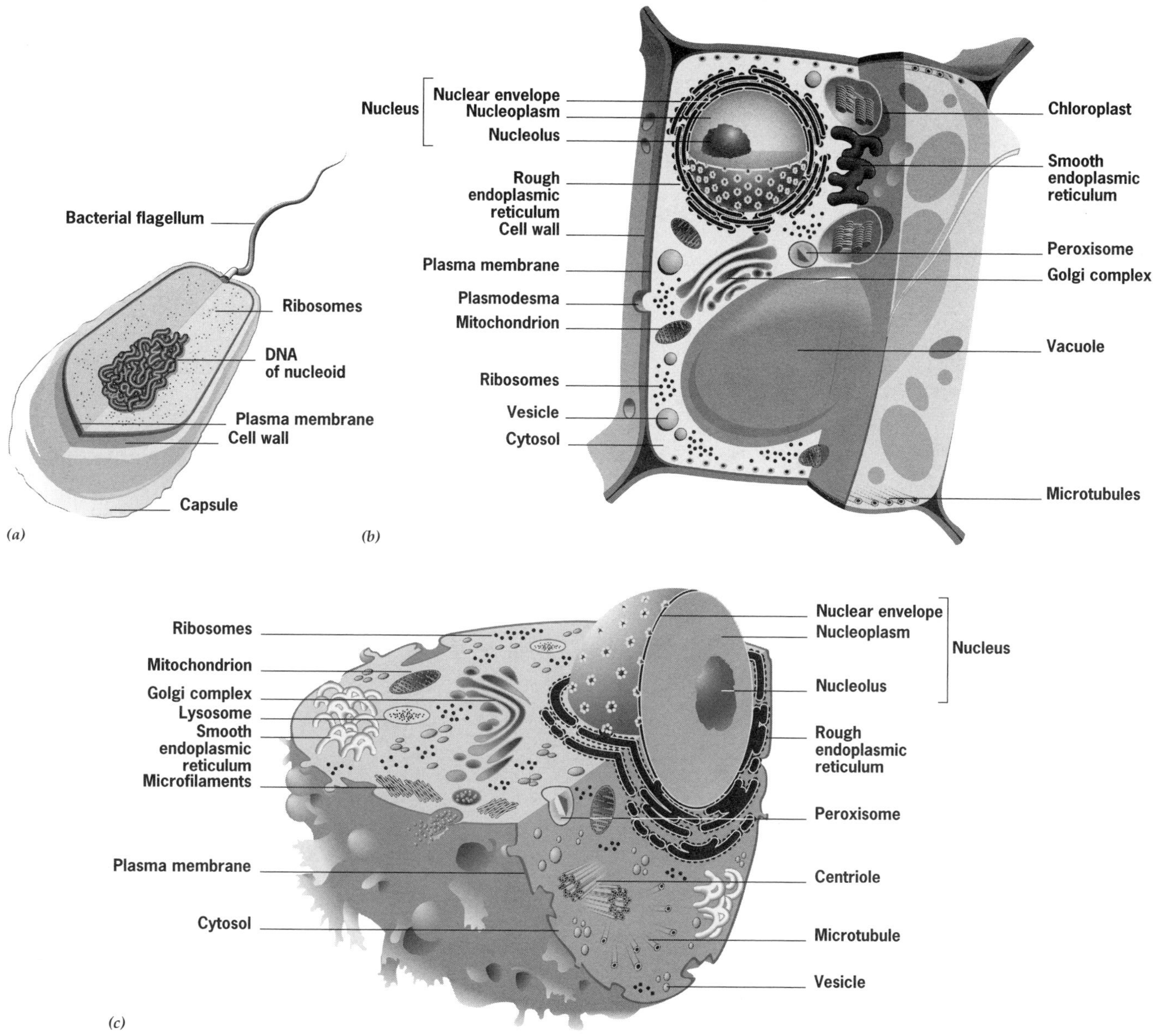

Figure 1.8 The structure of cells.

Figure 1.8

Copyright © 2005 John Wiley & Sons, Inc.

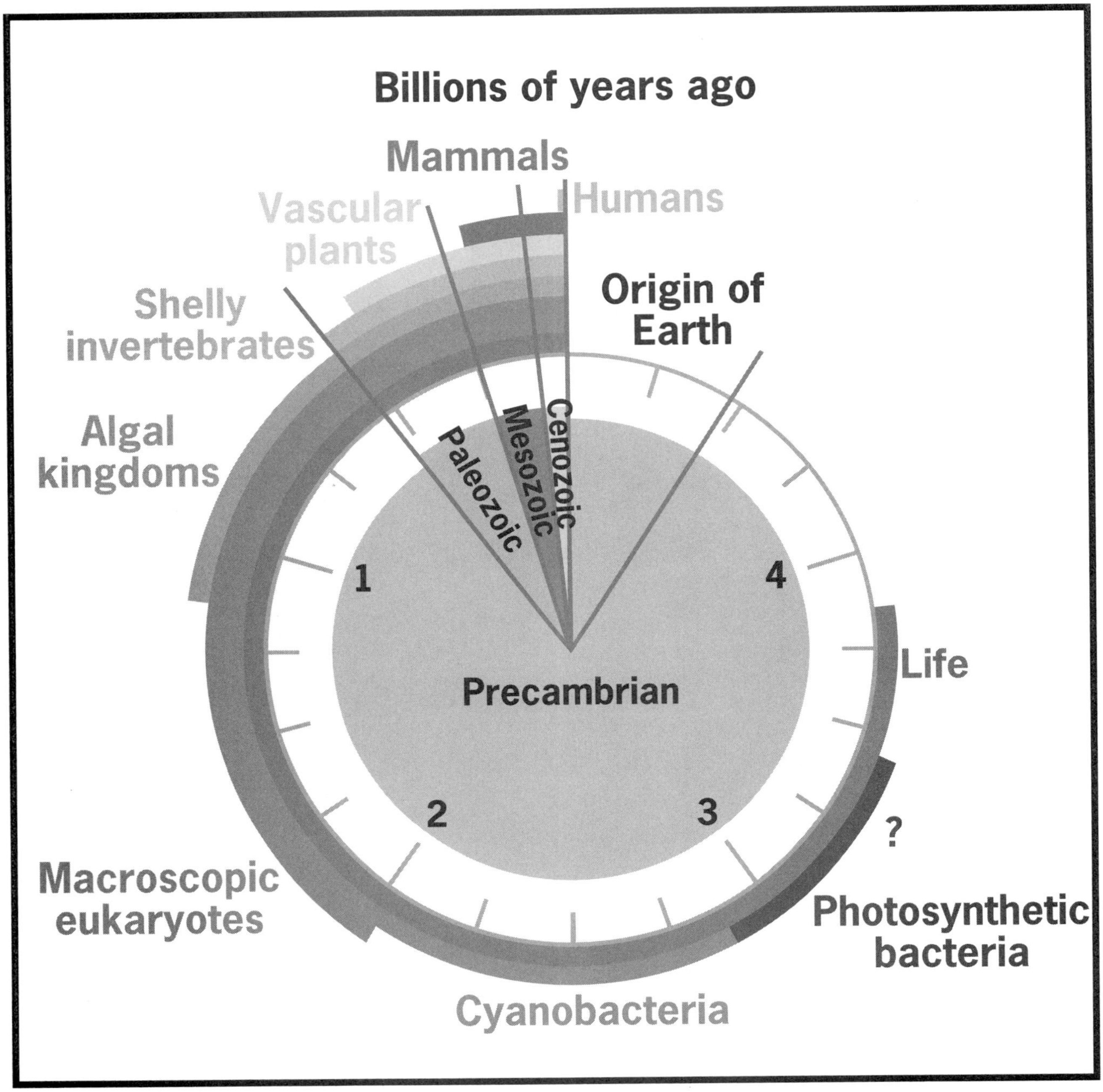

Figure 1.9 Earth's biogeologic clock.

Copyright © 2005 John Wiley & Sons, Inc.

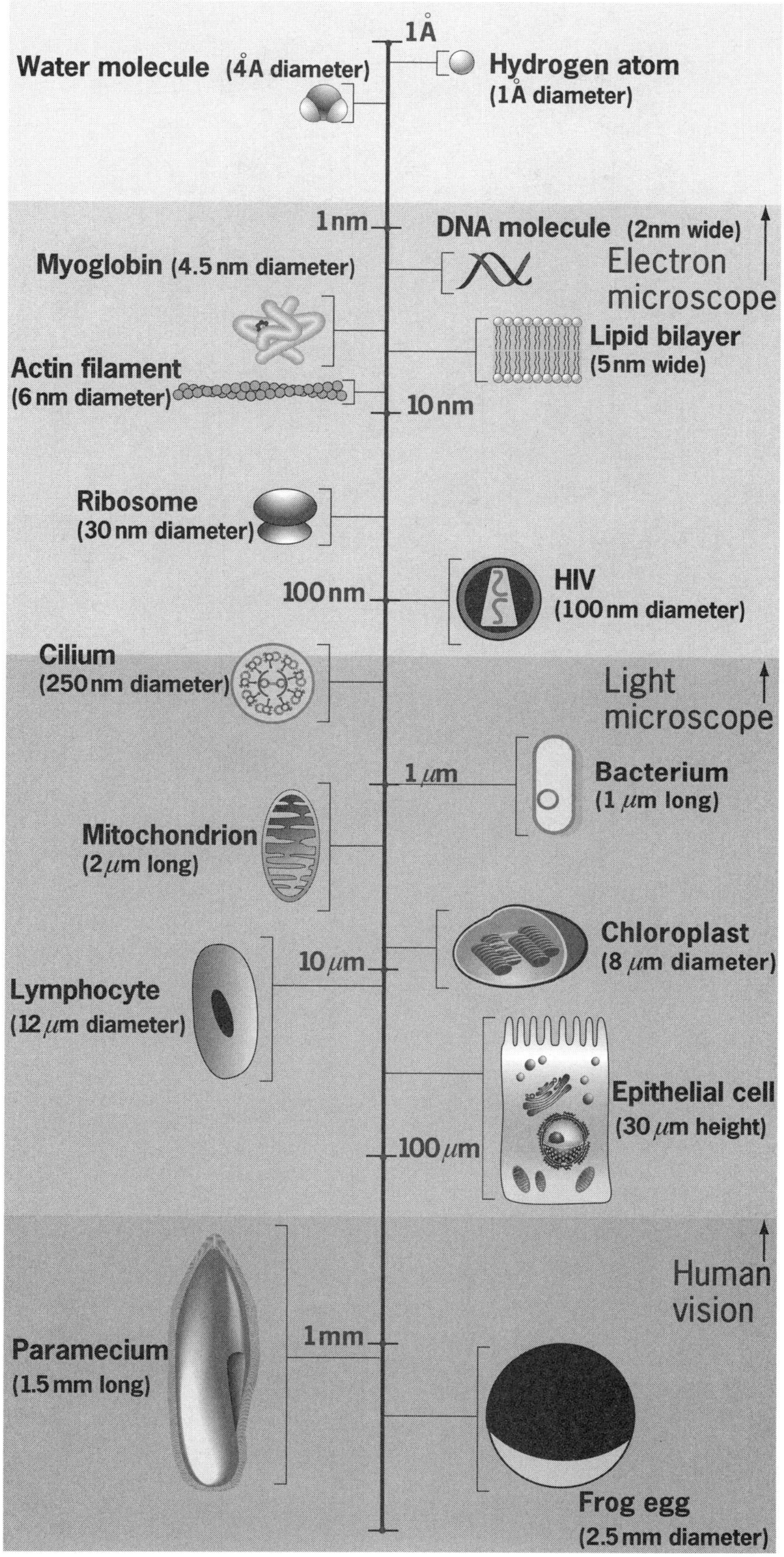

Figure 1.19 Relative sizes of cells and cell components.

Copyright © 2005 John Wiley & Sons, Inc.

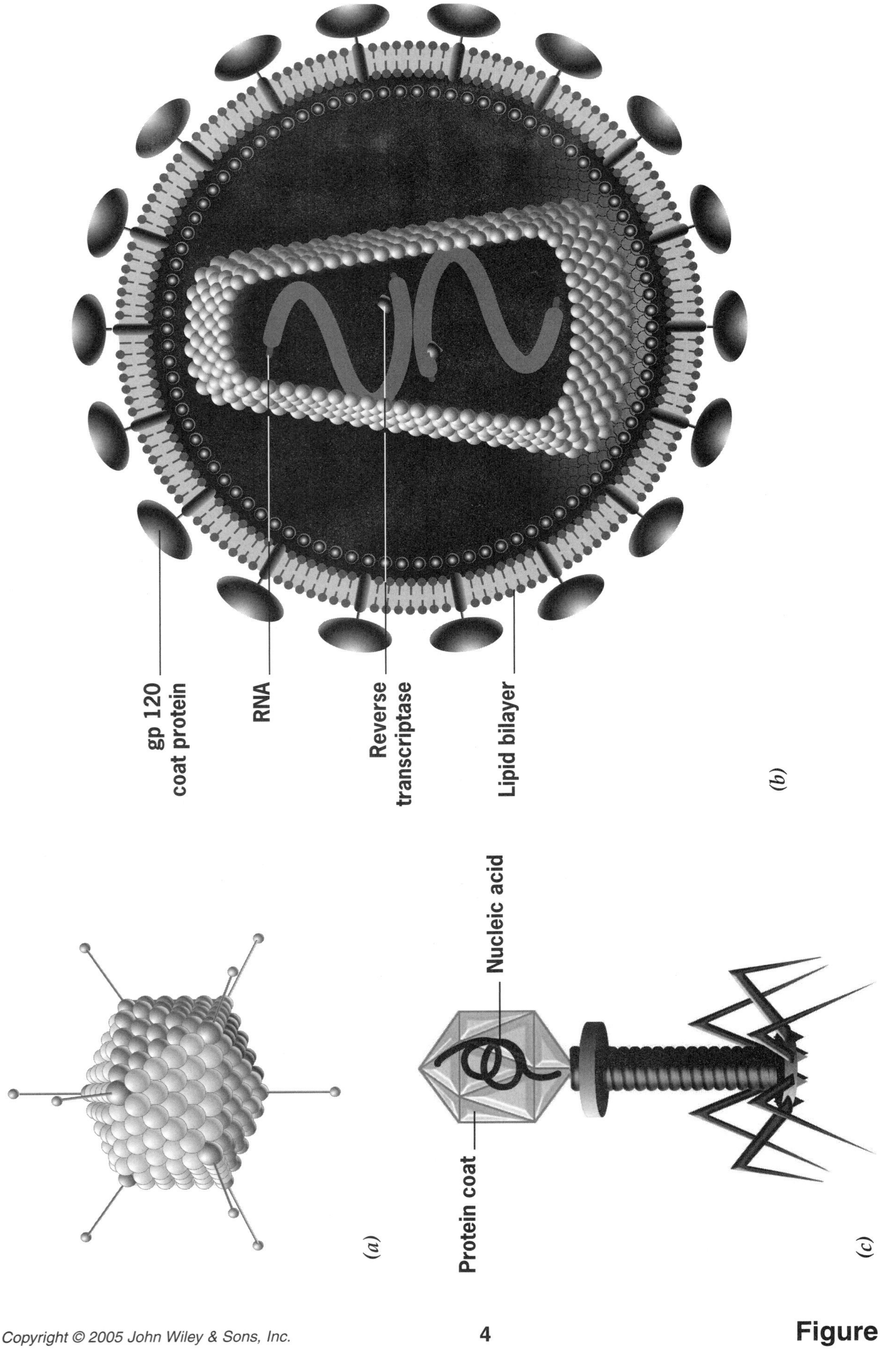

Figure 1.21 Virus diversity.

Copyright © 2005 John Wiley & Sons, Inc.

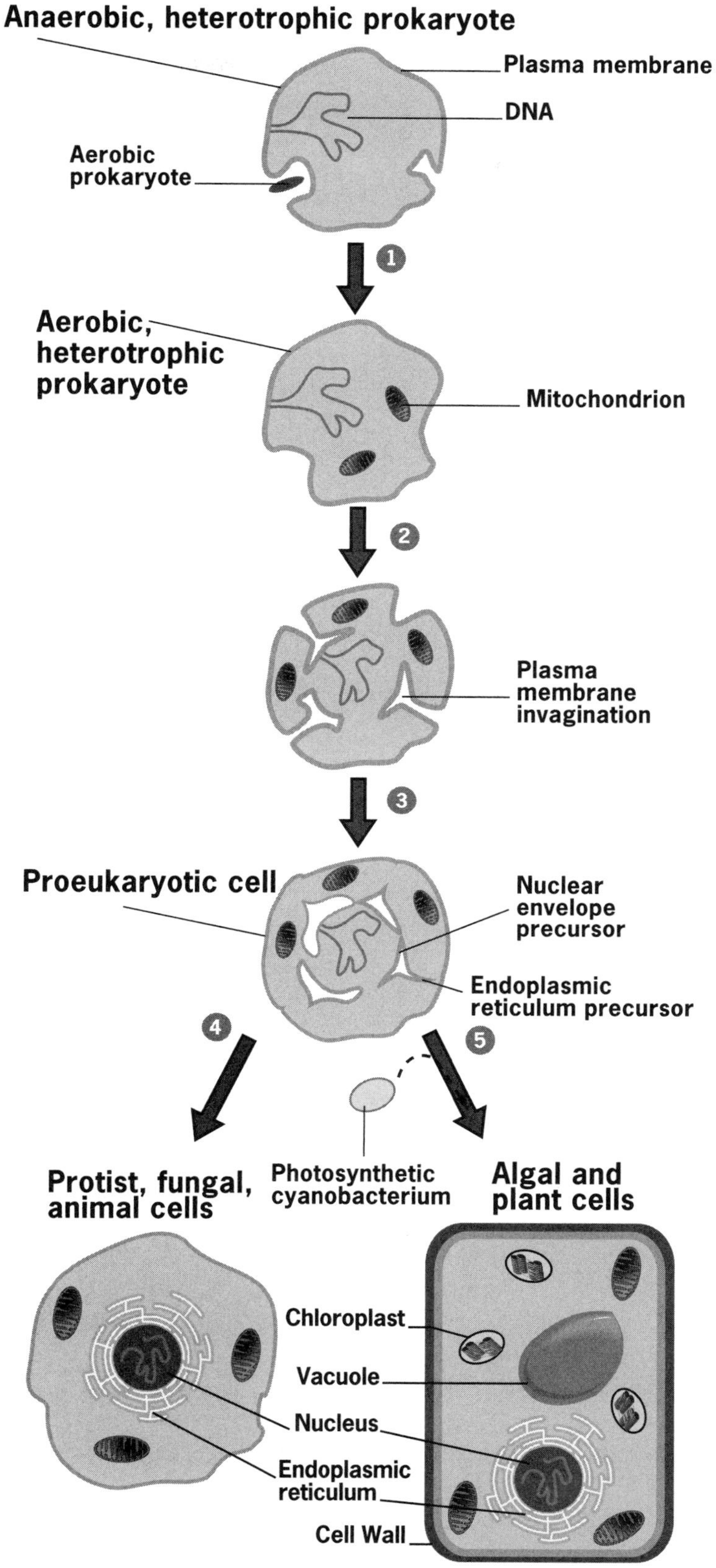

Figure 1 A model depicting possible steps in the evolution of eukaryotic cells, including the origin of mitochondria and chloroplasts by endosymbiosis.

Copyright © 2005 John Wiley & Sons, Inc.

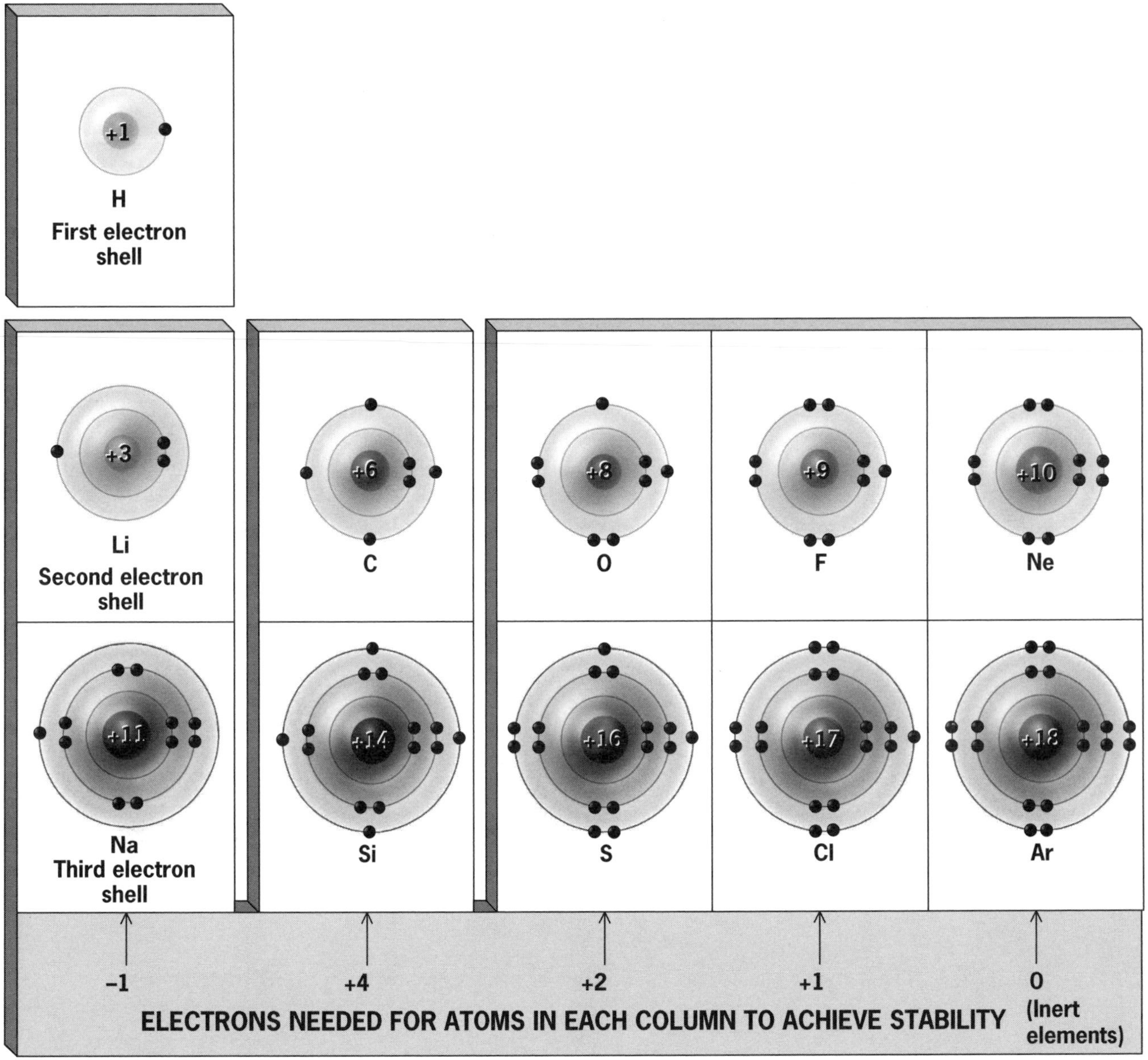

Figure 2.1 A representation of the arrangement of electrons in a number of common atoms.

Copyright © 2005 John Wiley & Sons, Inc.

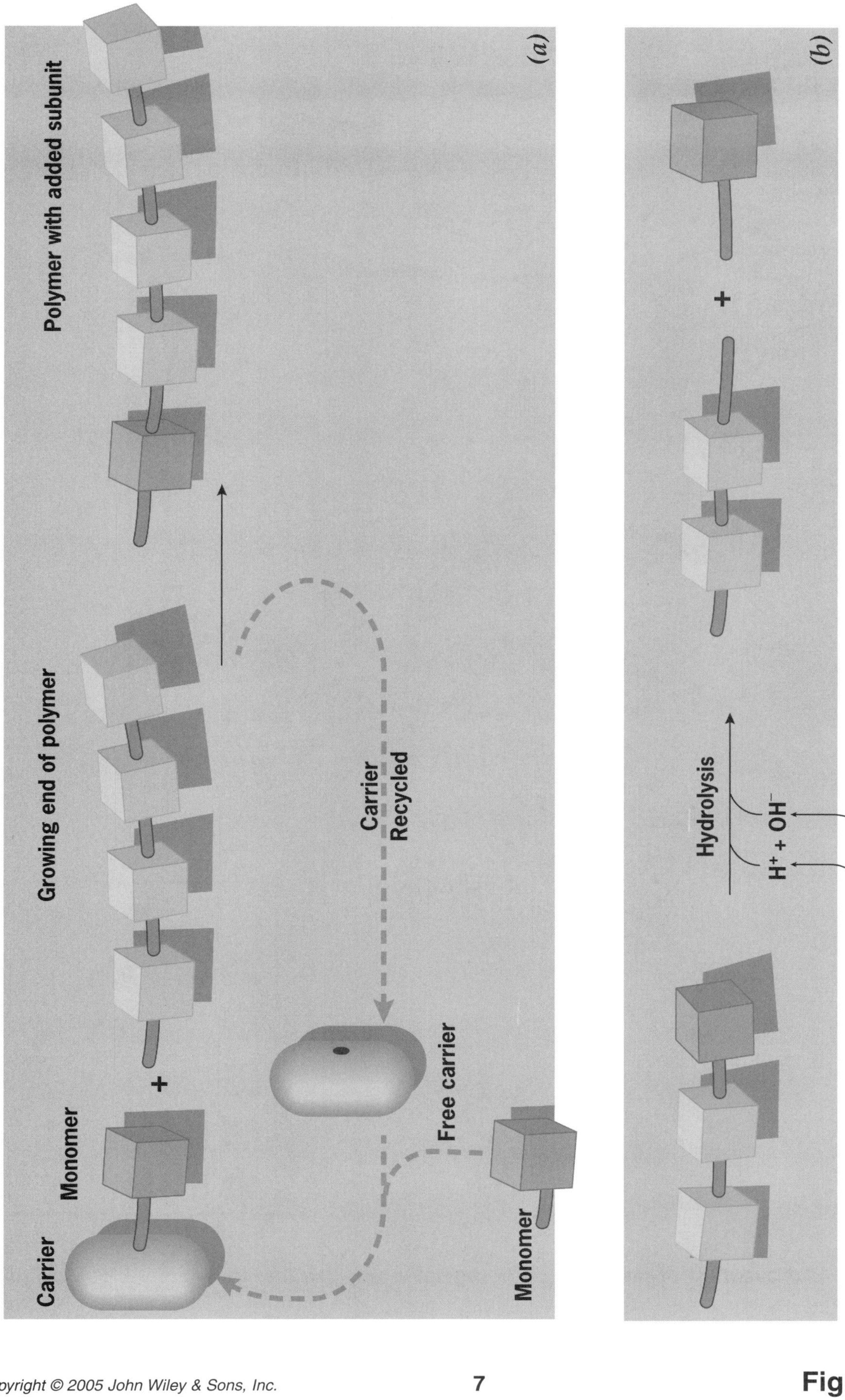

Figure 2.10 Monomers and polymers; polymerization and hydrolysis.

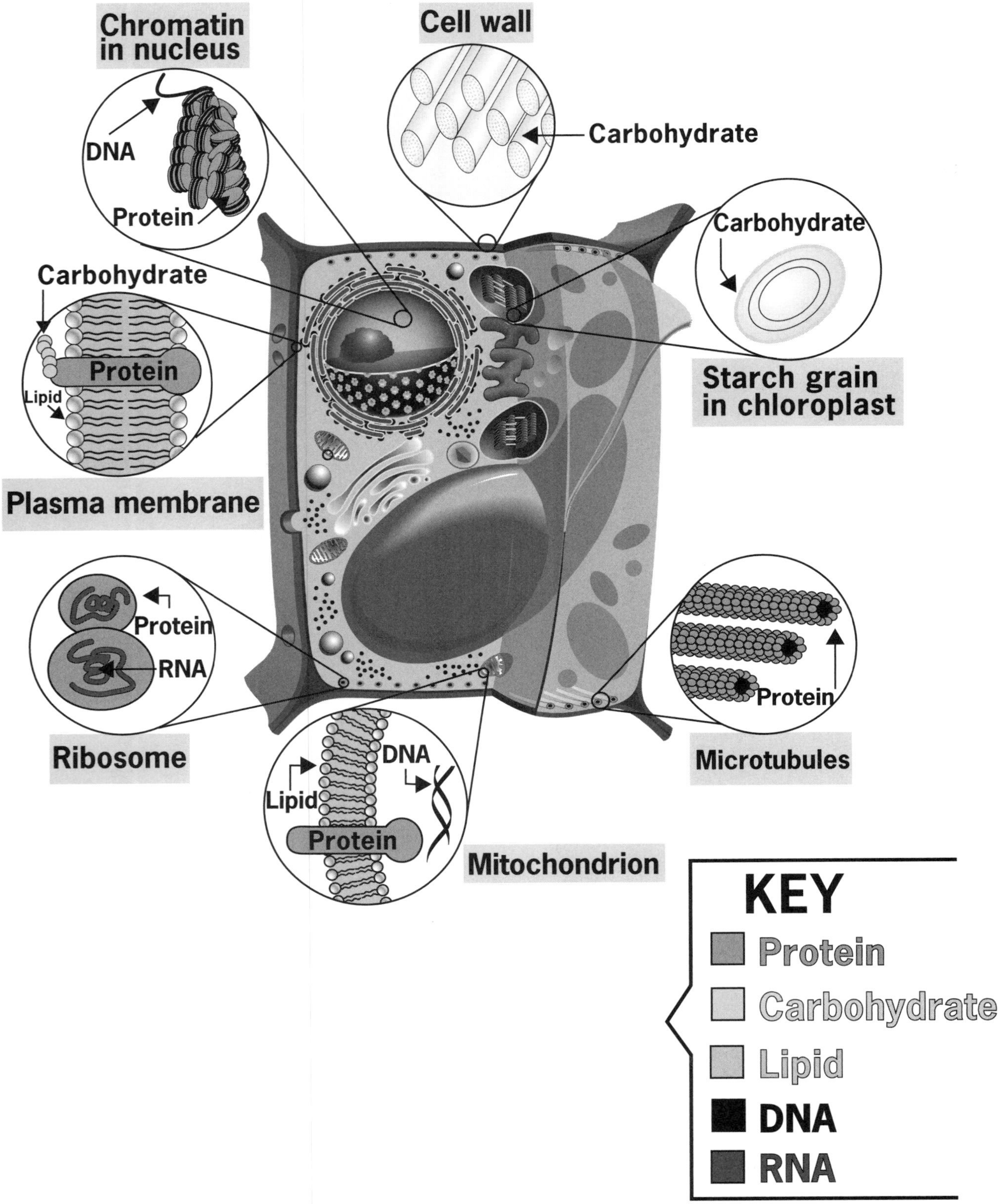

Figure 2.11 An overview of the types of biological molecules that make up various cellular structures.

Copyright © 2005 John Wiley & Sons, Inc.

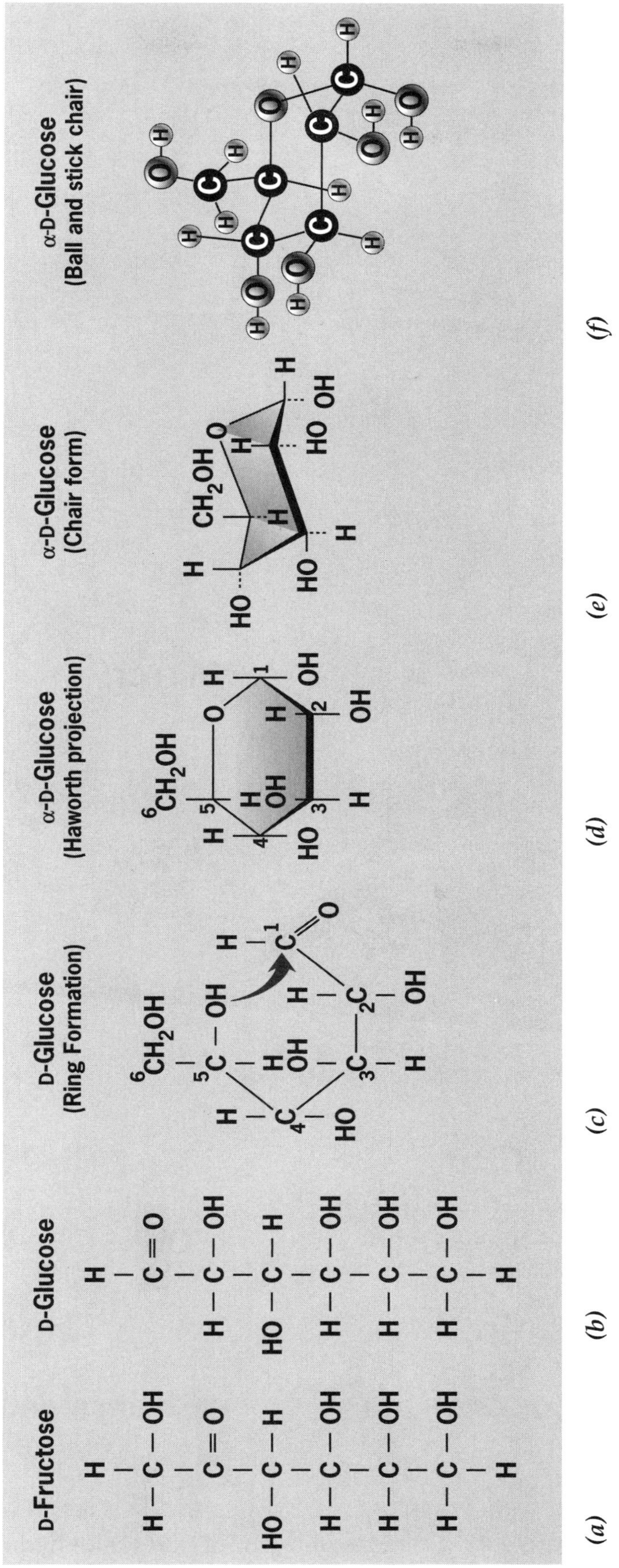

Figure 2.12 The structures of sugars.

Copyright © 2005 John Wiley & Sons, Inc.

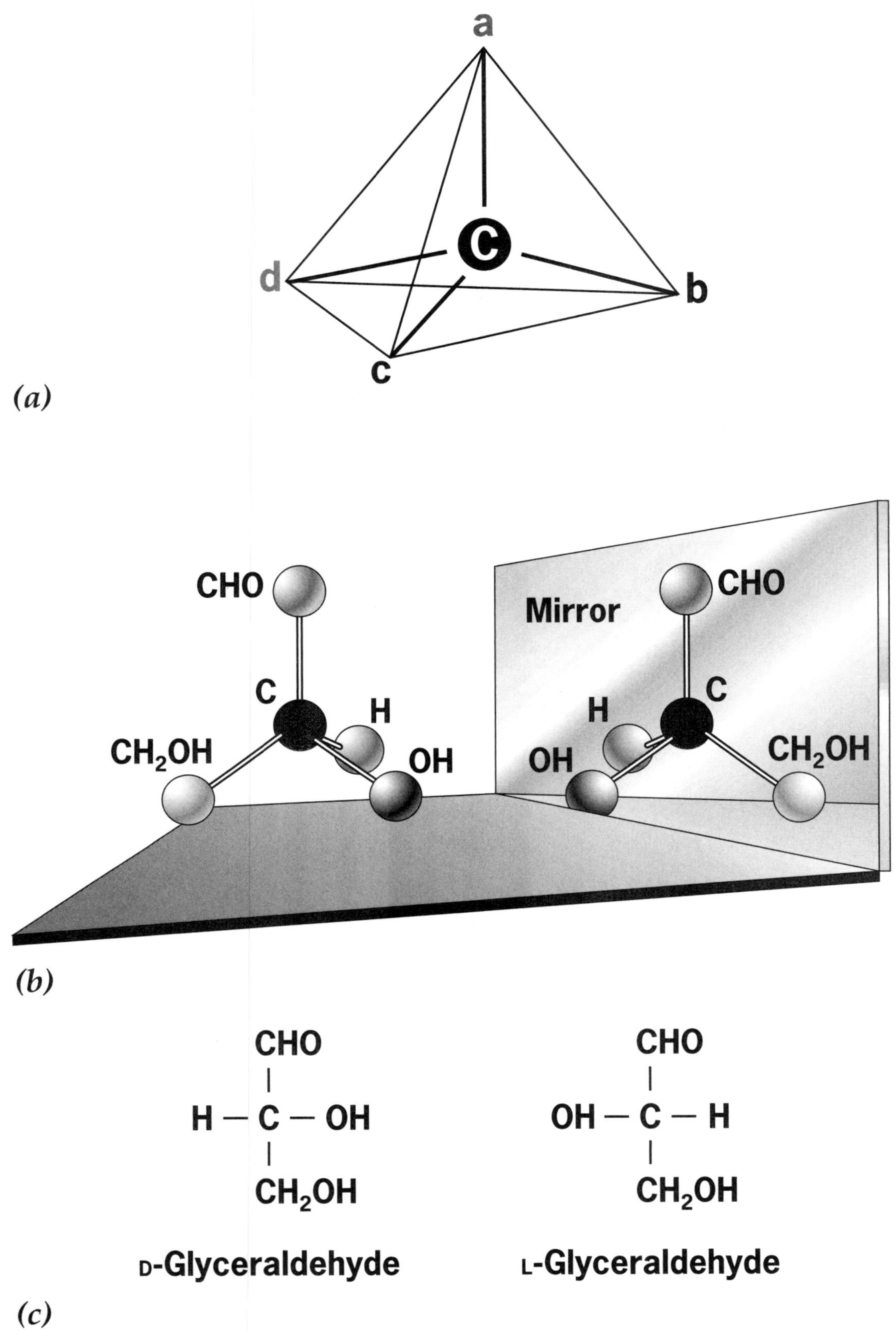

Figure 2.13 Stereoisomerism of glyceraldehyde.

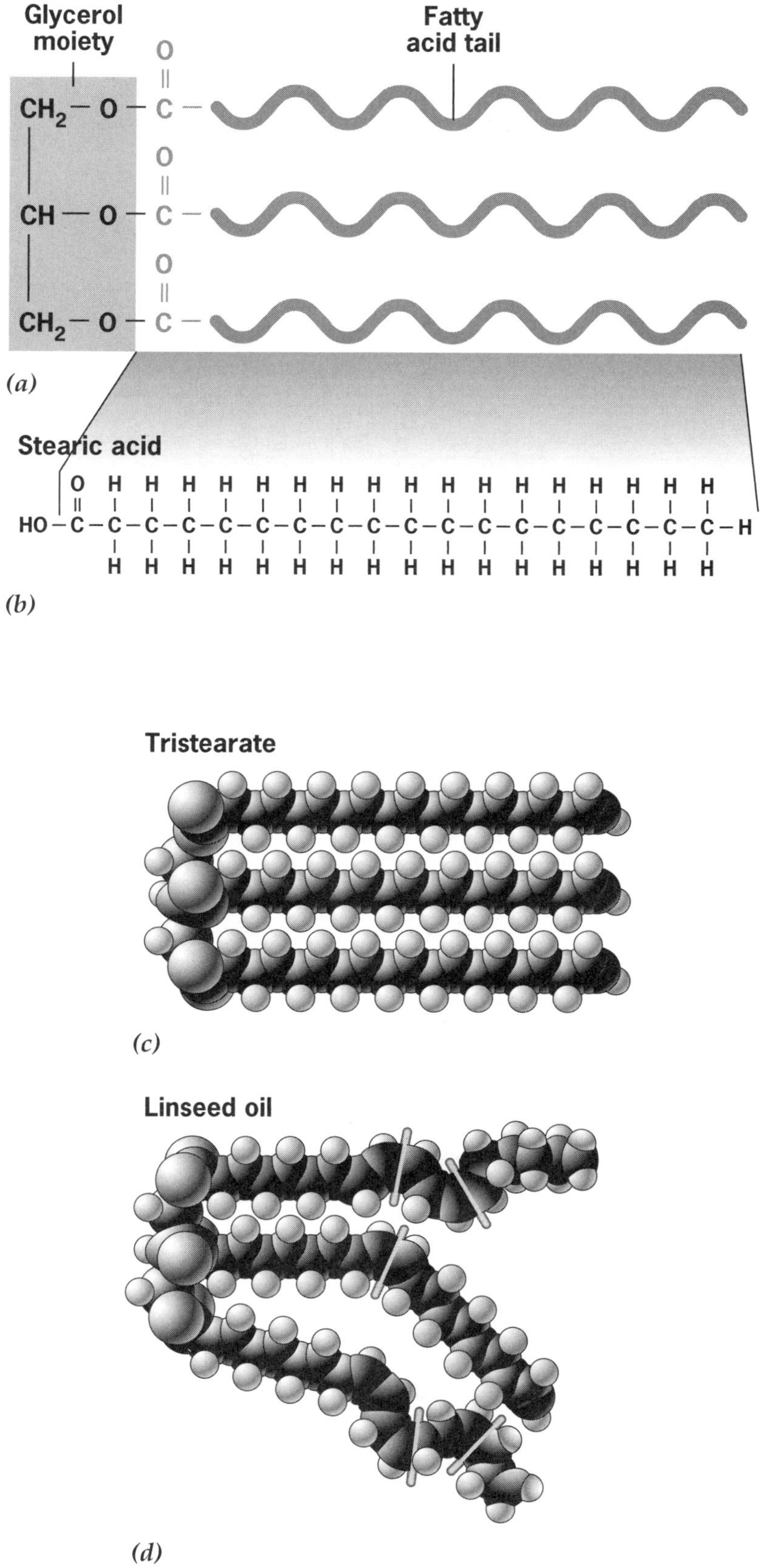

Figure 2.19 Fats and fatty acids.

Copyright © 2005 John Wiley & Sons, Inc.

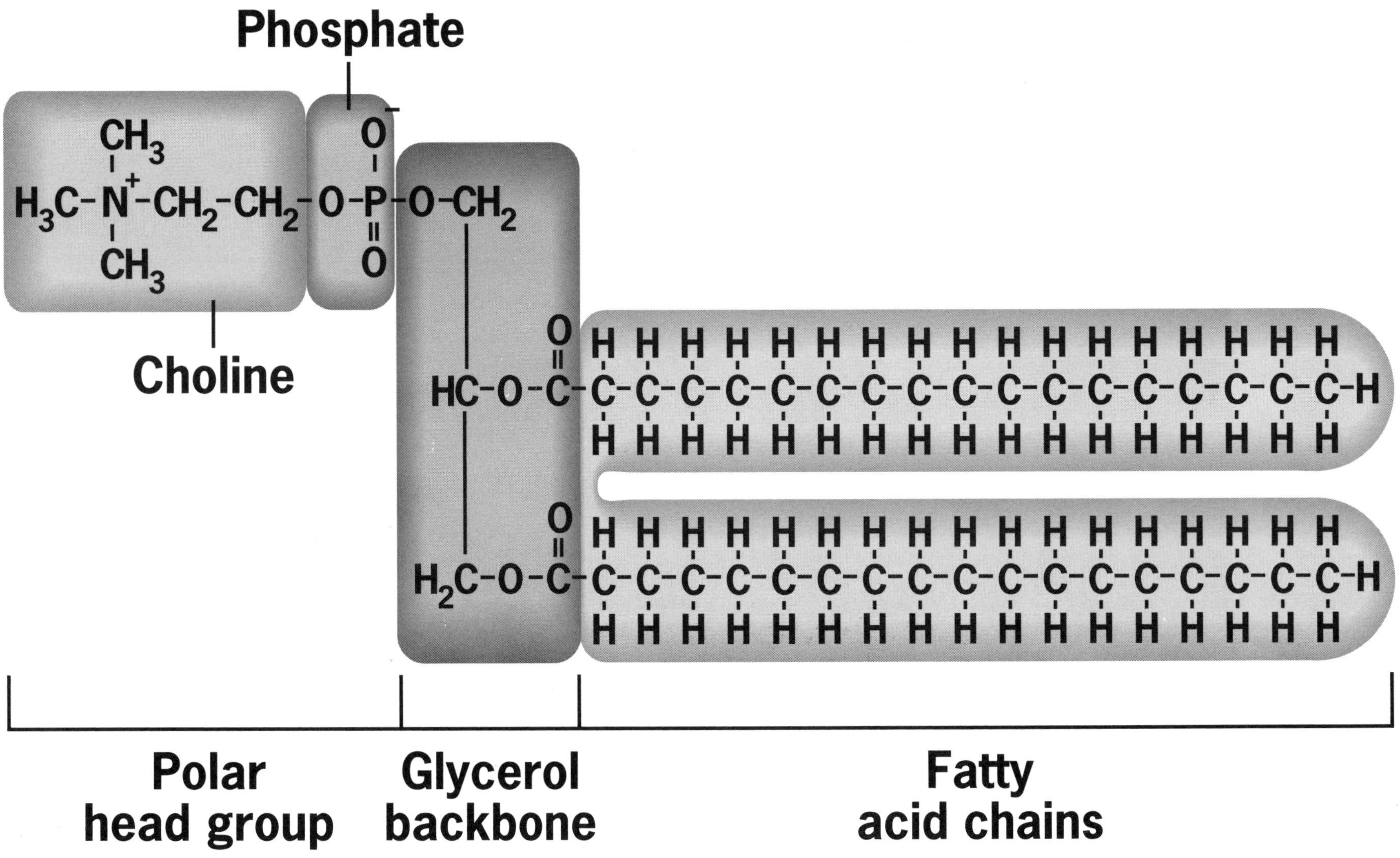

Figure 2.22 The phospholipid phosphatidylcholine.

Polar charged

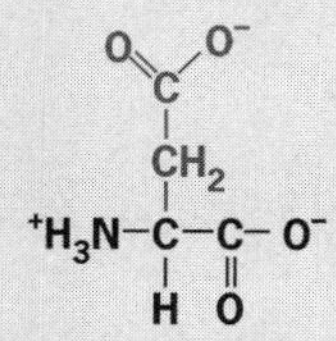

Aspartic acid
(Asp or D)

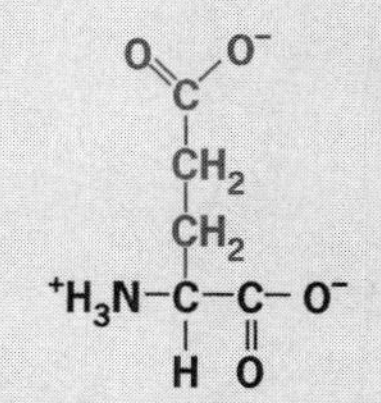

Glutamic acid
(Glu or E)

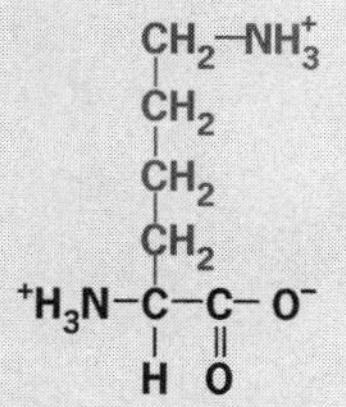

Lysine
(Lys or K)

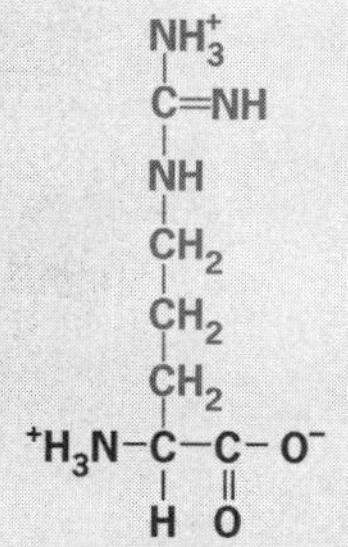

Arginine
(Arg or R)

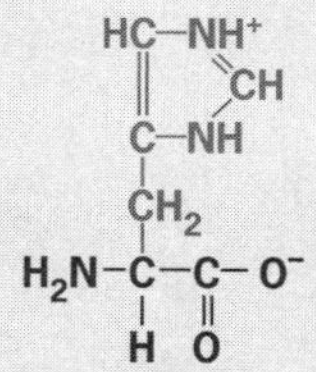

Histidine
(His or H)

Properties of side chains (R groups):

Hydrophilic side chains act as acids or bases which tend to be fully charged (+ or –) under physiologic conditions. Side chains form ionic bonds and are often involved in chemical reactions.

Polar uncharged

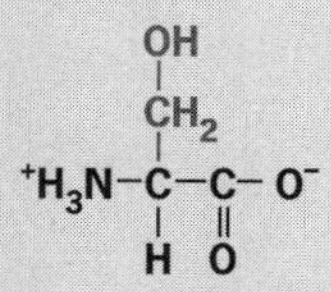

Serine
(Ser or S)

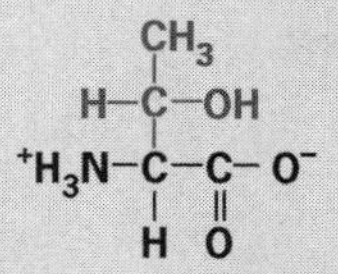

Threonine
(Thr or T)

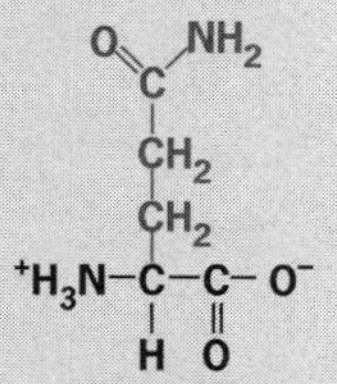

Glutamine
(Gln or Q)

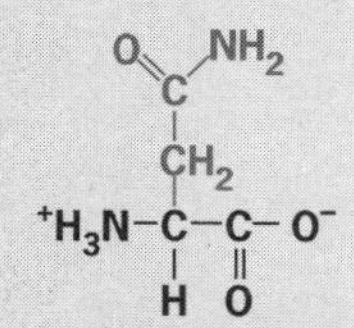

Asparagine
(Asn or N)

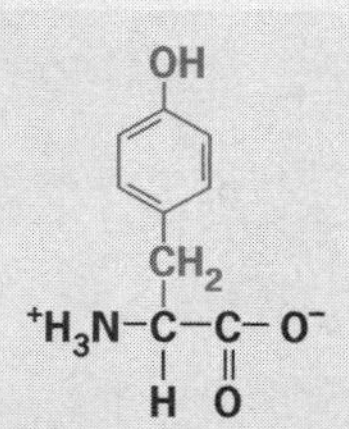

Tyrosine
(Tyr or Y)

Properties of side chain:

Hydrophilic side chains tend to have partial + or – charge allowing them to participate in chemical reactions, form H-bonds, and associate with water.

Nonpolar

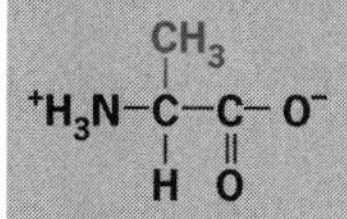

Alanine
(Ala or A)

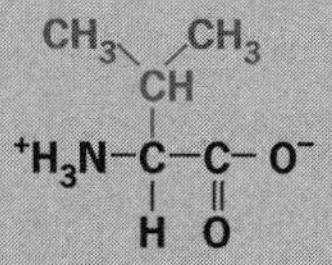

Valine
(Val or V)

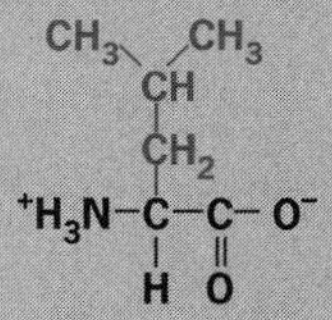

Leucine
(Leu or L)

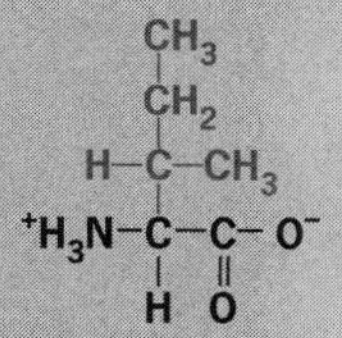

Isoleucine
(Ile or I)

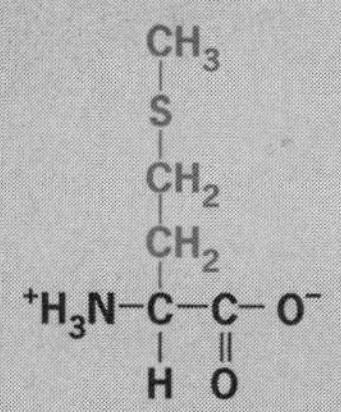

Methionine
(Met or M)

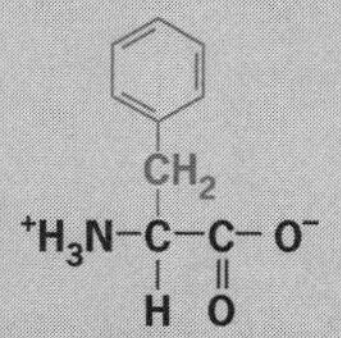

Phenylalanine
(Phe or F)

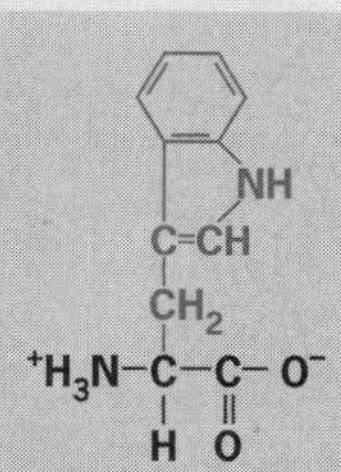

Tryptophan
(Trp or W)

Properties of side chain:

Hydrophobic side chain consists almost entirely of C and H atoms. These amino acids tend to form the inner core of soluble proteins, buried away from the aqueous medium. They play an important role in membranes by associating with the lipid bilayer.

Side chains with unique properties

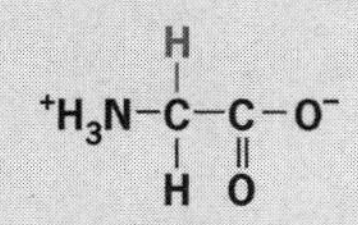

Glycine
(Gly or G)

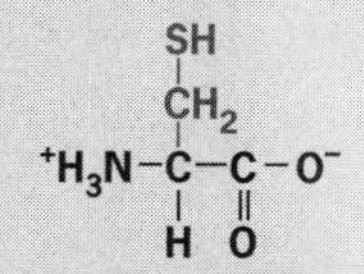

Cysteine
(Cys or C)

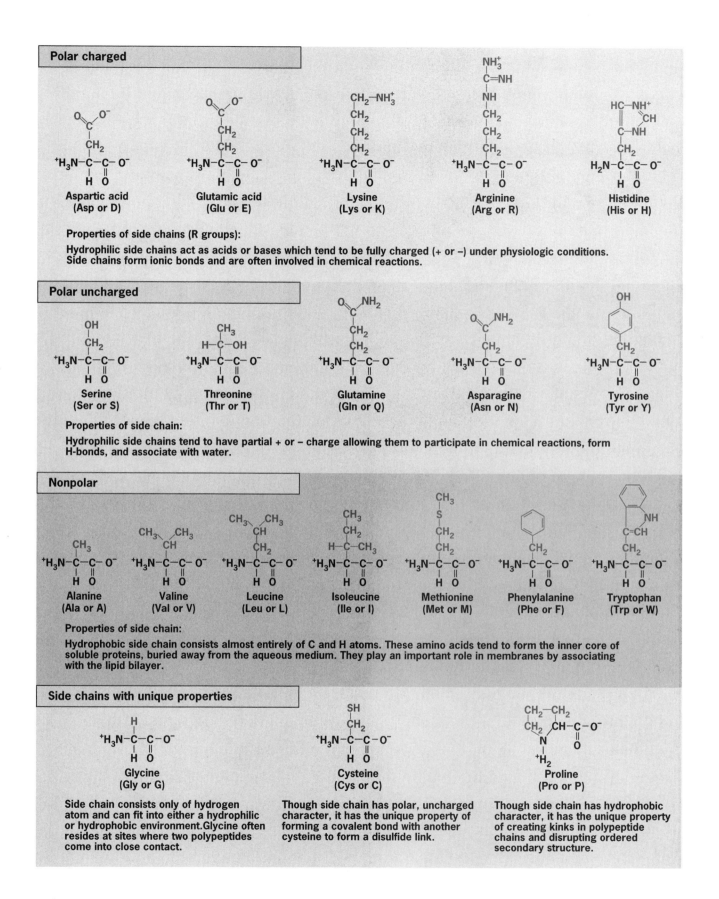

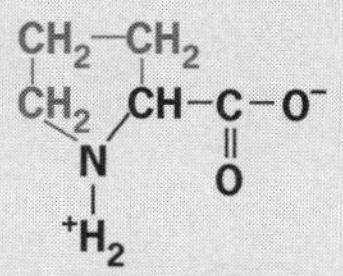

Proline
(Pro or P)

Side chain consists only of hydrogen atom and can fit into either a hydrophilic or hydrophobic environment.Glycine often resides at sites where two polypeptides come into close contact.

Though side chain has polar, uncharged character, it has the unique property of forming a covalent bond with another cysteine to form a disulfide link.

Though side chain has hydrophobic character, it has the unique property of creating kinks in polypeptide chains and disrupting ordered secondary structure.

Figure 2.26 The chemical structure of amino acids.

Copyright © 2005 John Wiley & Sons, Inc.

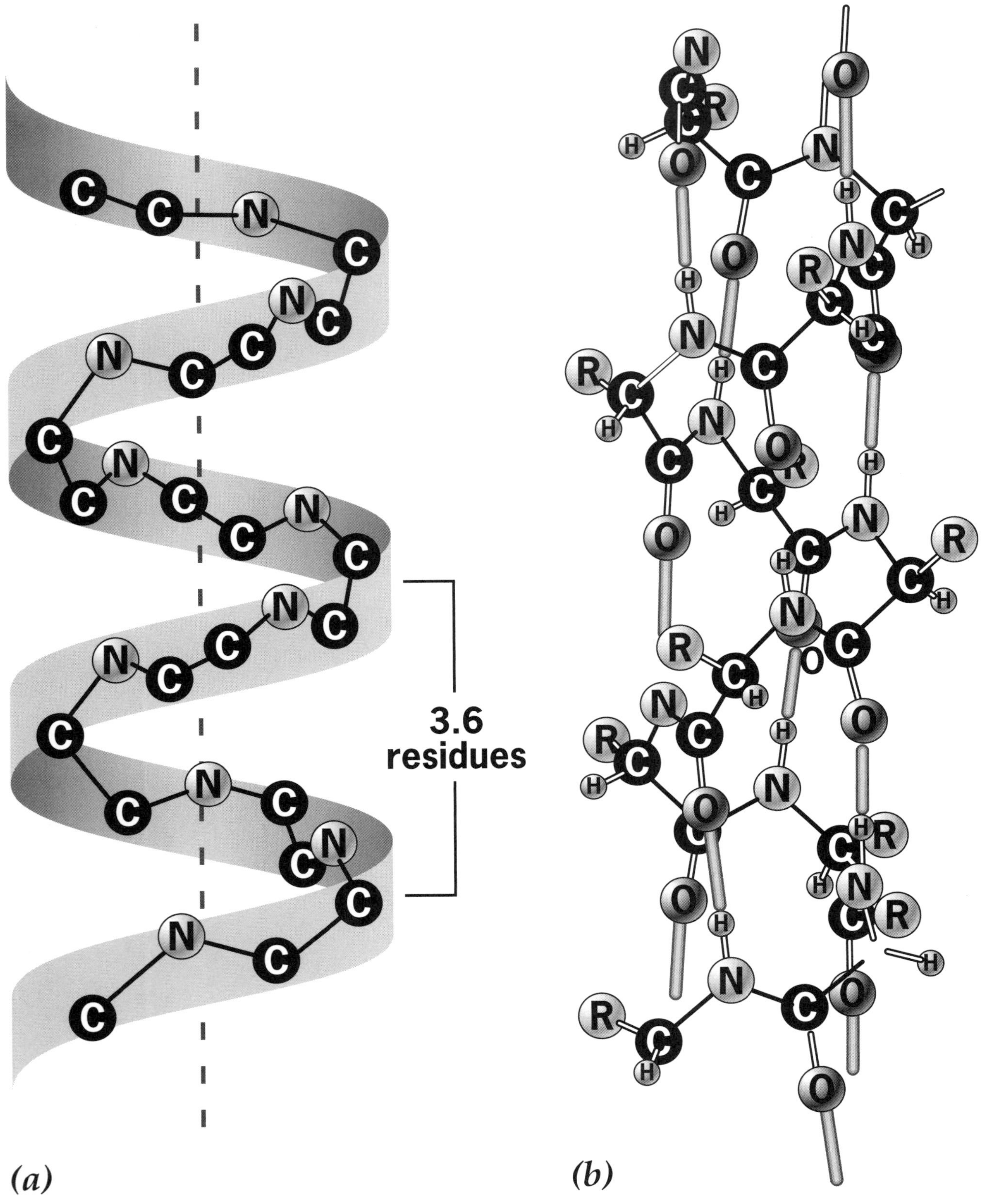

Figure 2.30 The alpha helix.

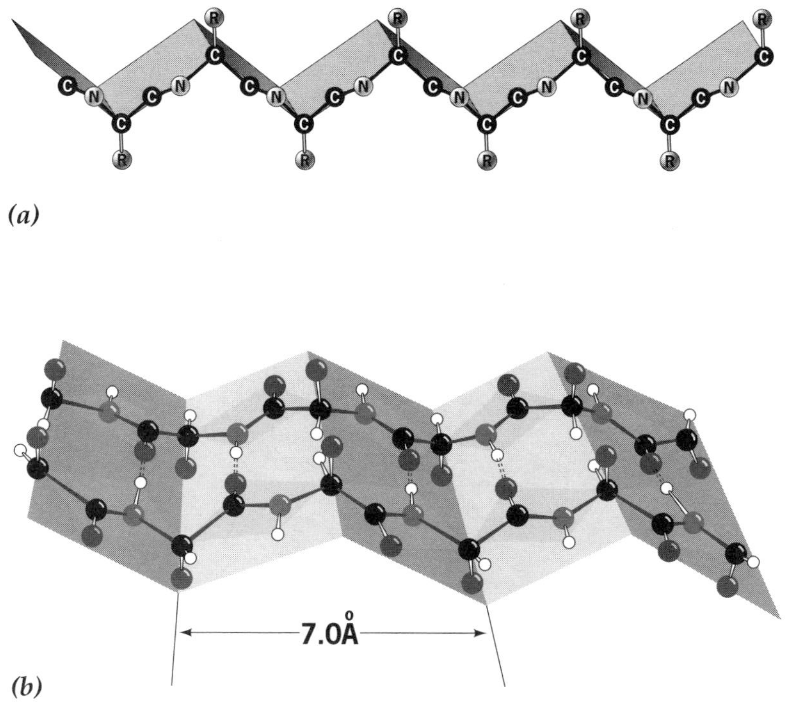

Figure 2.31 The b-pleated sheet.

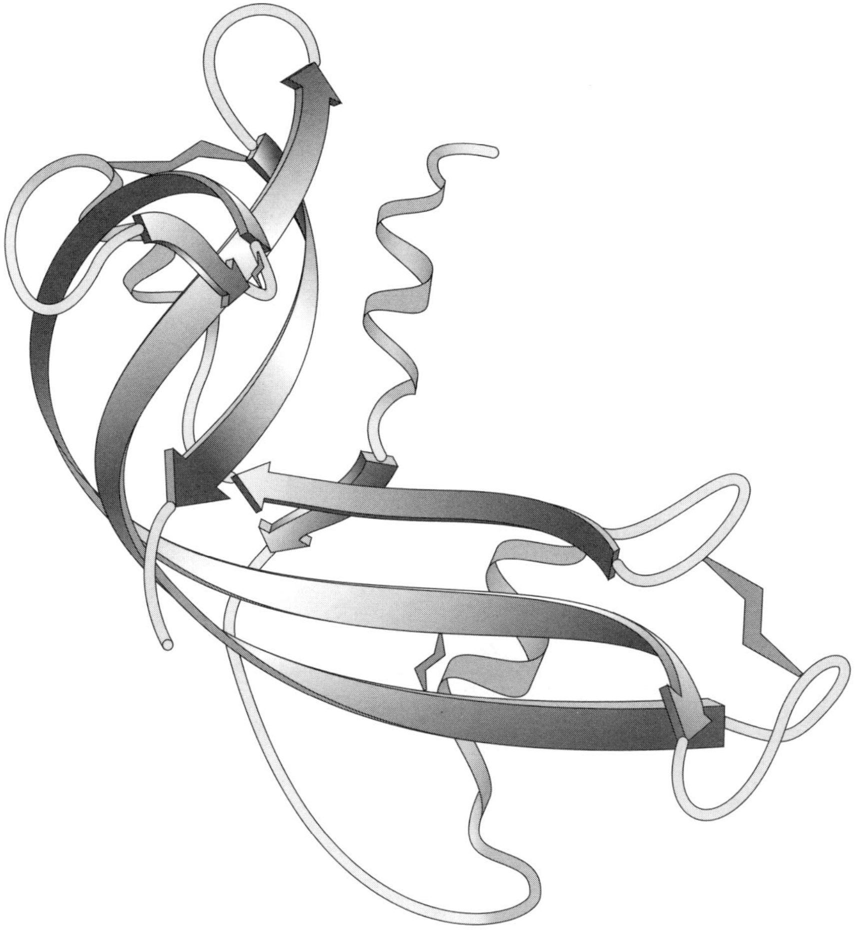

Figure 2.32 A ribbon model of ribonuclease.

Copyright © 2005 John Wiley & Sons, Inc.

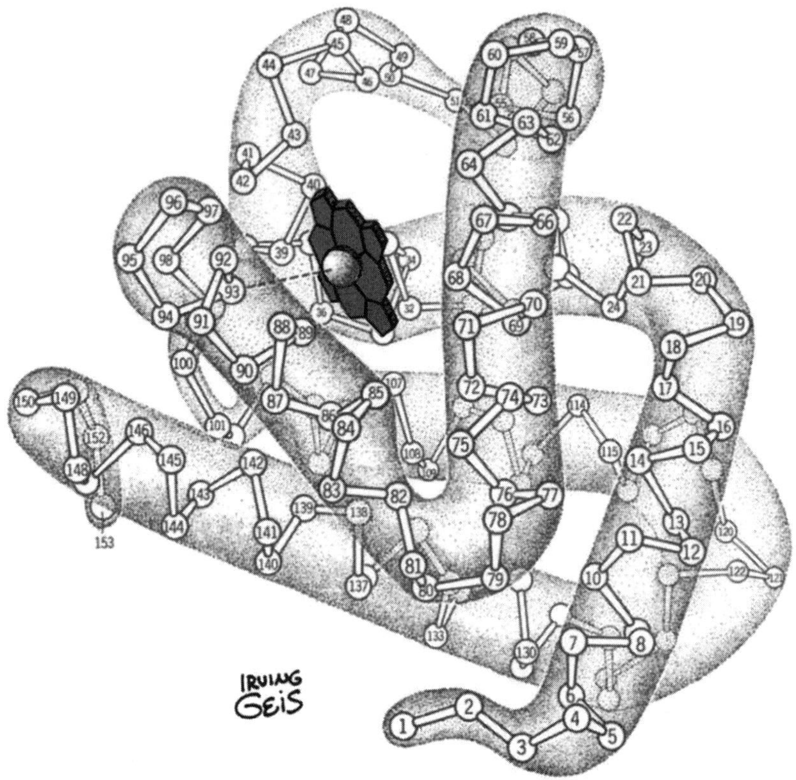

Figure 2.34a The three-dimensional structure of myoglobin.

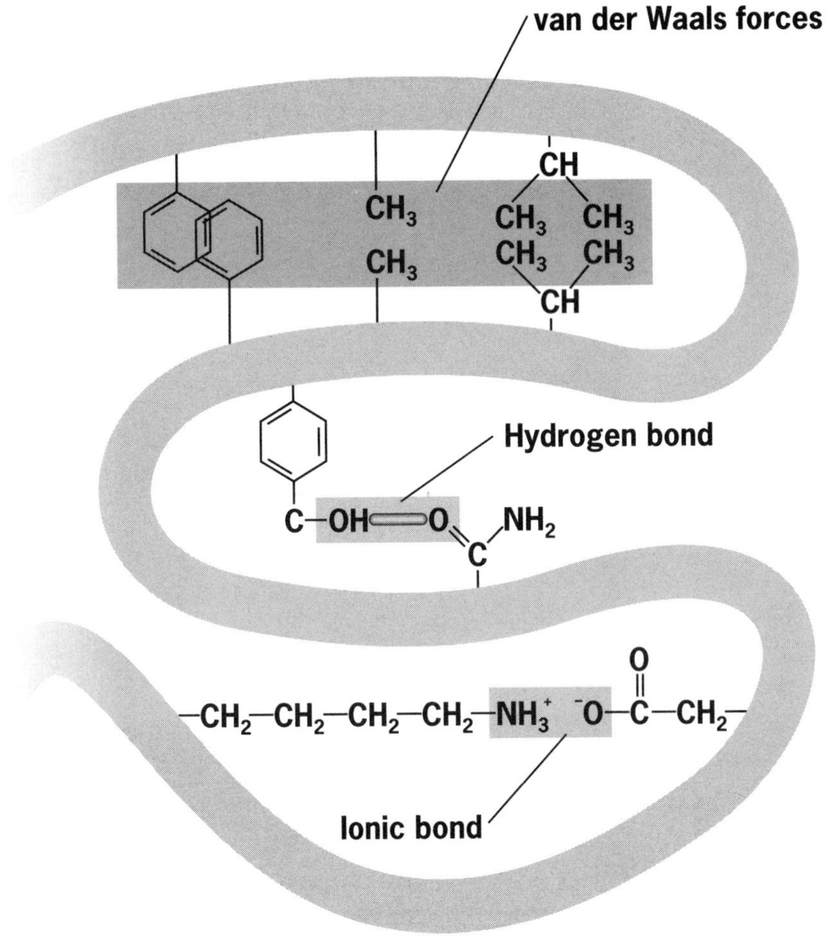

Figure 2.35 Types of noncovalent bonds maintaining the conformation of proteins.

Copyright © 2005 John Wiley & Sons, Inc.

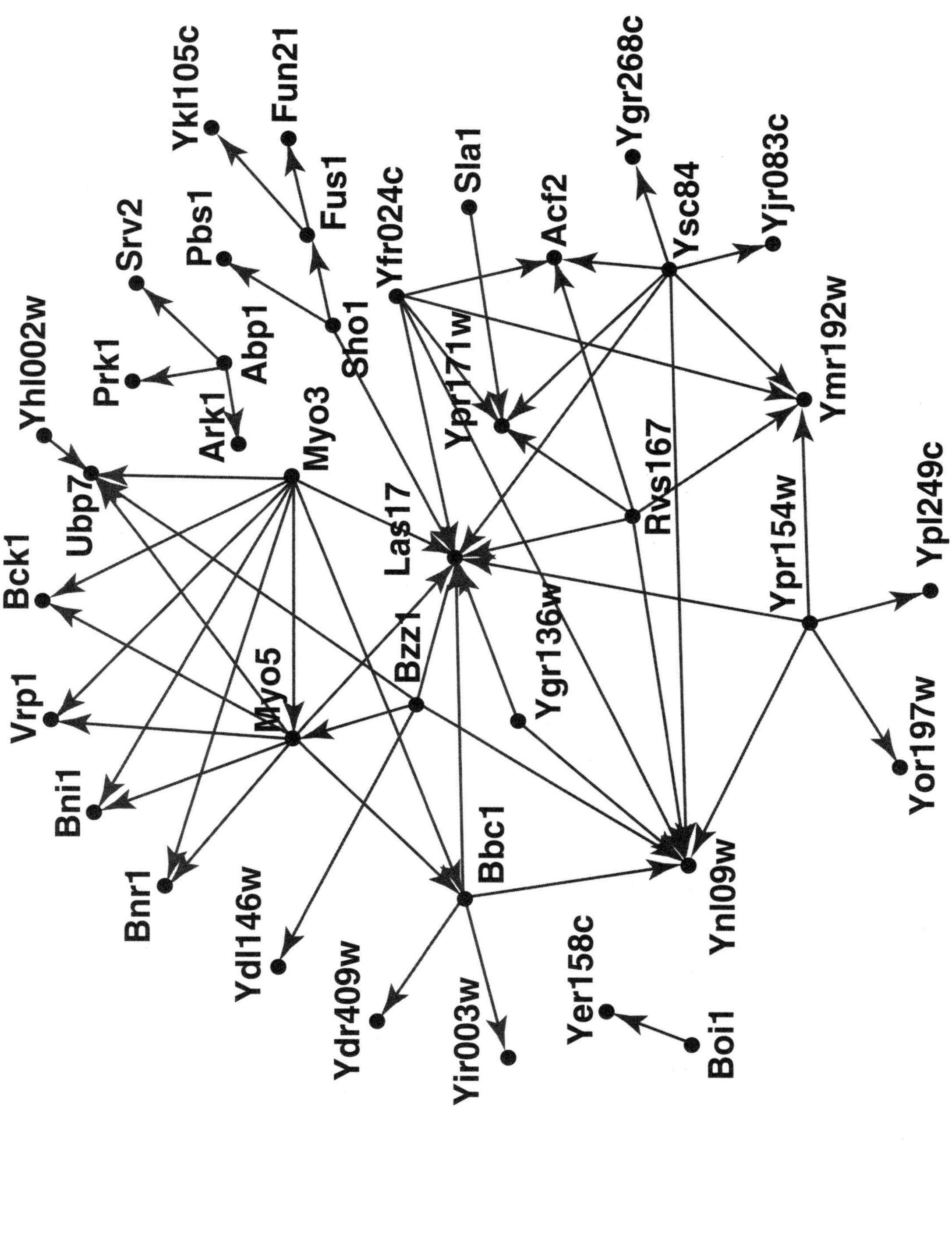

Figure 2.42 A network of protein–protein interactions.

Copyright © 2005 John Wiley & Sons, Inc.

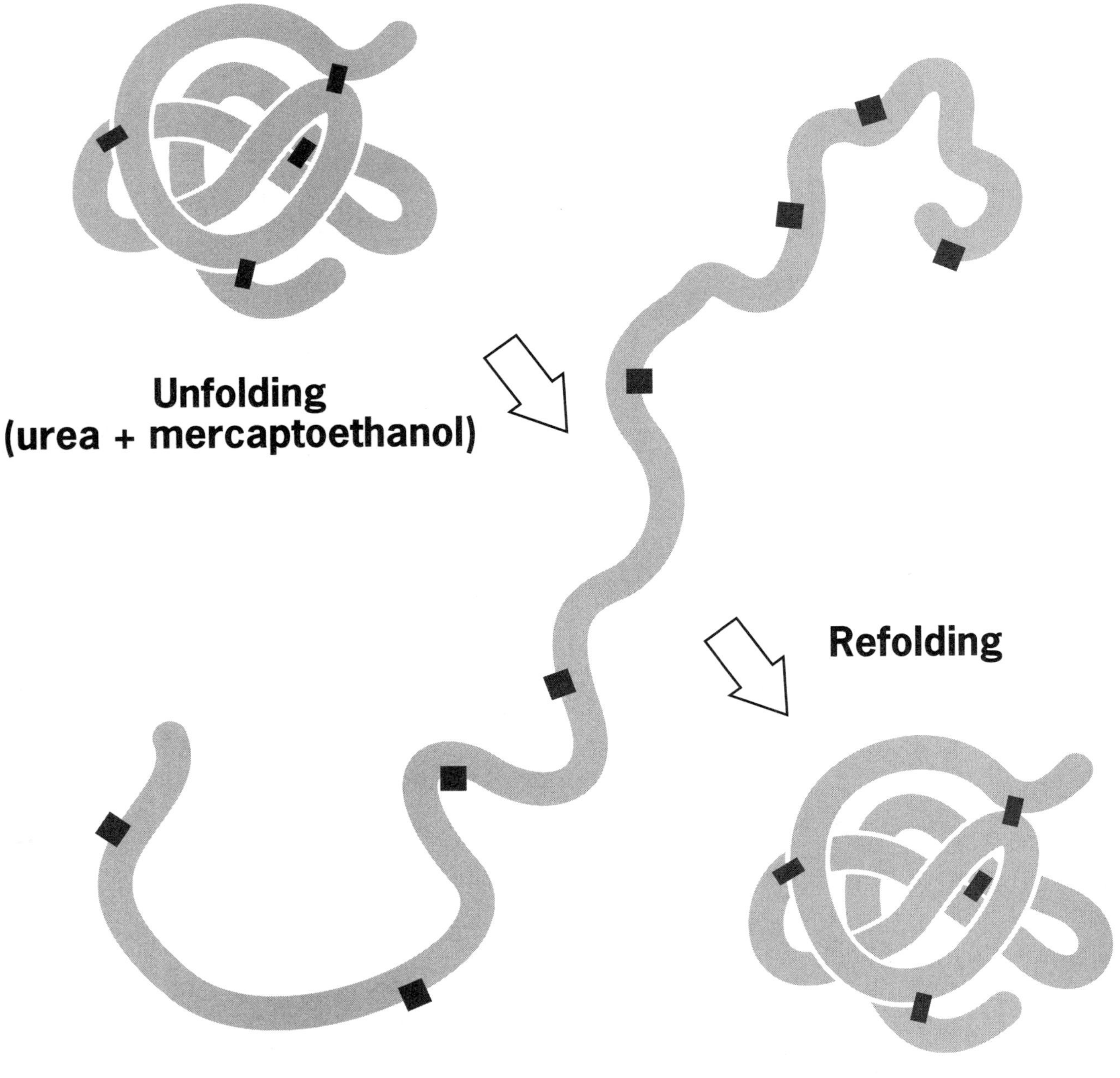

Figure 2.43 Denaturation and refolding of ribonuclease.

Copyright © 2005 John Wiley & Sons, Inc.

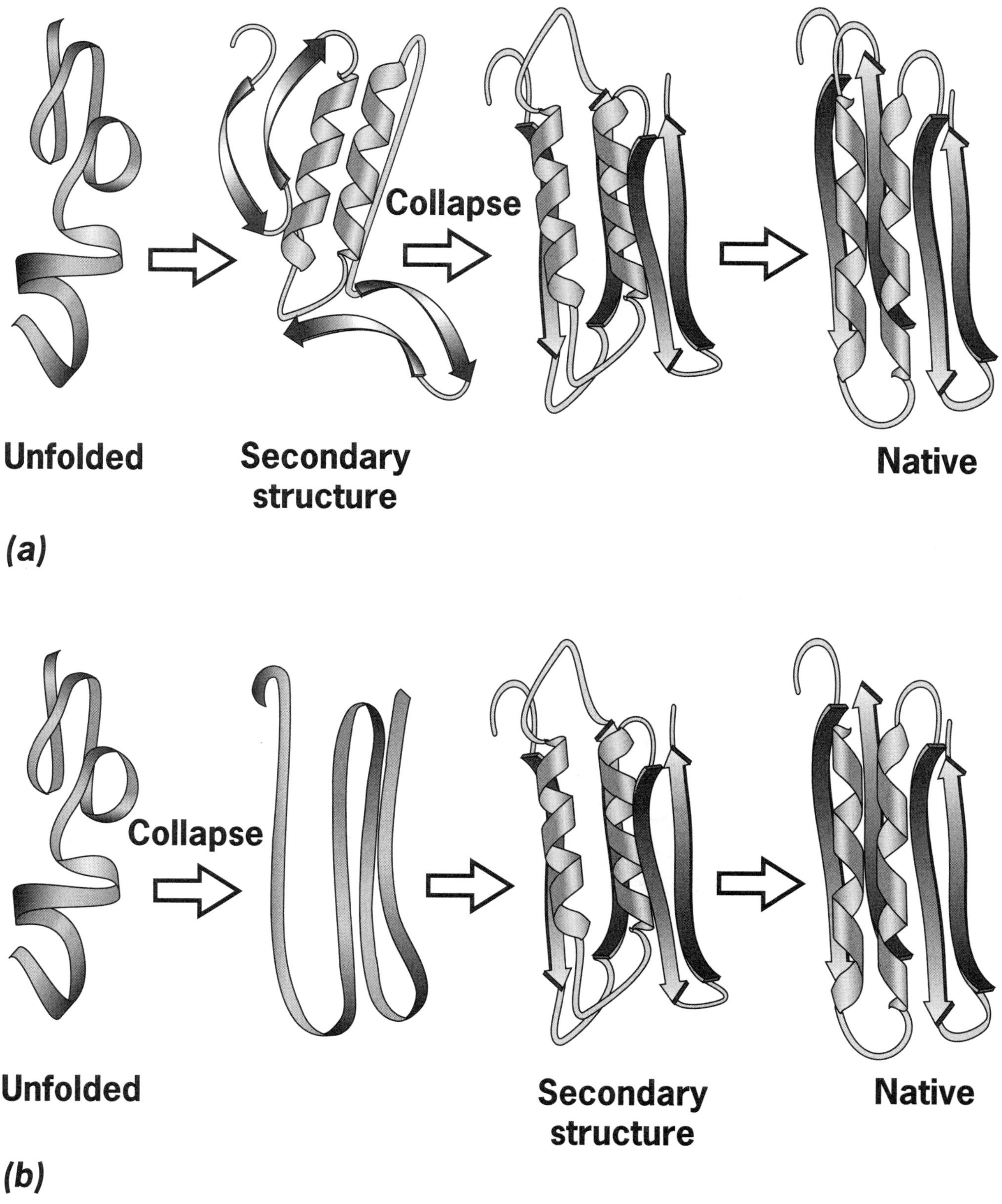

Figure 2.44 Two alternate pathways.

Copyright © 2005 John Wiley & Sons, Inc.

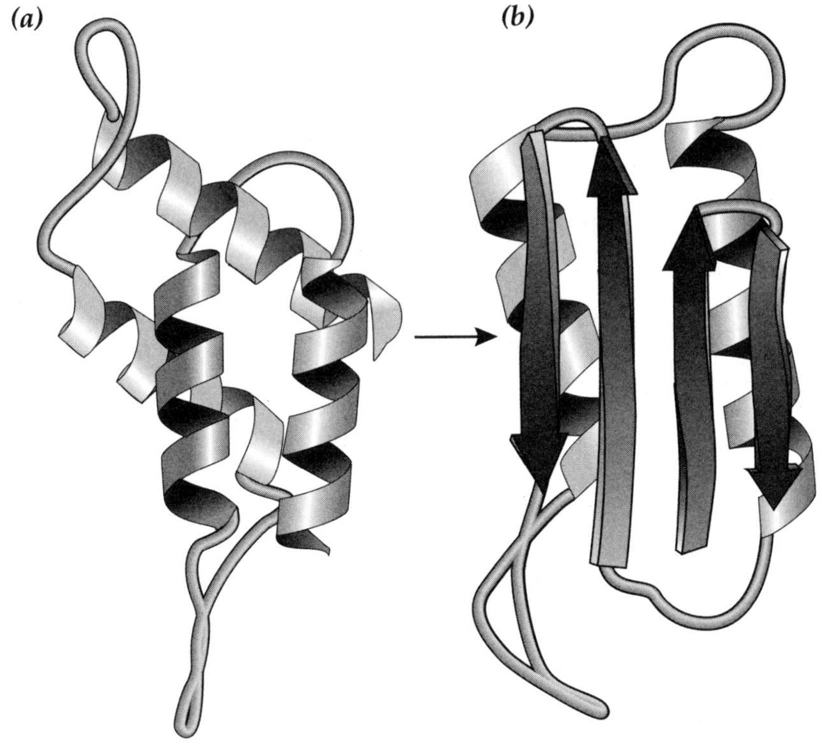

Figure 1 A comparison of the normal (PrPC) and abnormal (PrPSc) prion proteins.

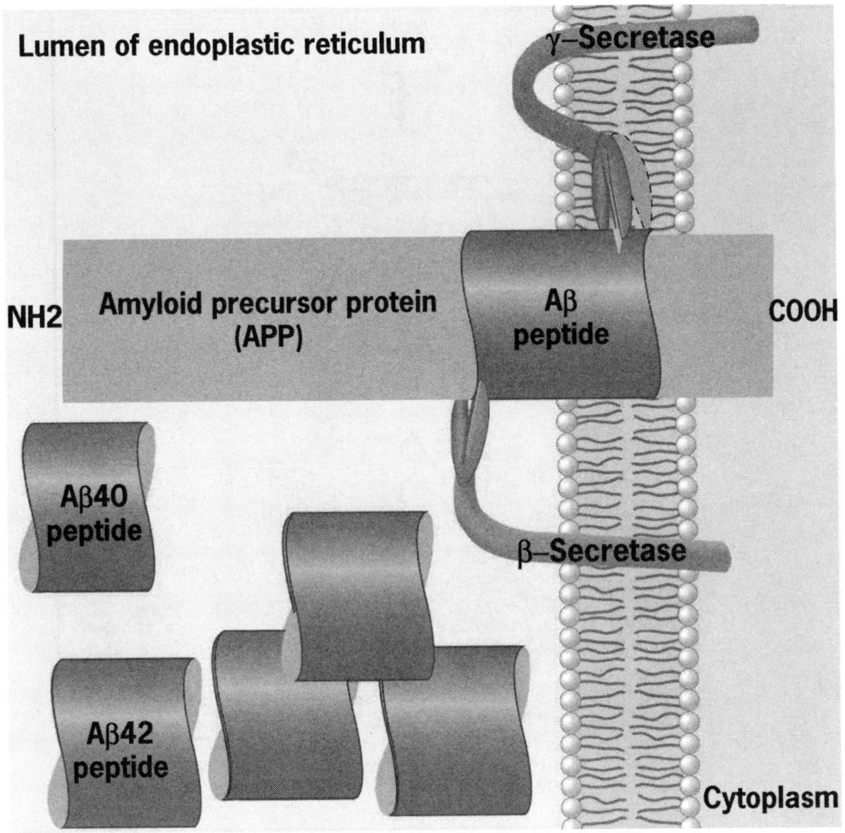

Figure 3 Formation of the Aβ peptide.

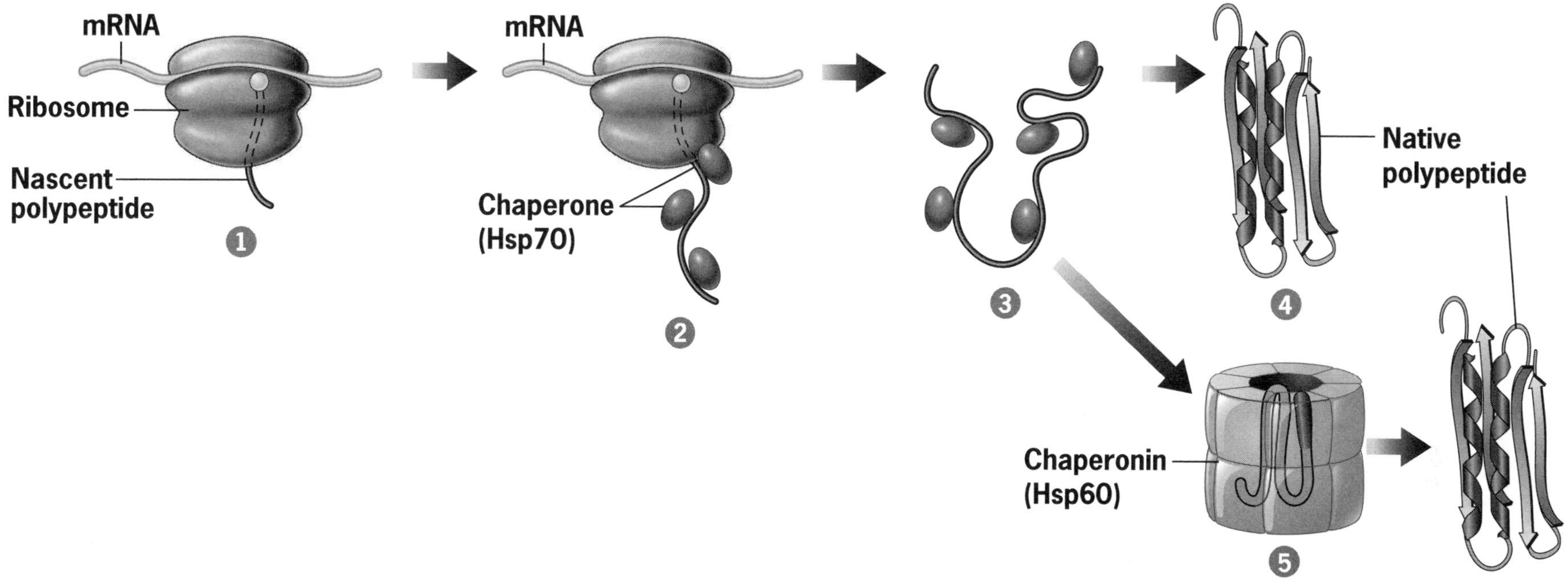

Figure 2.45 The role of molecular chaperones in encouraging protein folding.

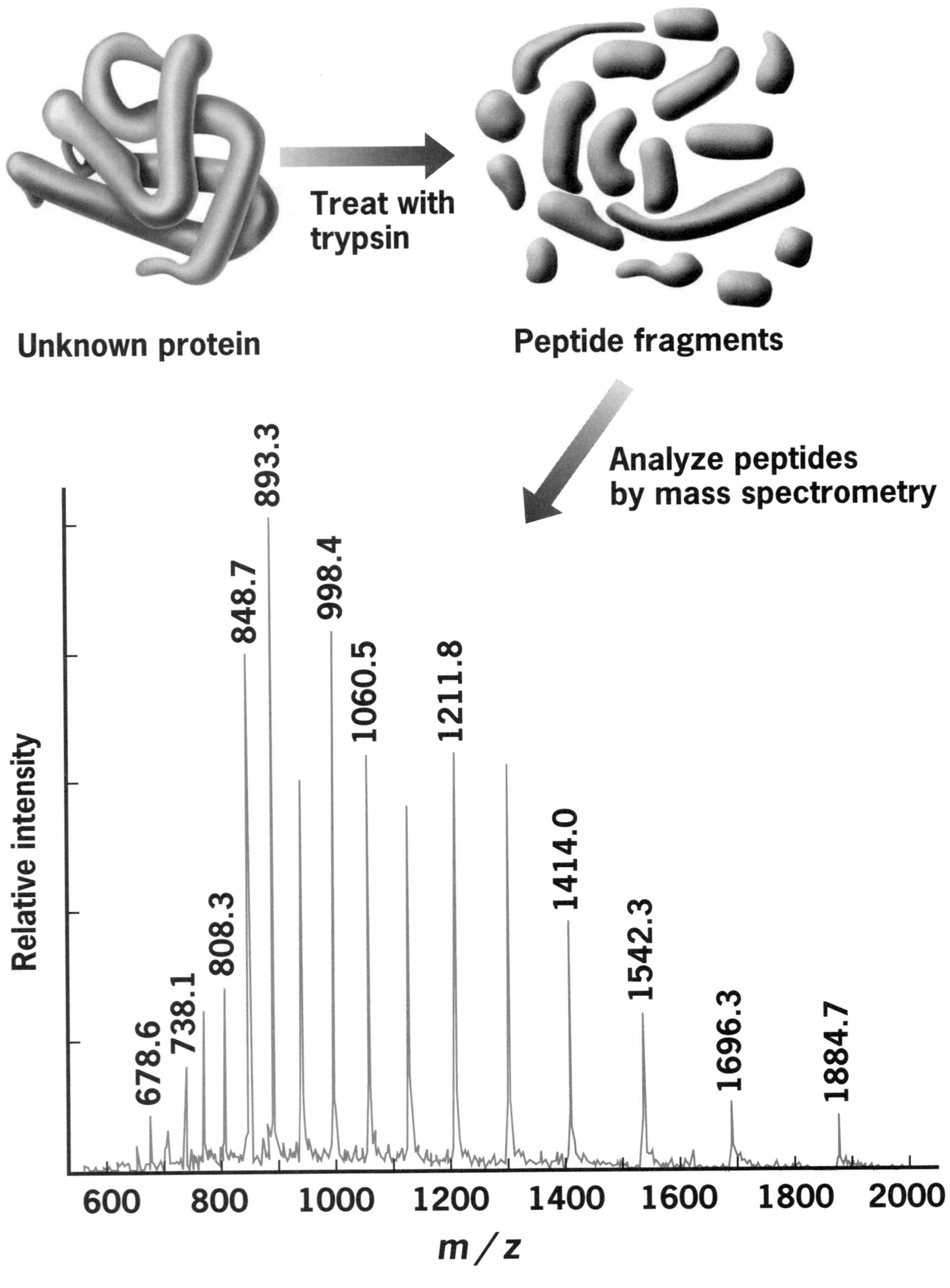

Figure 2.48 Identifying proteins by mass spectrometry.

Copyright © 2005 John Wiley & Sons, Inc.

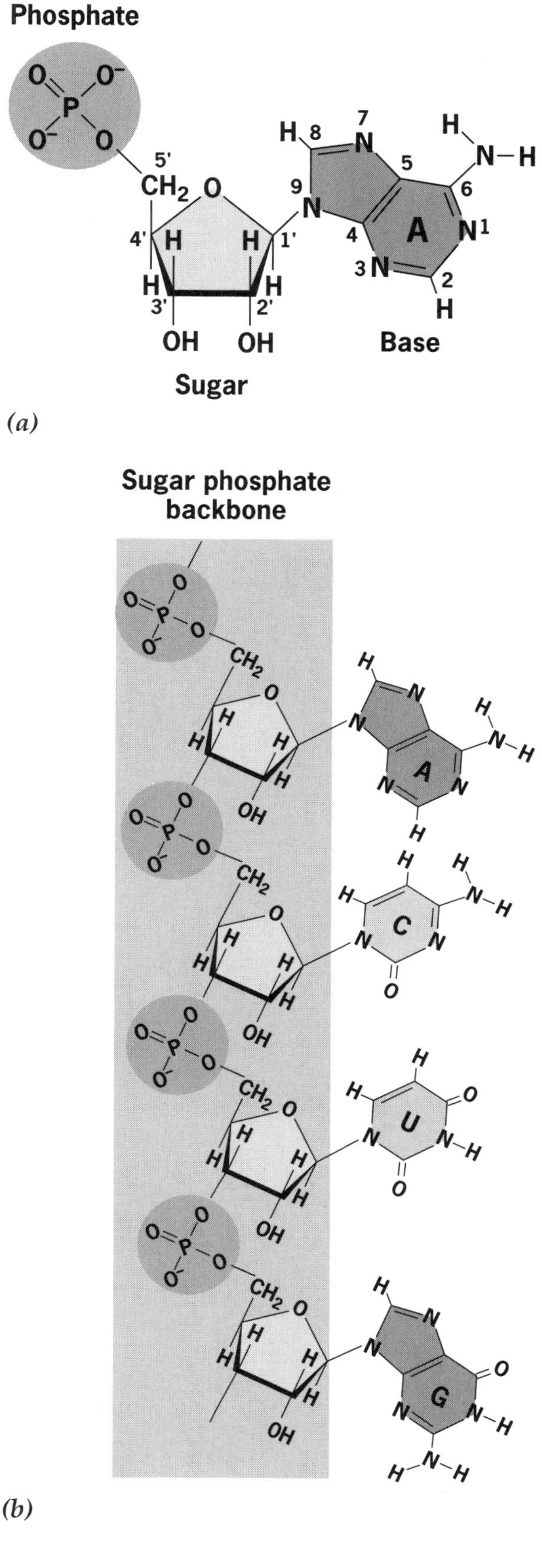

Figure 2.52 Nucleotides and nucleotide strands of RNA.

Copyright © 2005 John Wiley & Sons, Inc.

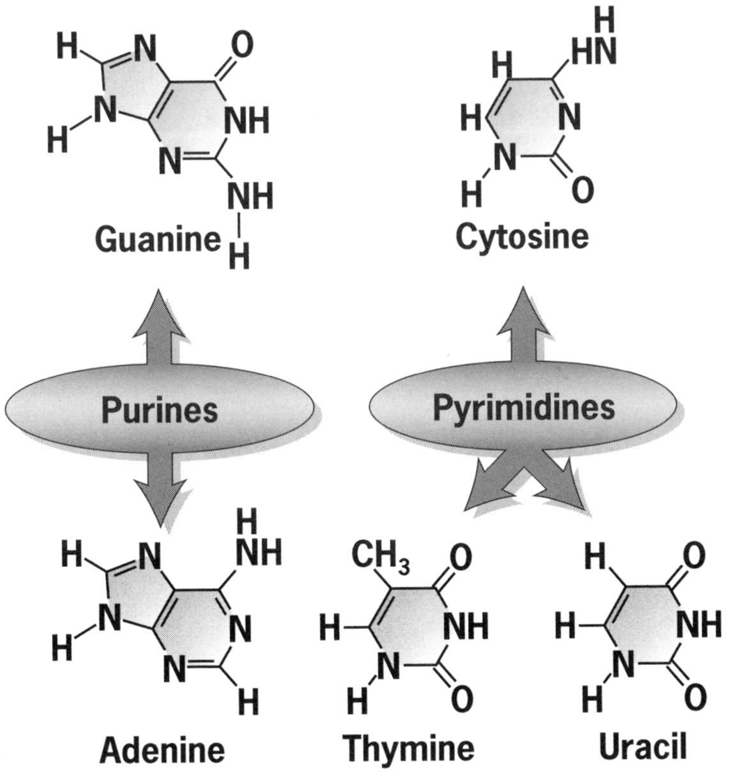

Figure 2.53 Nitrogenous bases in nucleic acids.

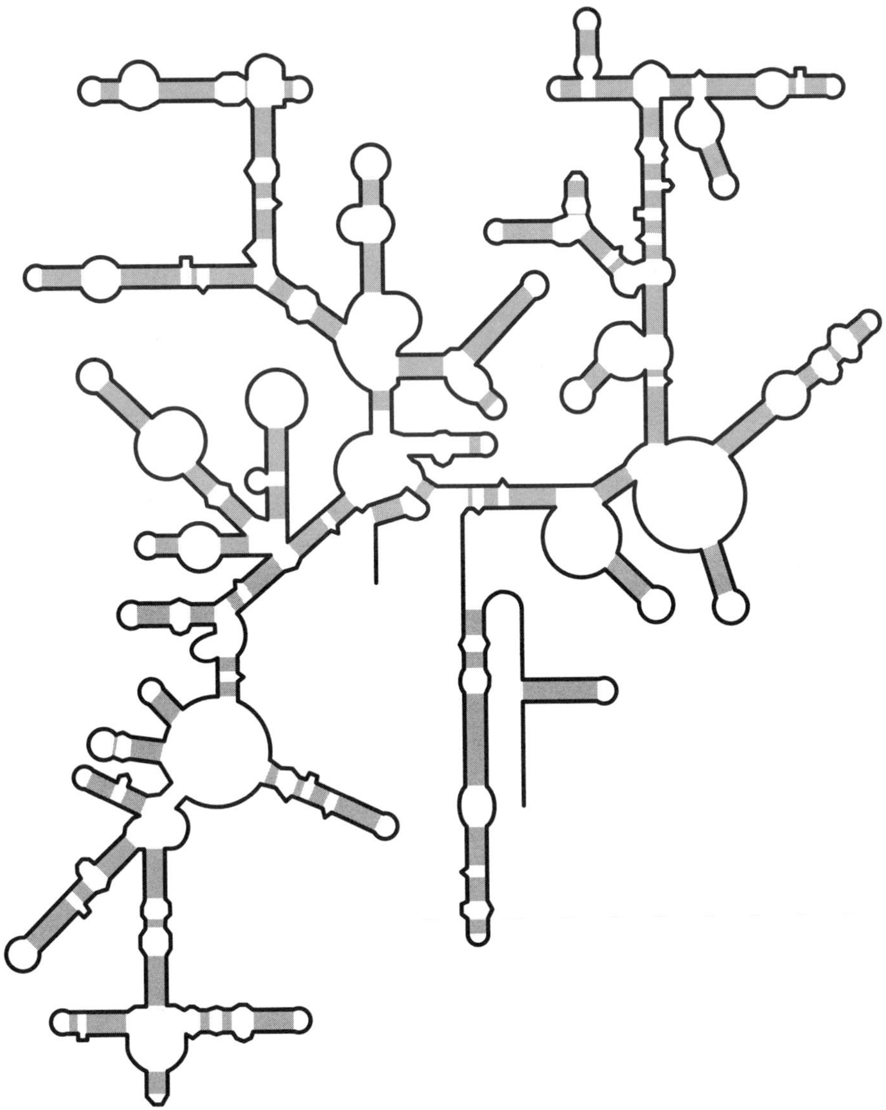

(a)

Figure 2.54a RNAs can assume complex shapes.

Copyright © 2005 John Wiley & Sons, Inc.

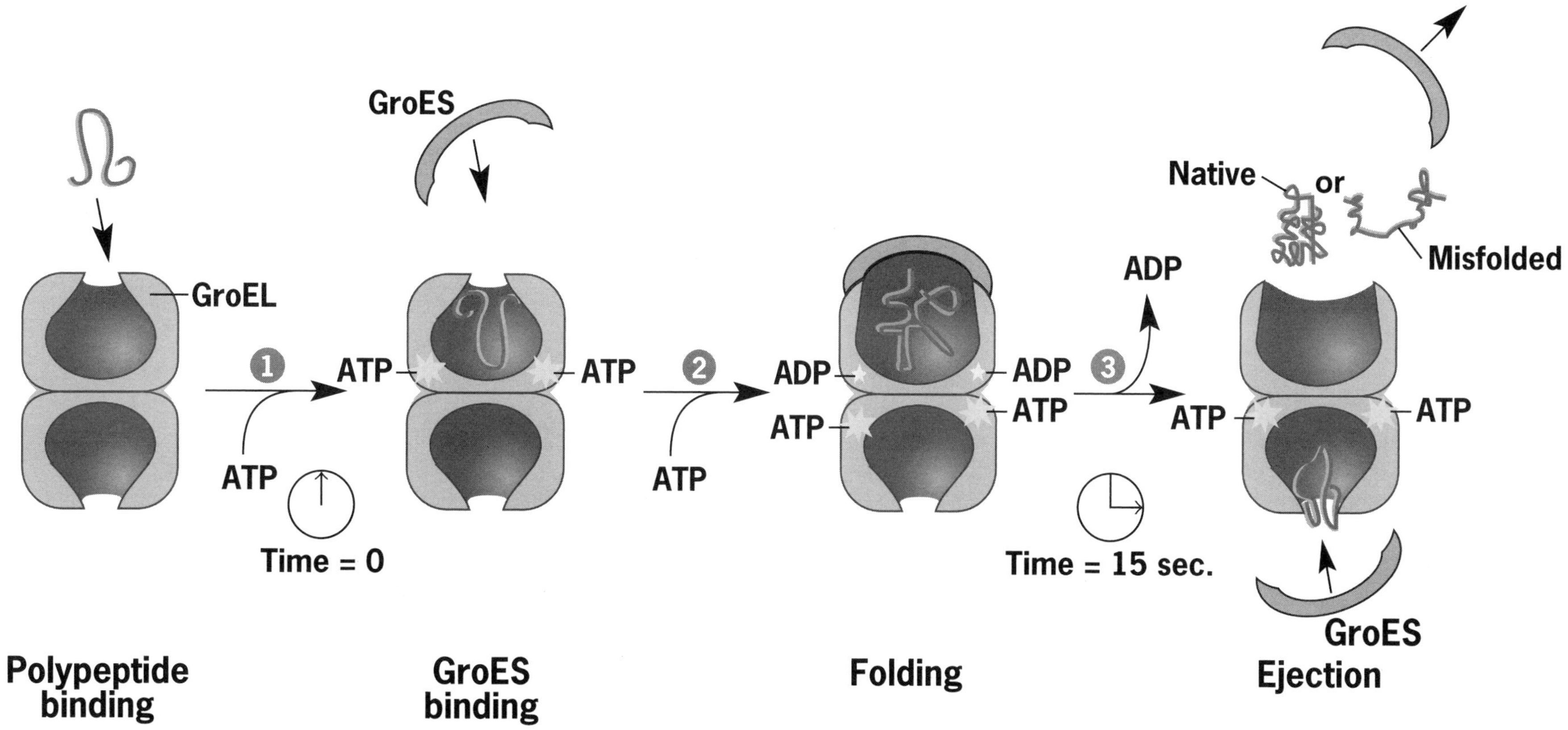

Figure 4 A schematic illustration of the proposed steps that occur during the GroEL-GroES-assisted folding of a polypeptide.

Copyright © 2005 John Wiley & Sons, Inc.

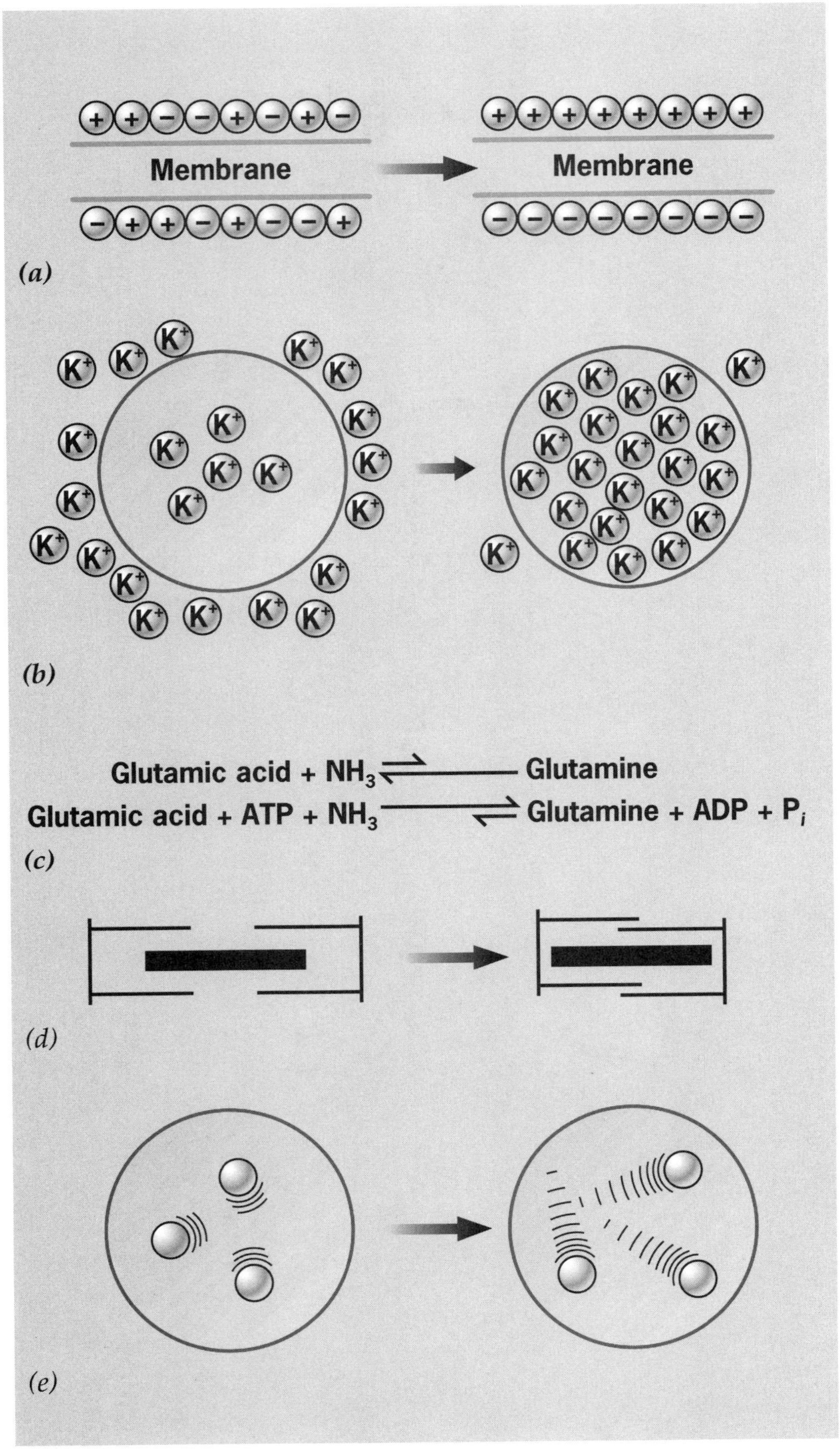

Figure 3.6 A few roles for ATP hydrolysis.

Copyright © 2005 John Wiley & Sons, Inc.

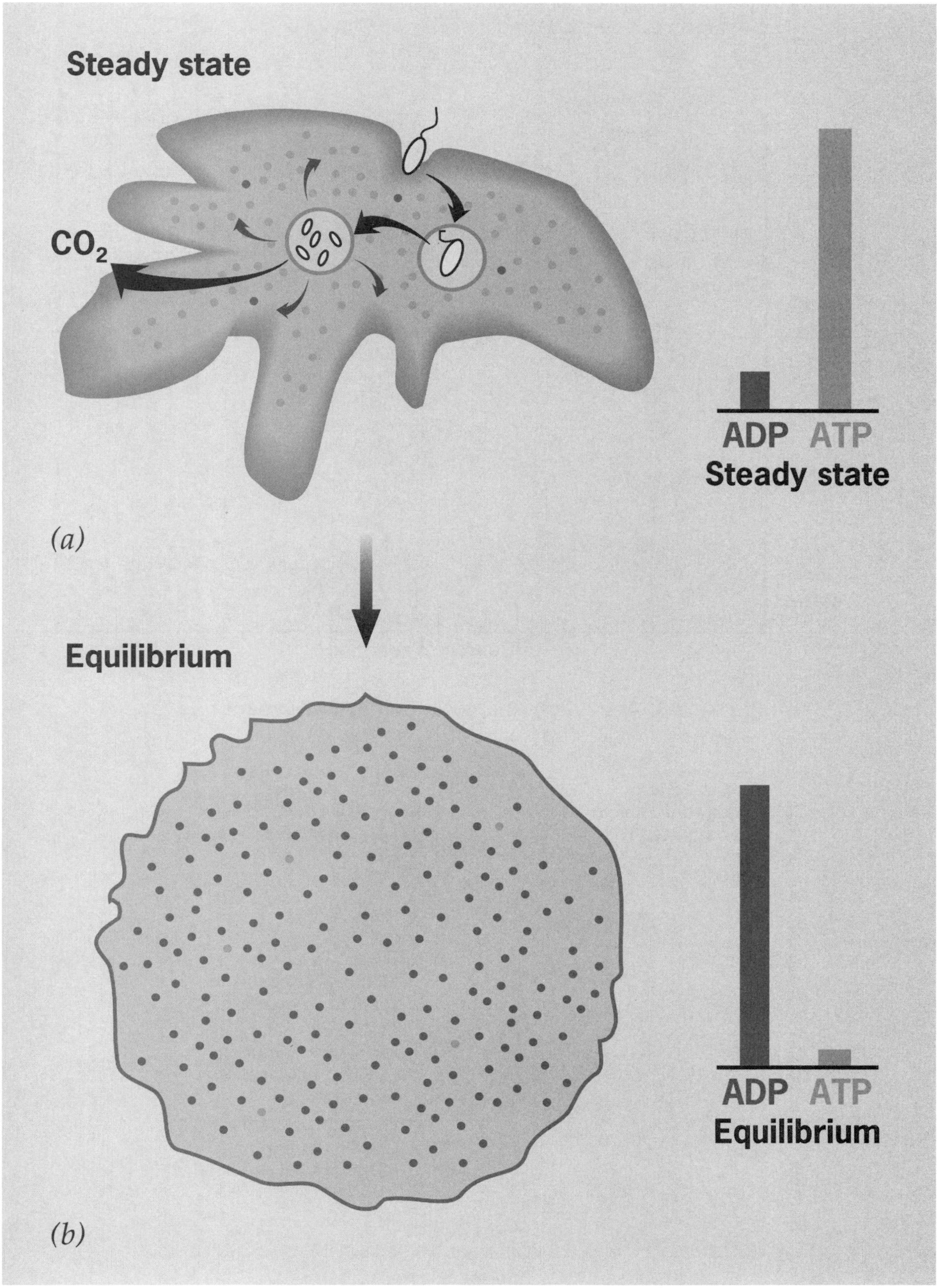

Figure 3.7 Steady state versus equilibrium.

Copyright © 2005 John Wiley & Sons, Inc.

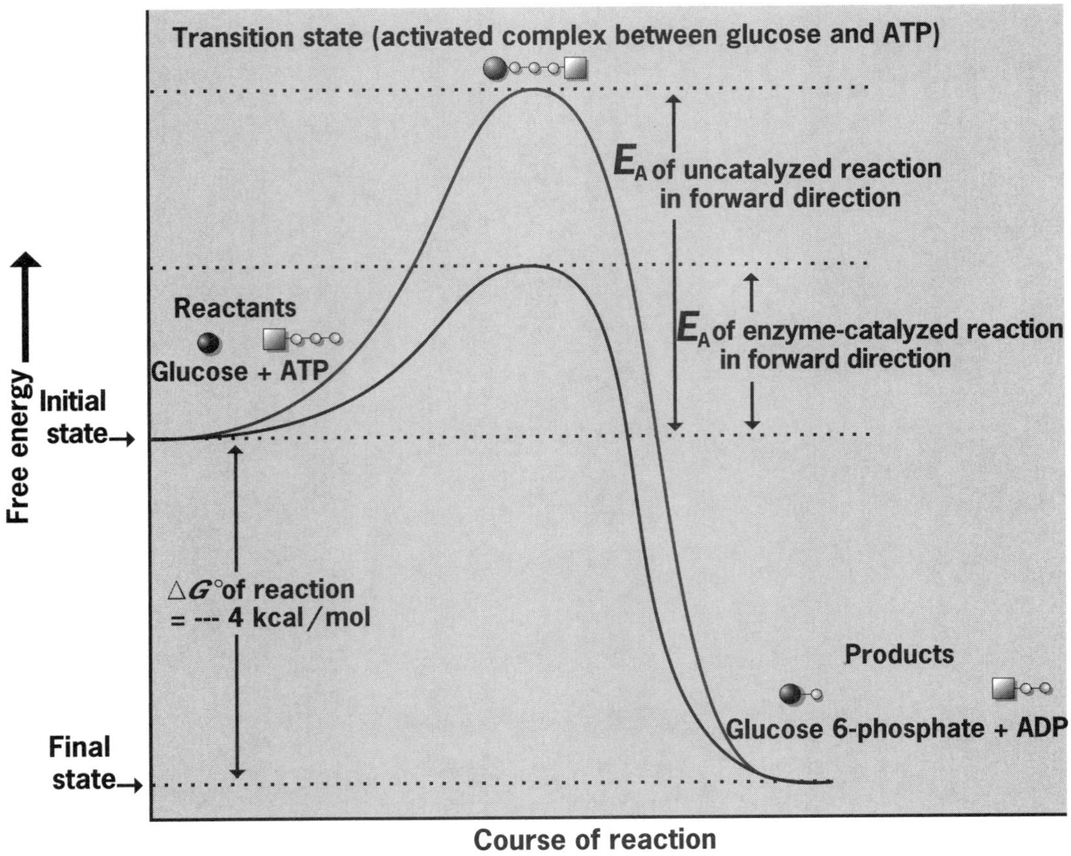

Figure 3.8 Activation energy and enzymatic reactions.

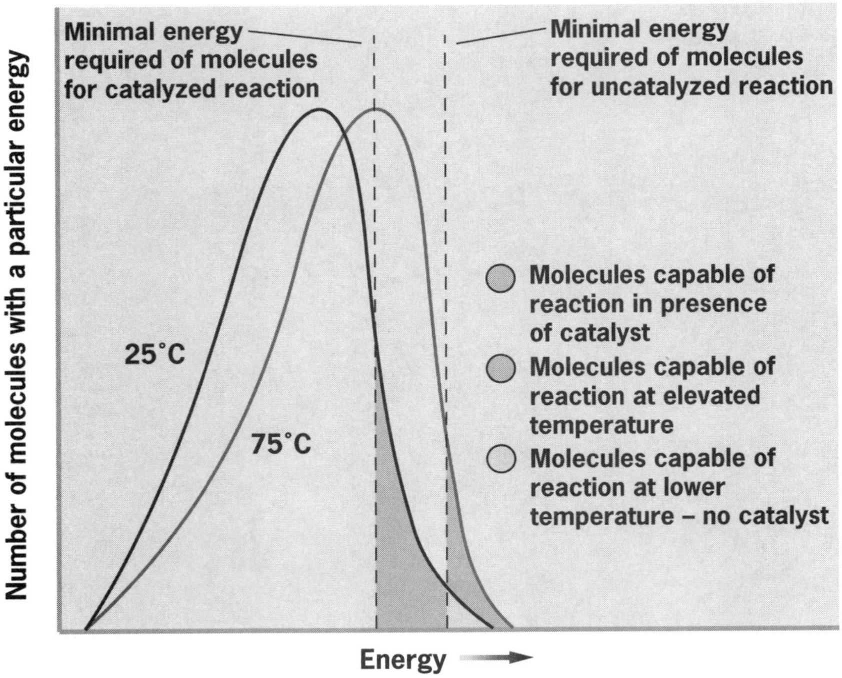

Figure 3.9 The effect of lowering activation energy on the rate of a reaction.

Copyright © 2005 John Wiley & Sons, Inc.

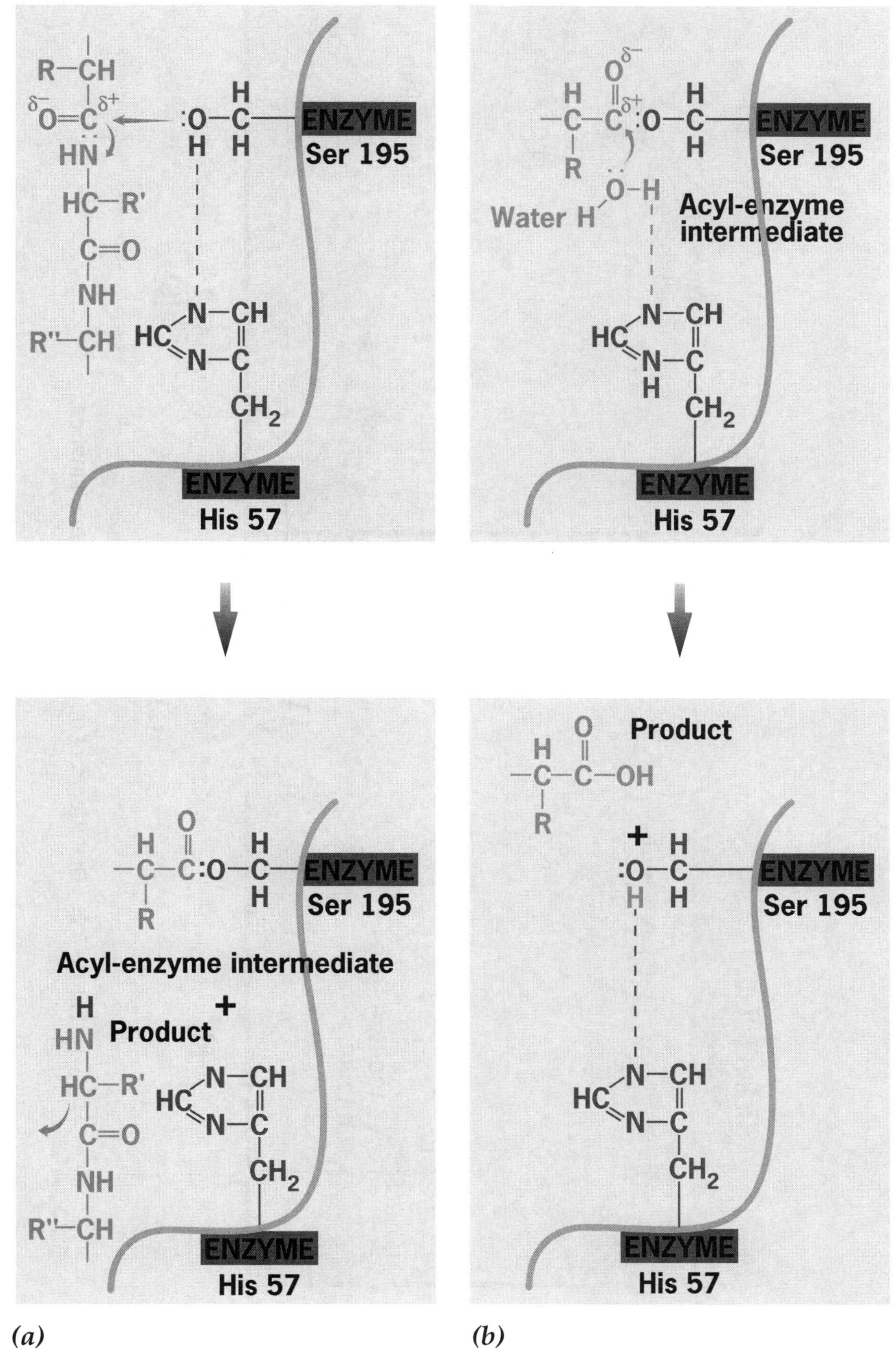

Figure 3.13 Diagrammatic representation of the catalytic mechanism of chymotrypsin.

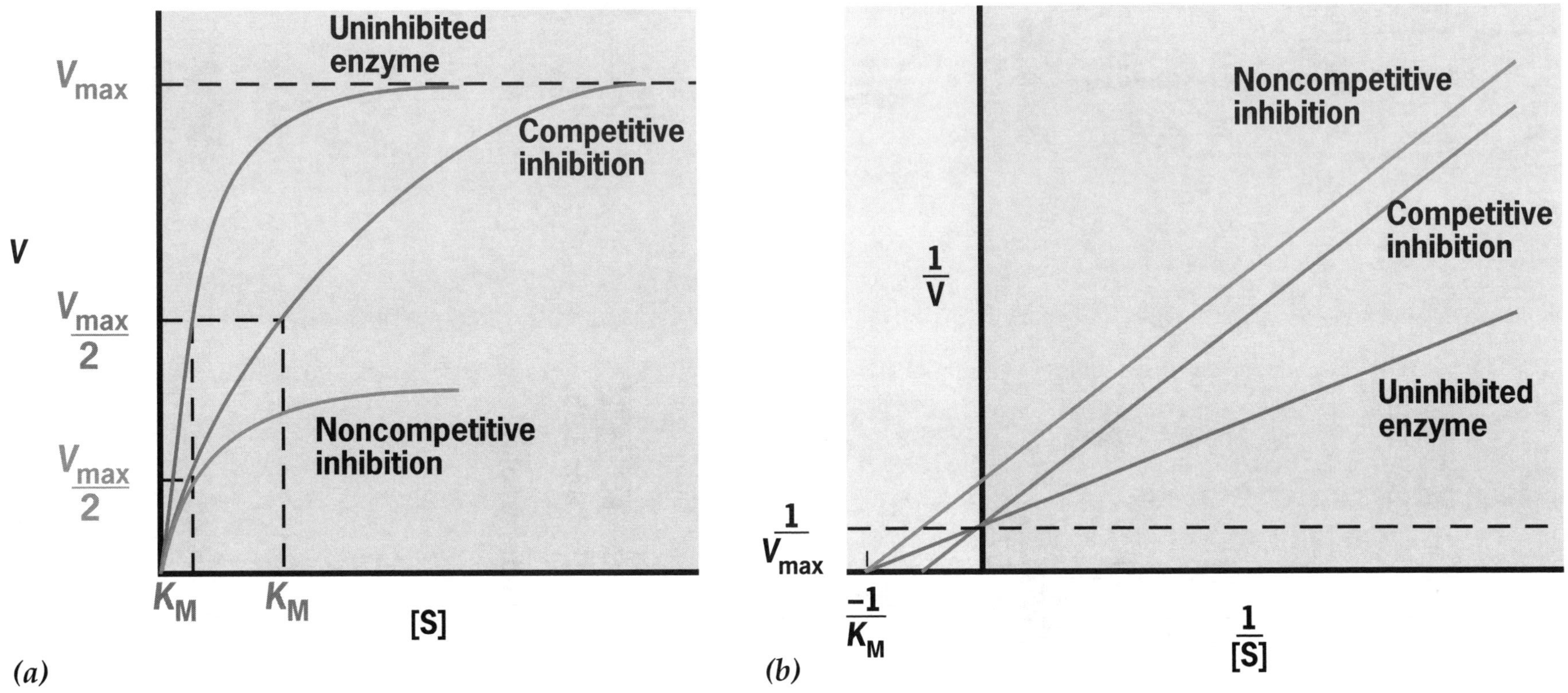

Figure 3.20 The effects of inhibitors on enzyme kinetics.

Copyright © 2005 John Wiley & Sons, Inc.

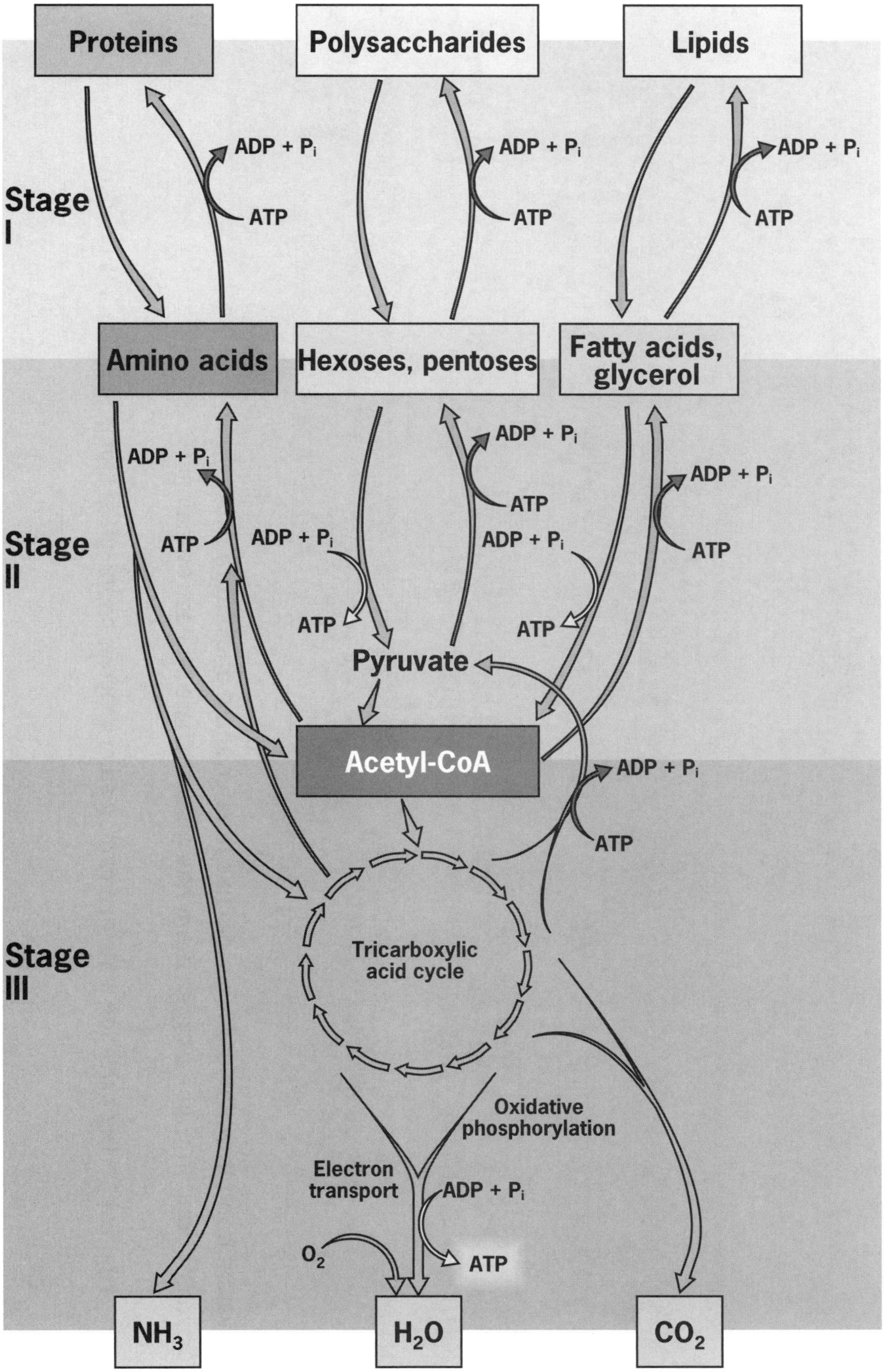

Proteins
Polysaccharides
Lipids
ADP + Pi
ATP
ADP + Pi
ATP
ADP + Pi
ATP
Stage I
Amino acids
Hexoses, pentoses
Fatty acids, glycerol
ADP + Pi
ATP
ADP + Pi
ATP
ADP + Pi
ATP
Stage II
ADP + Pi
ATP
ADP + Pi
ATP
Pyruvate
Acetyl-CoA
ADP + Pi
ATP
Stage III
Tricarboxylic acid cycle
Oxidative phosphorylation
Electron transport
ADP + Pi
O2
ATP
NH3
H2O
CO2

Figure 3.21 Three stages of metabolism.

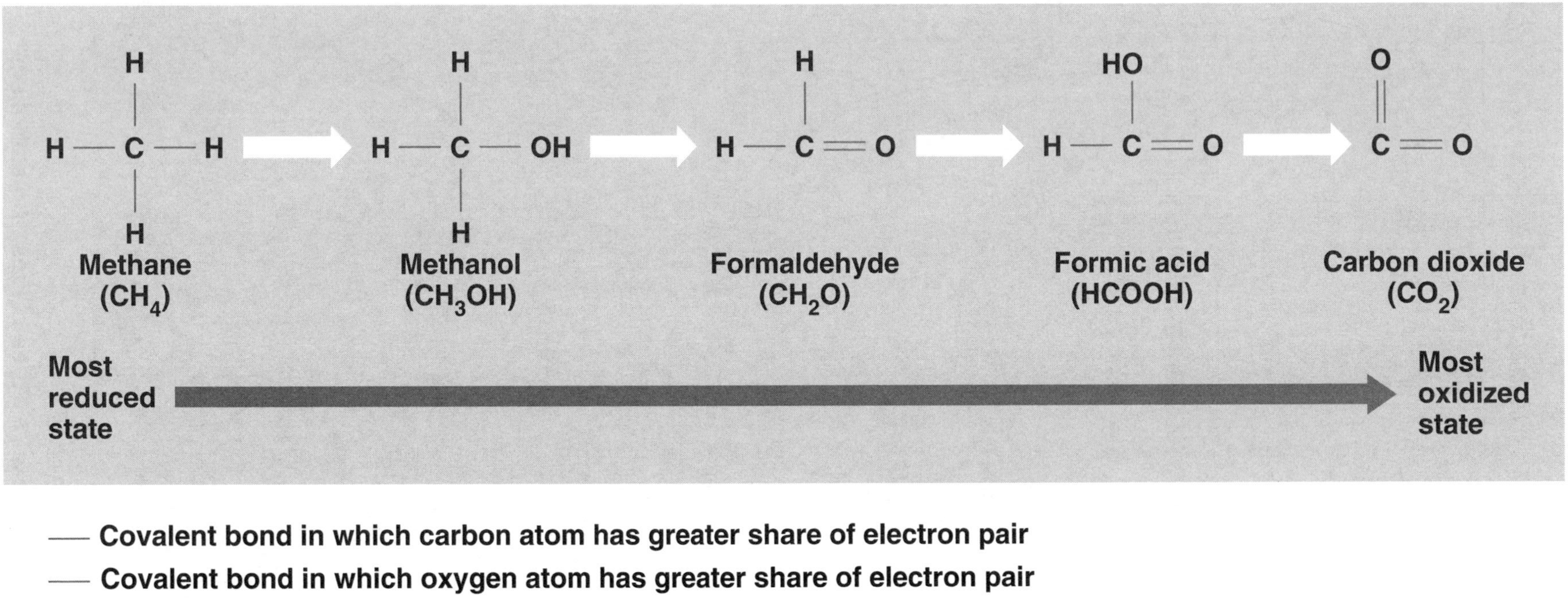

Figure 3.22 The oxidation state of a carbon atom depends on the other atoms to which it is bonded.

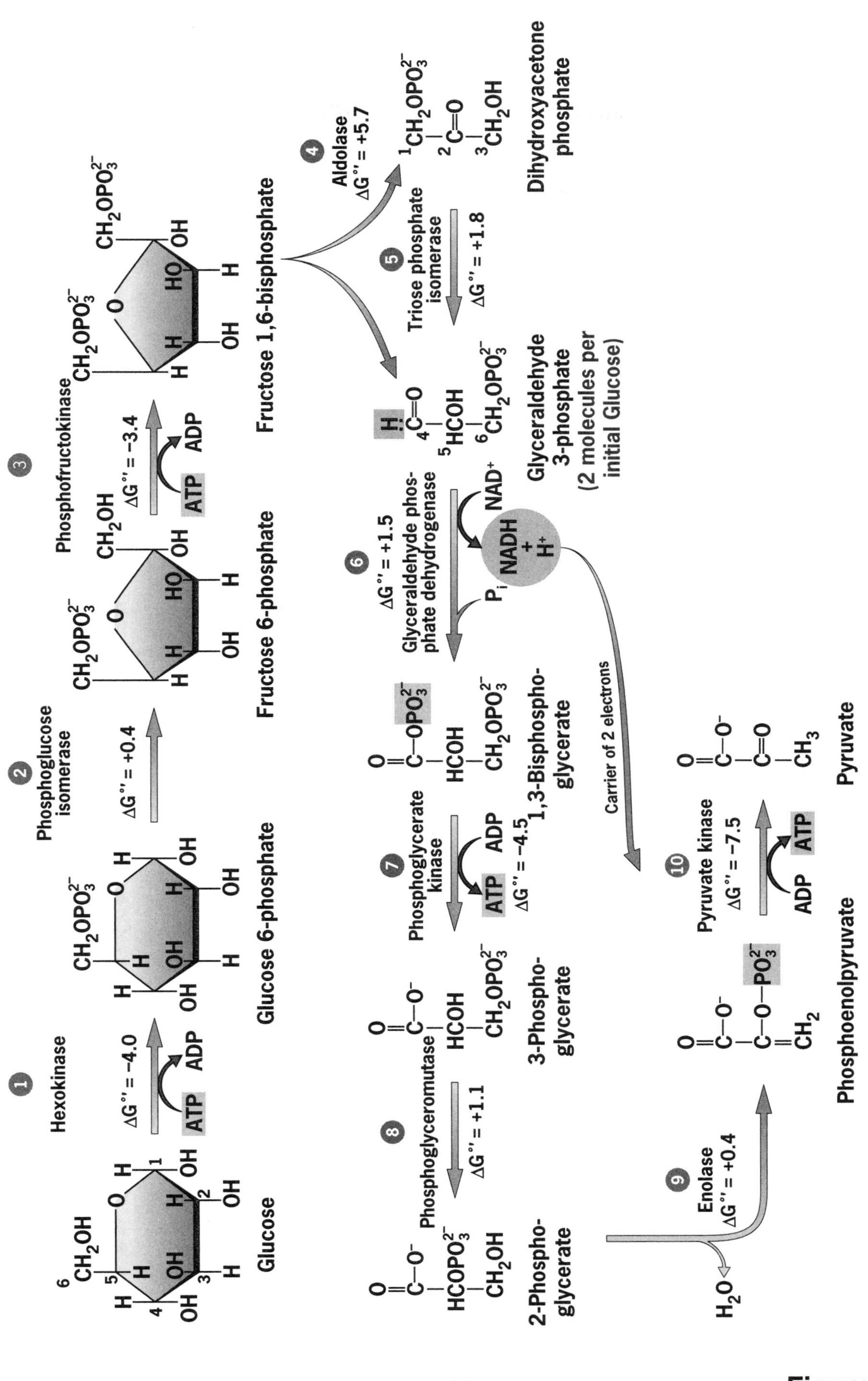

Figure 3.23 The steps of glycolysis.

Copyright © 2005 John Wiley & Sons, Inc.

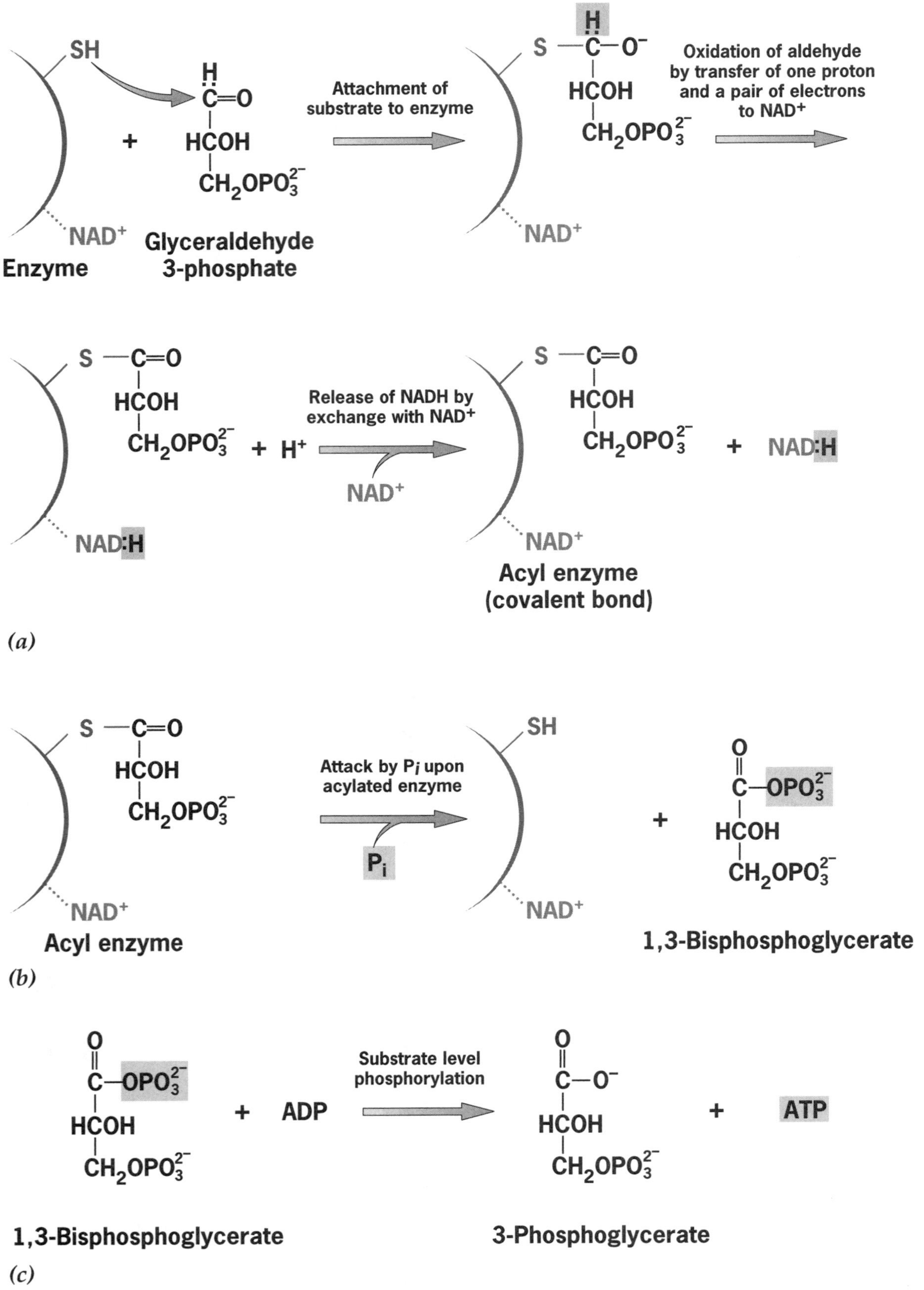

Figure 3.25 The transfer of energy during a chemical oxidation.

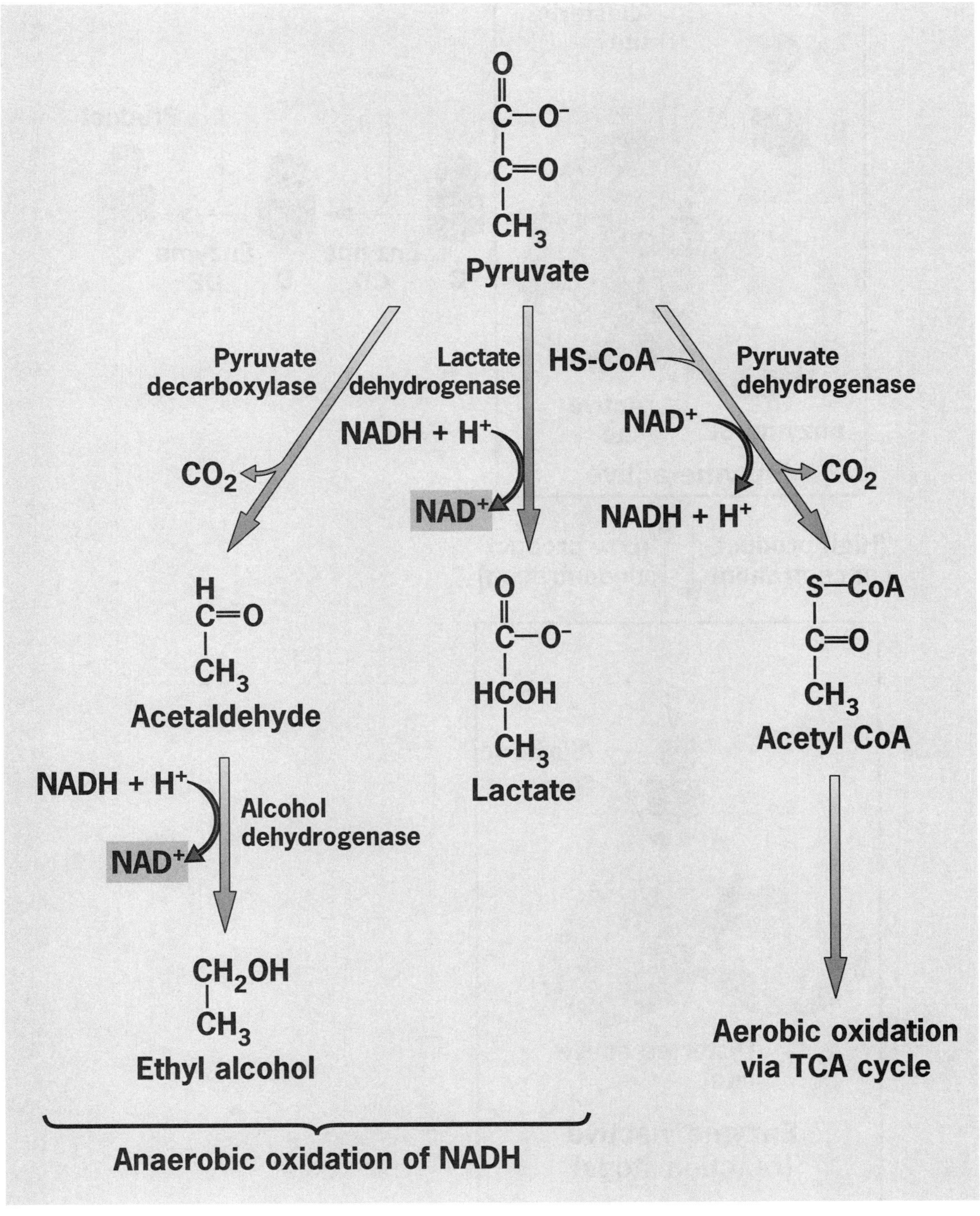

Figure 3.28 Fermentation.

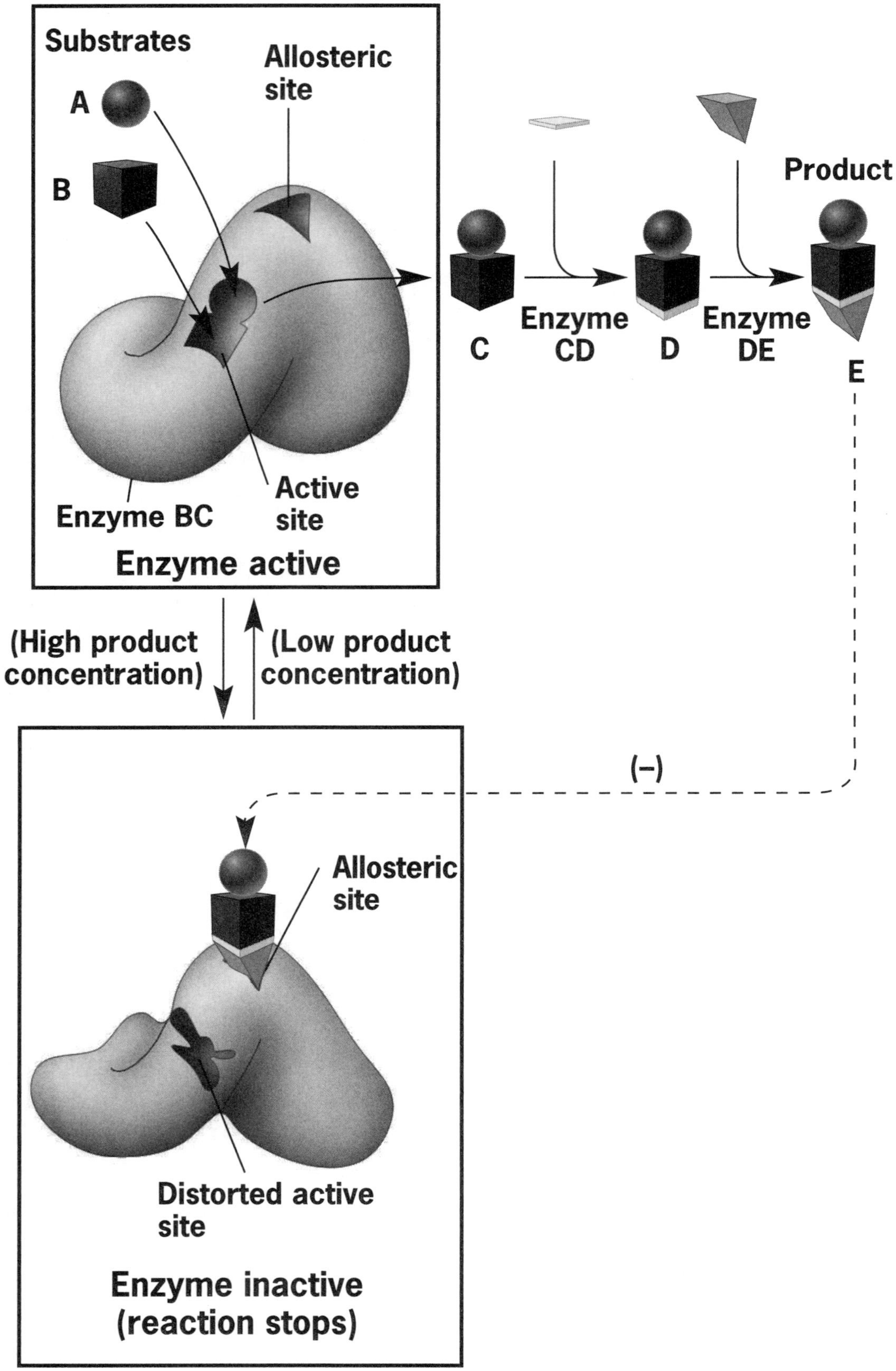

Figure 3.29 Feedback inhibition.

Copyright © 2005 John Wiley & Sons, Inc.

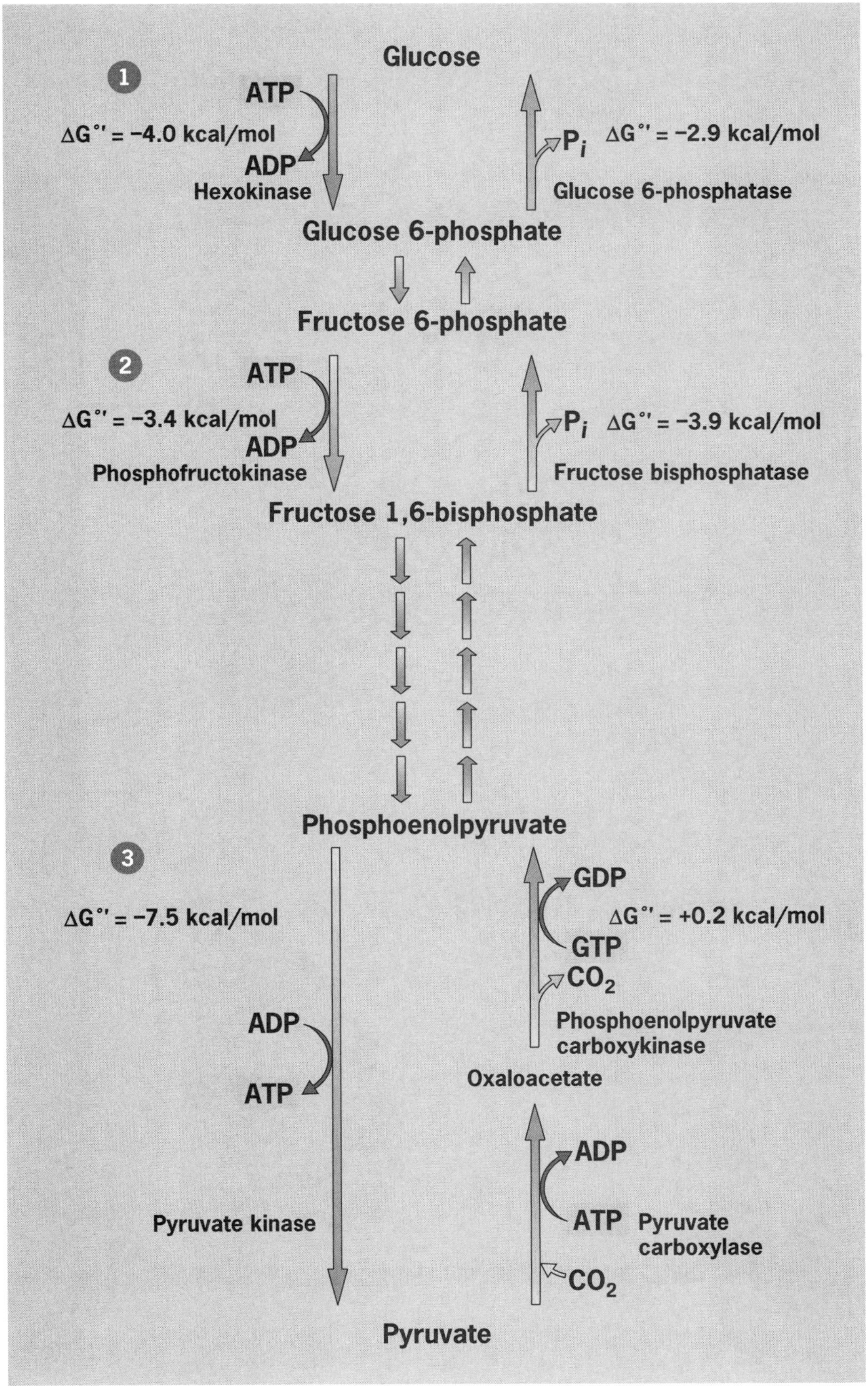

Figure 3.30 Glycolysis versus gluconeogenesis.

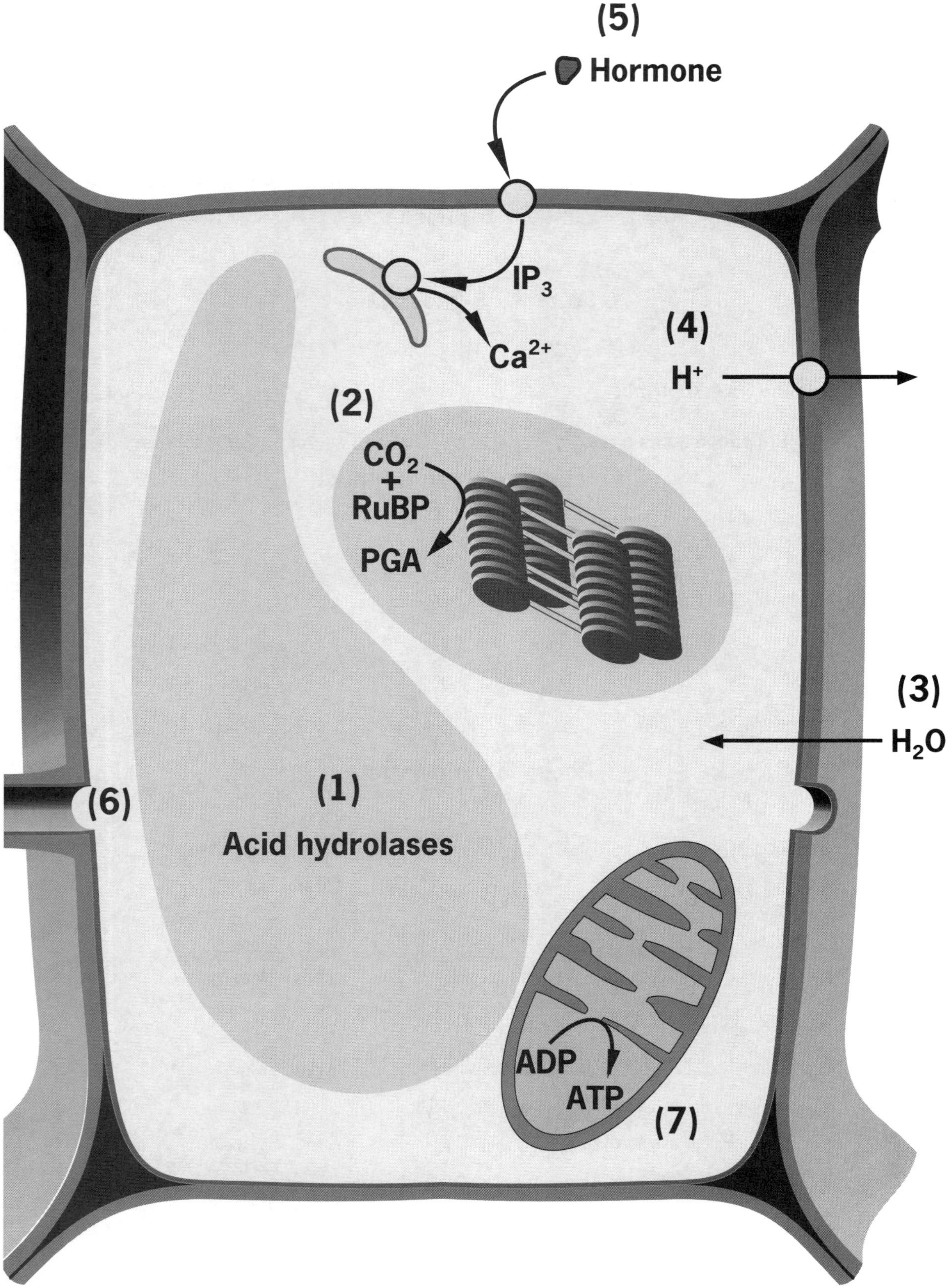

Figure 4.2 A summary of membrane functions in a plant cell.

Copyright © 2005 John Wiley & Sons, Inc.

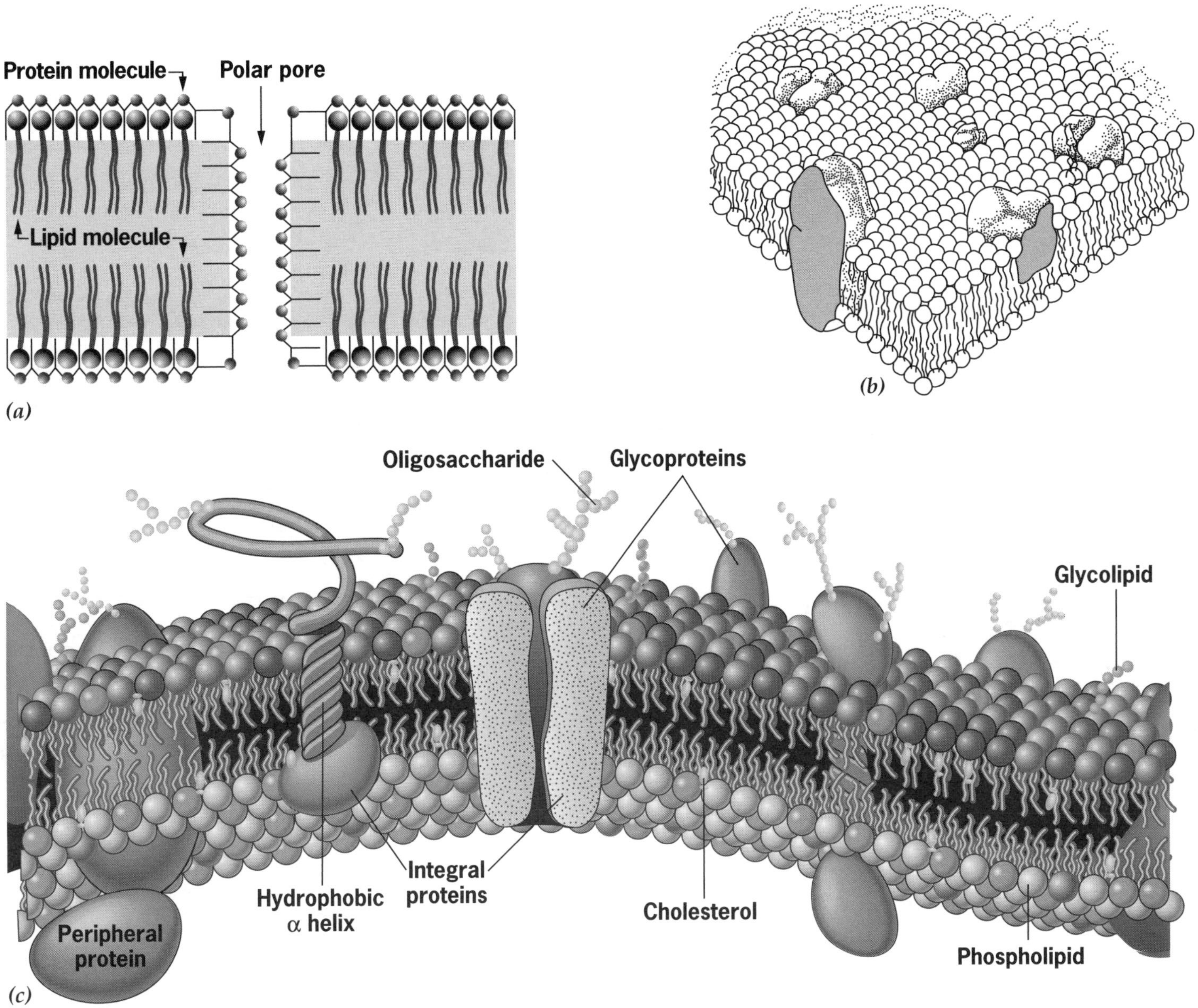

Figure 4.4 A brief history of the structure of the plasma membrane.

Copyright © 2005 John Wiley & Sons, Inc.

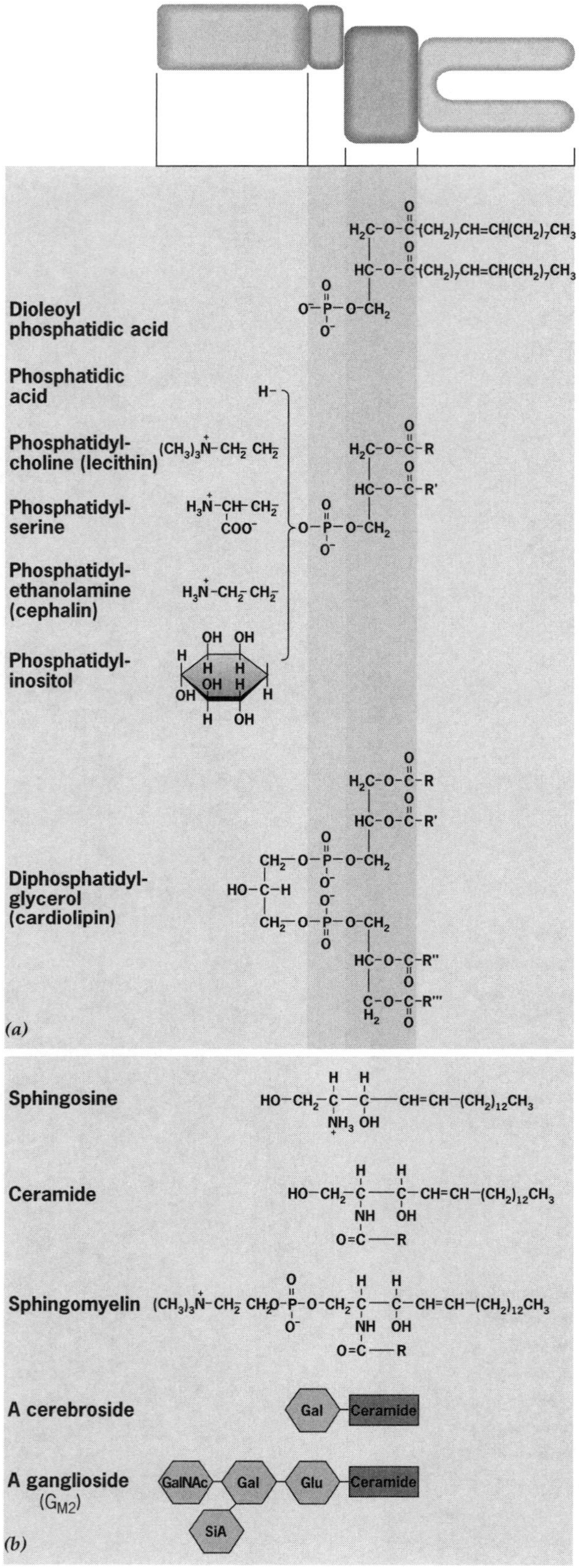

Figure 4.6 The chemical structure of membrane lipids.

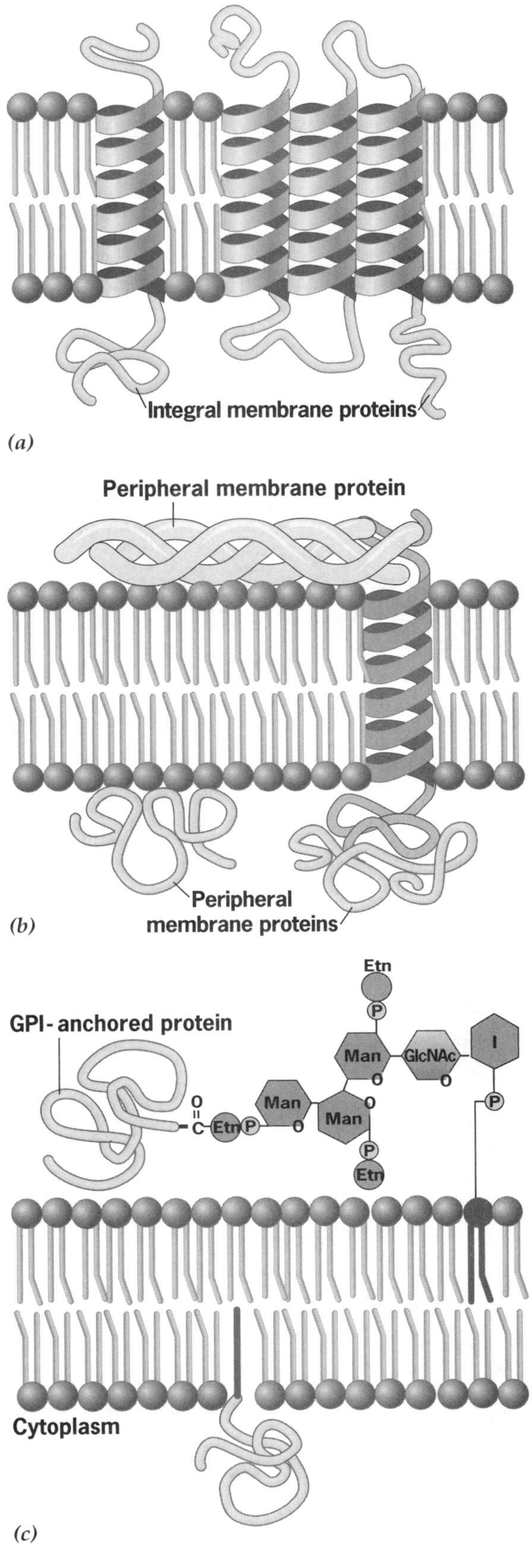

Figure 4.12 Three classes of membrane protein.

Copyright © 2005 John Wiley & Sons, Inc.

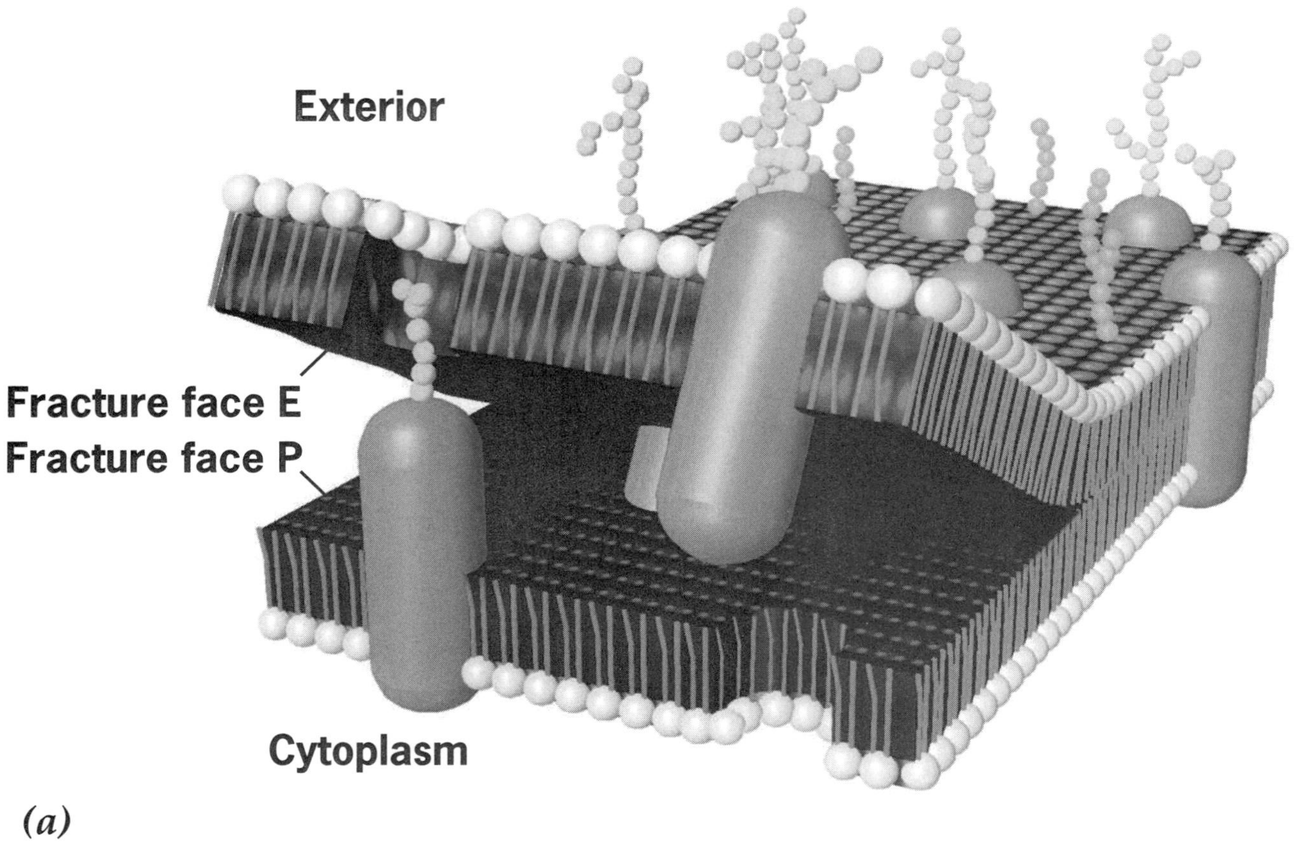

Figure 4.13a Freeze fracture: a technique for investigating cell membrane structure.

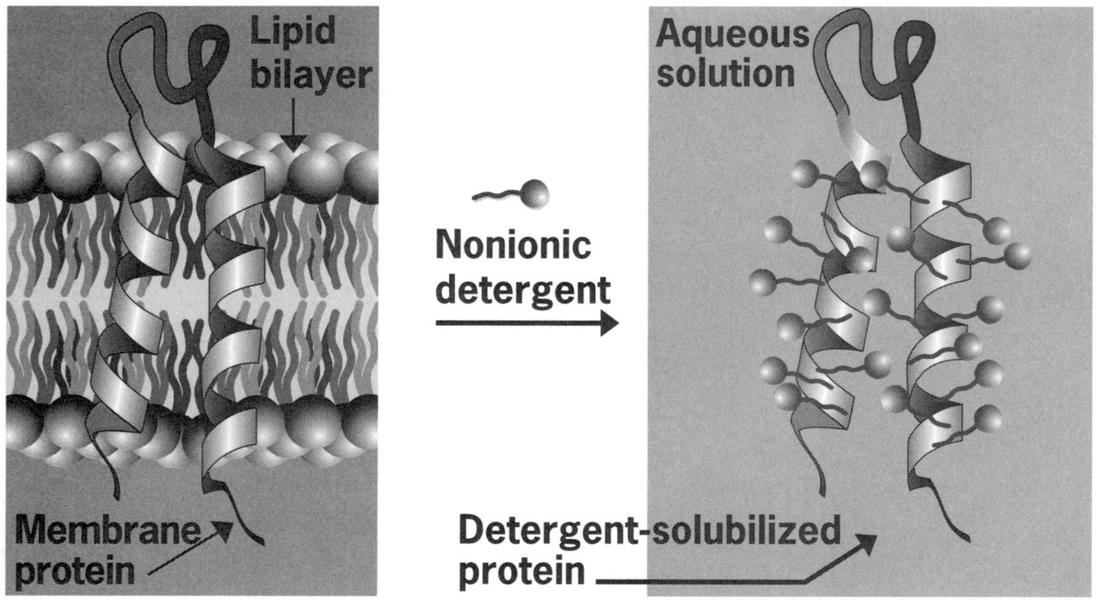

Figure 4.14 Solubilization of membrane proteins with detergents.

Copyright © 2005 John Wiley & Sons, Inc.

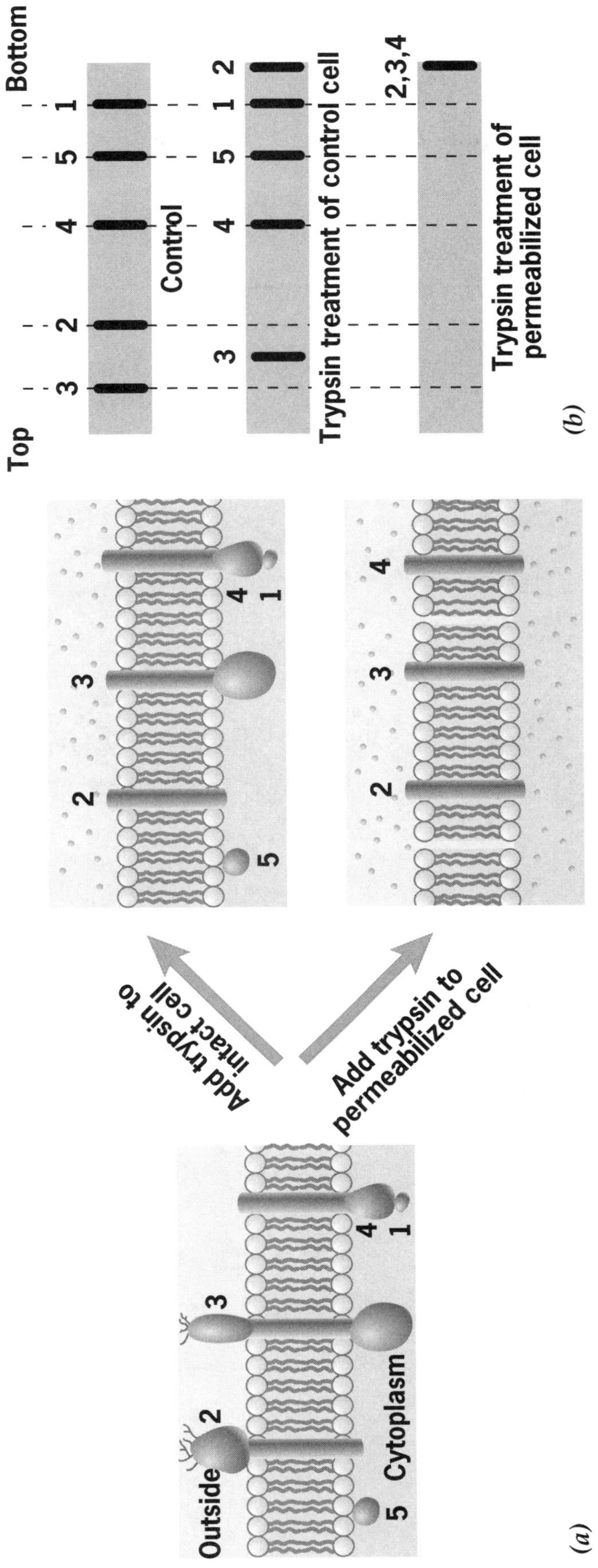

Figure 4.16 Experimental procedure for determining the orientation of proteins within a plasma membrane.

Copyright © 2005 John Wiley & Sons, Inc.

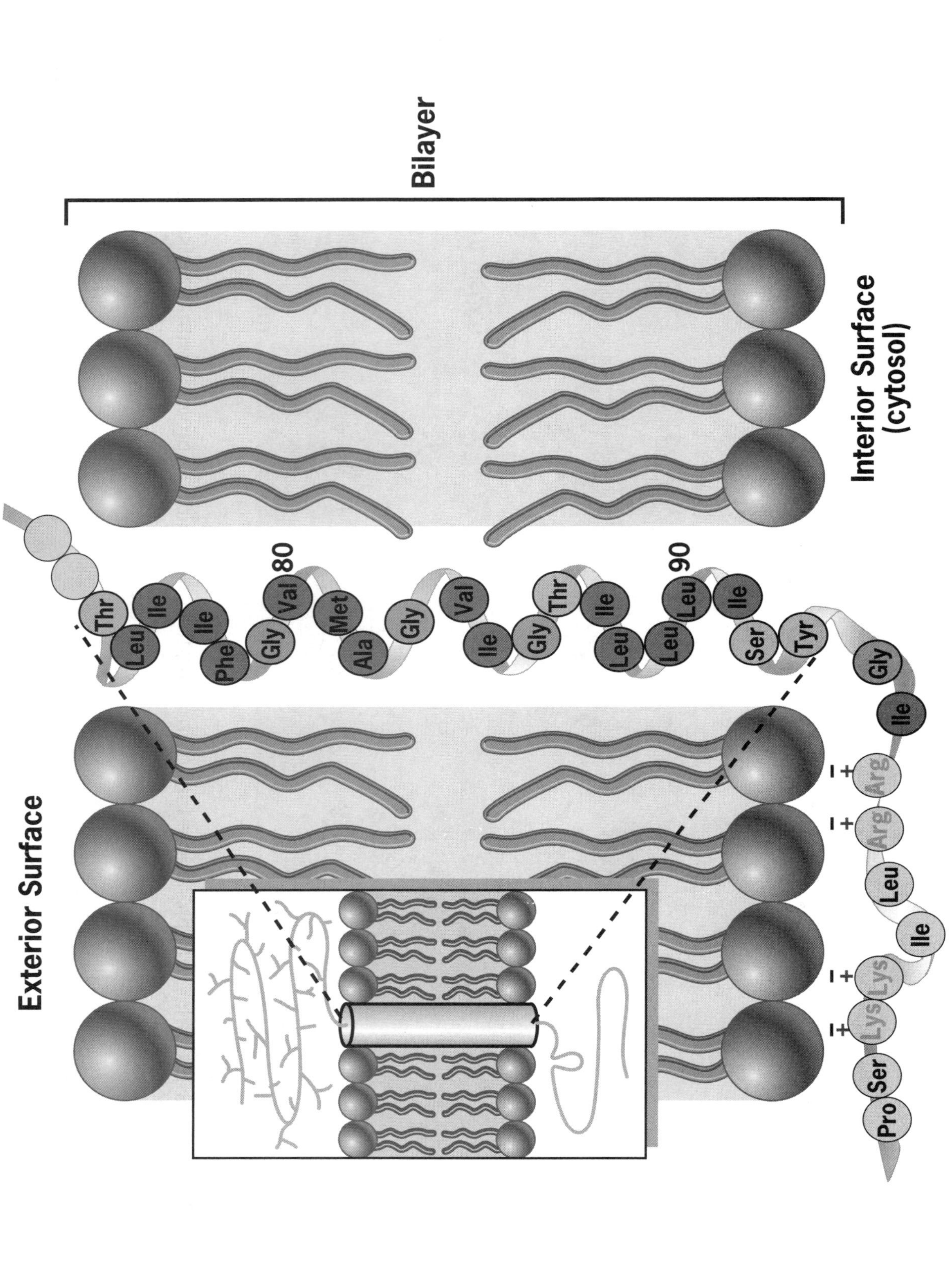

Figure 4.17 Glycophorin A, an integral protein with a single transmembrane domain.

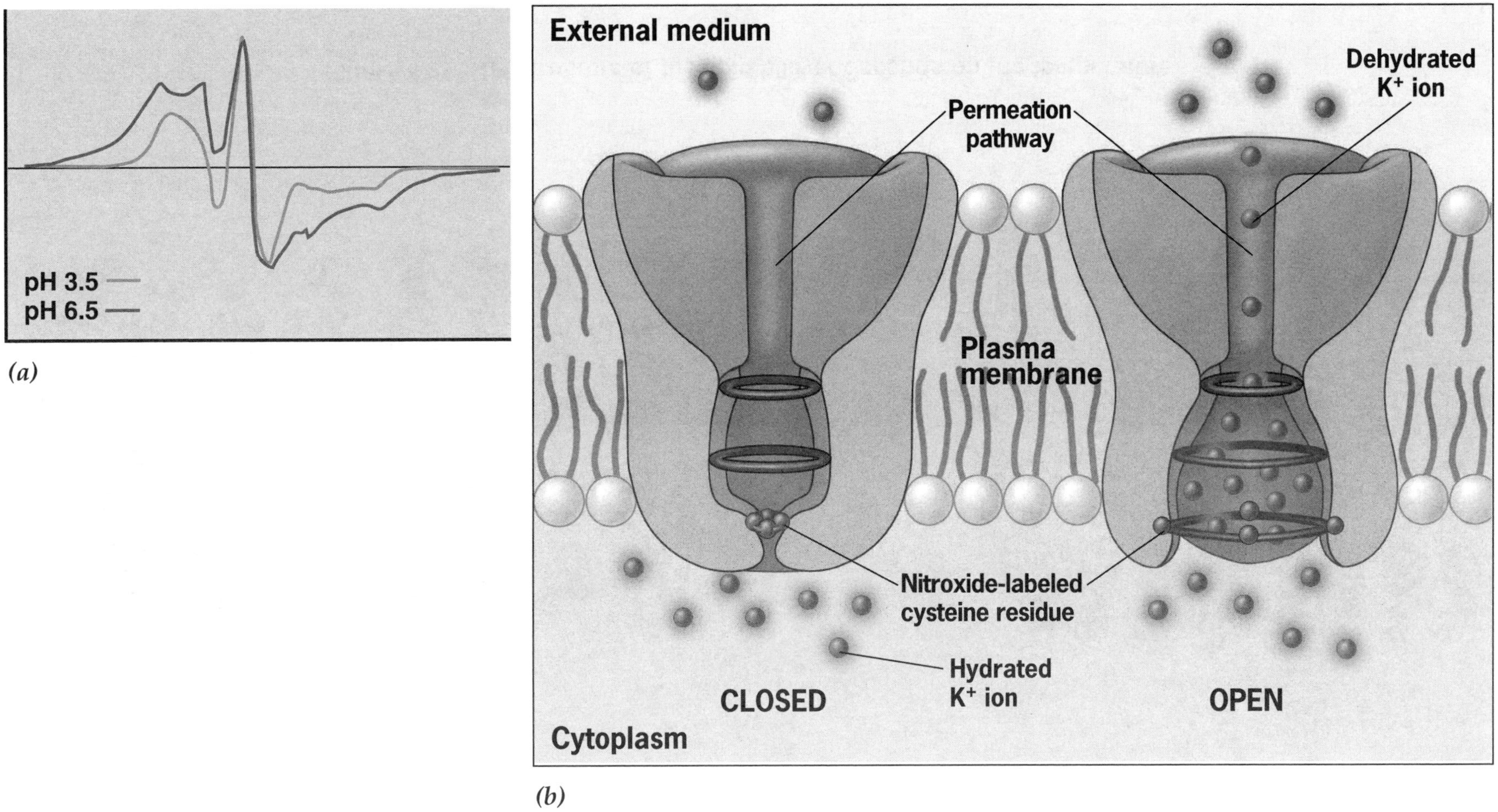

Figure 4.20 Use of EPR spectroscopy to monitor changes in conformation of a bacterial K^+ ion channel as it opens and closes.

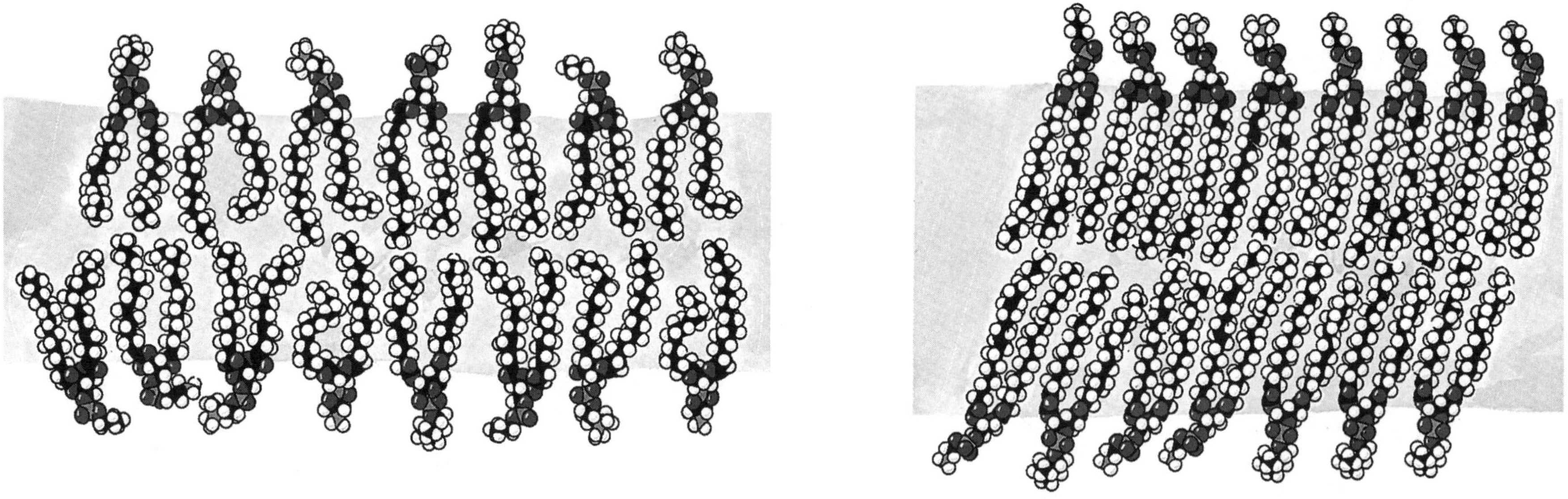

(a) *(b)*

Figure 4.21 The structure of the lipid bilayer depends on the temperature.

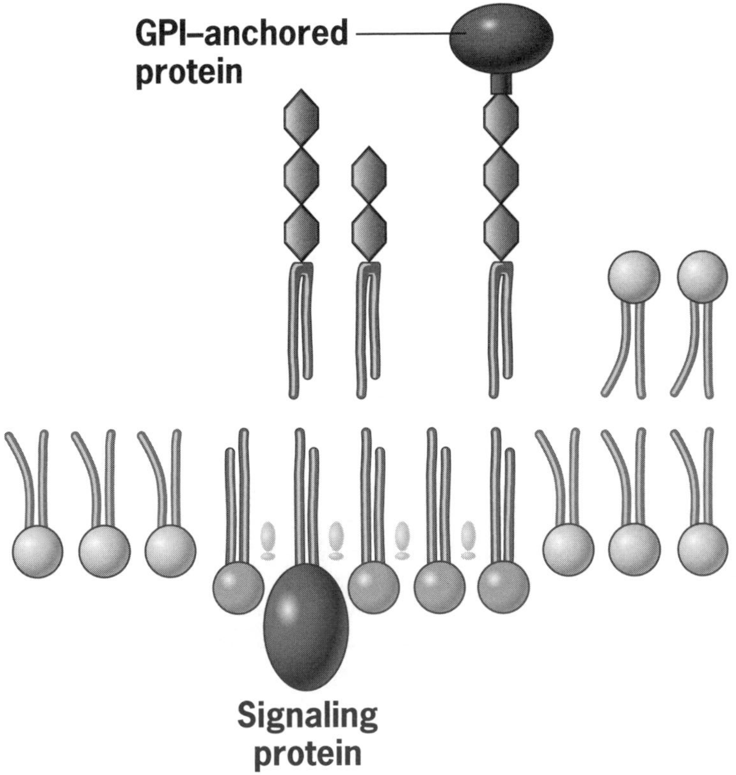

Figure 4.23b Lipid rafts.

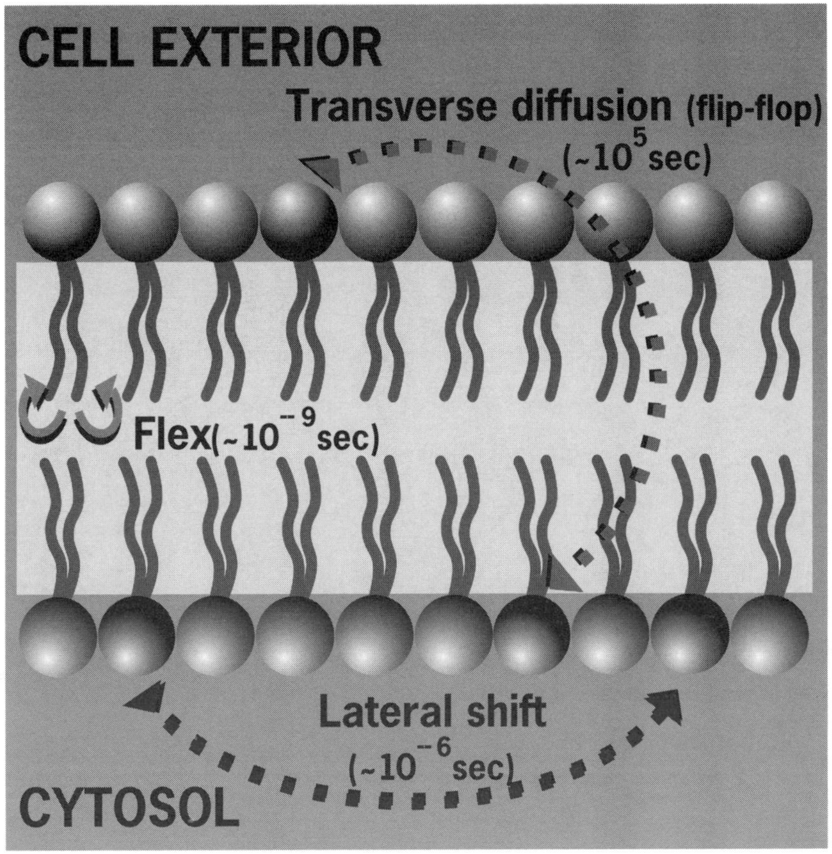

Figure 4.24 The possible movements of phospholipids in a membrane.

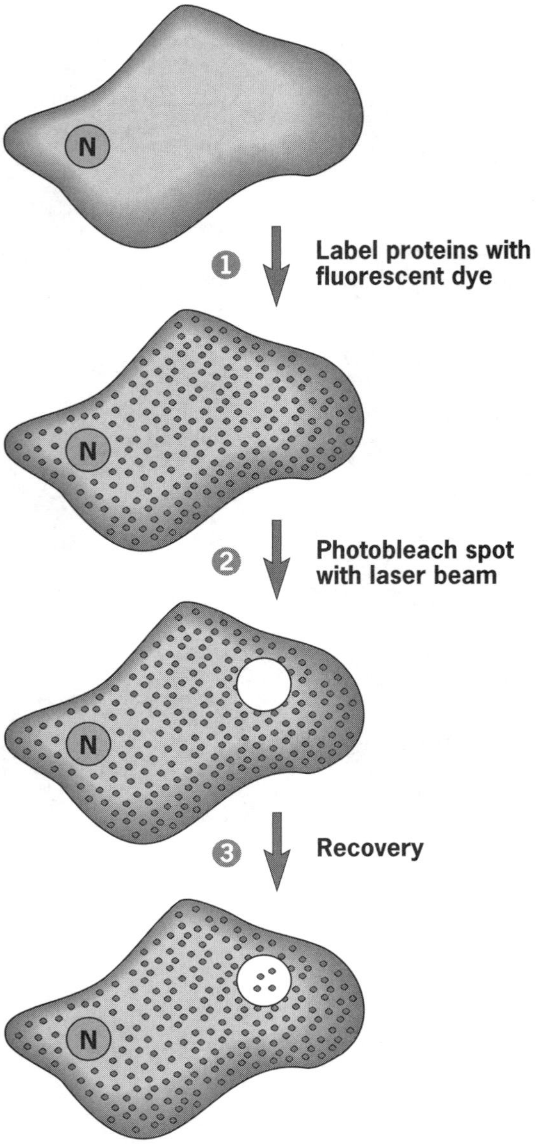

(a)

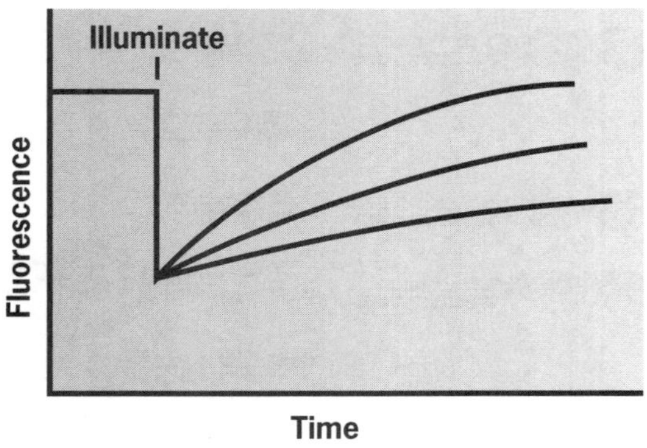

(b)

Figure 4.26 Measuring the diffusion rates of membrane proteins by fluorescence recovery after photobleaching (FRAP).

Copyright © 2005 John Wiley & Sons, Inc.

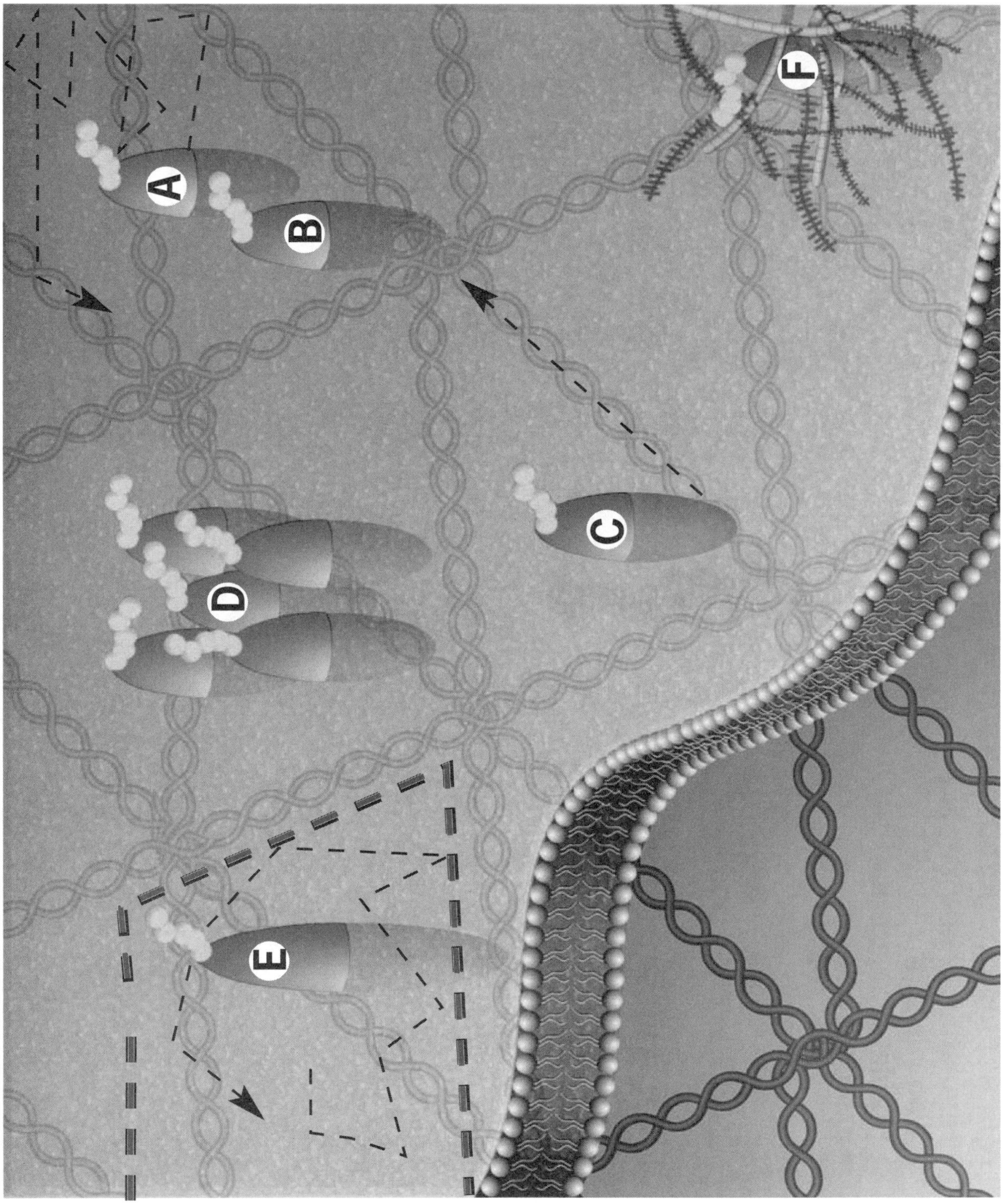

Figure 4.27 Patterns of movement of integral membrane proteins.

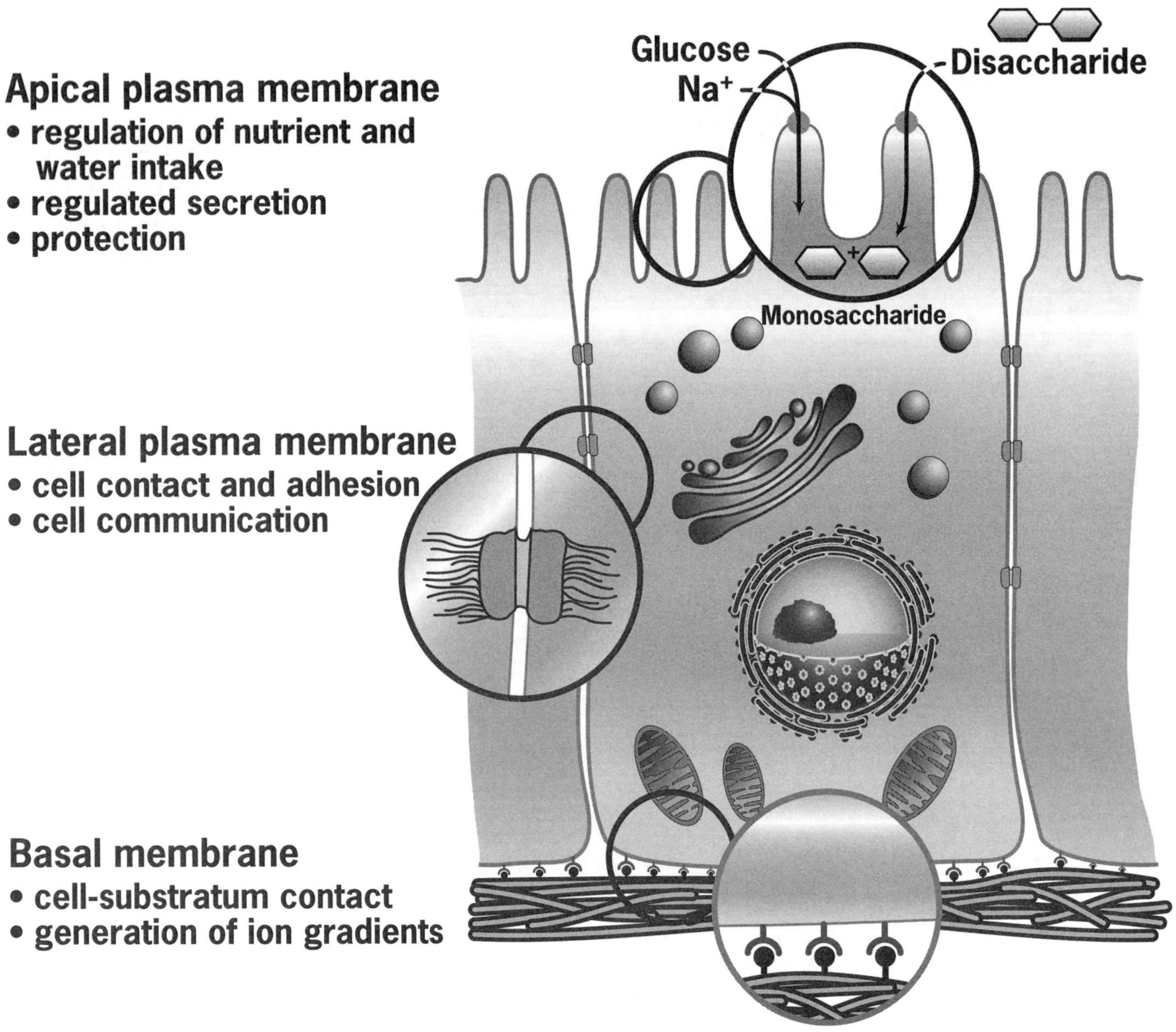

Figure 4.29 Differentiated functions of the plasma membrane of an epithelial cell.

Copyright © 2005 John Wiley & Sons, Inc.

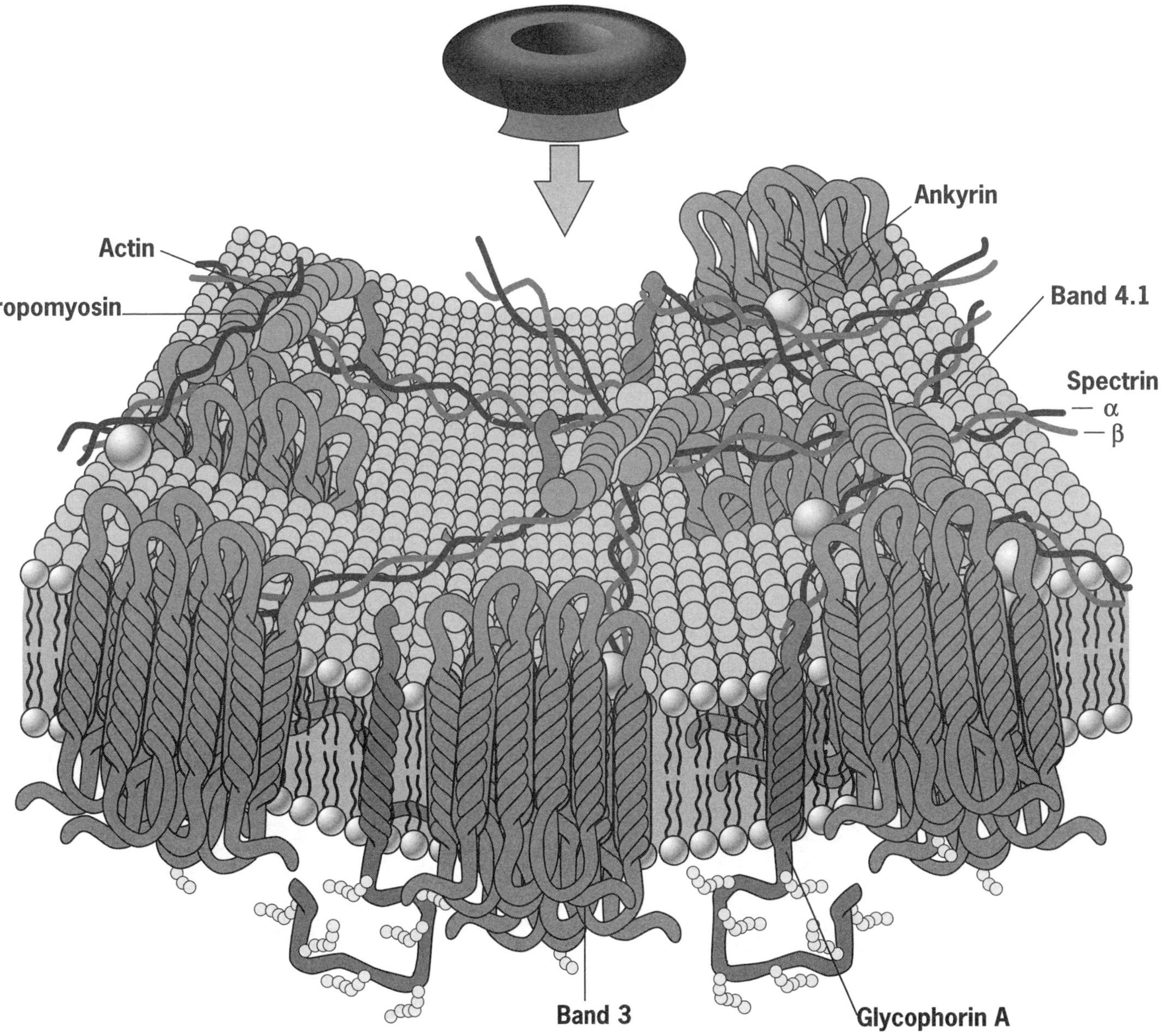

Figure 4.31d The plasma membrane of the human erythrocyte.

Copyright © 2005 John Wiley & Sons, Inc.

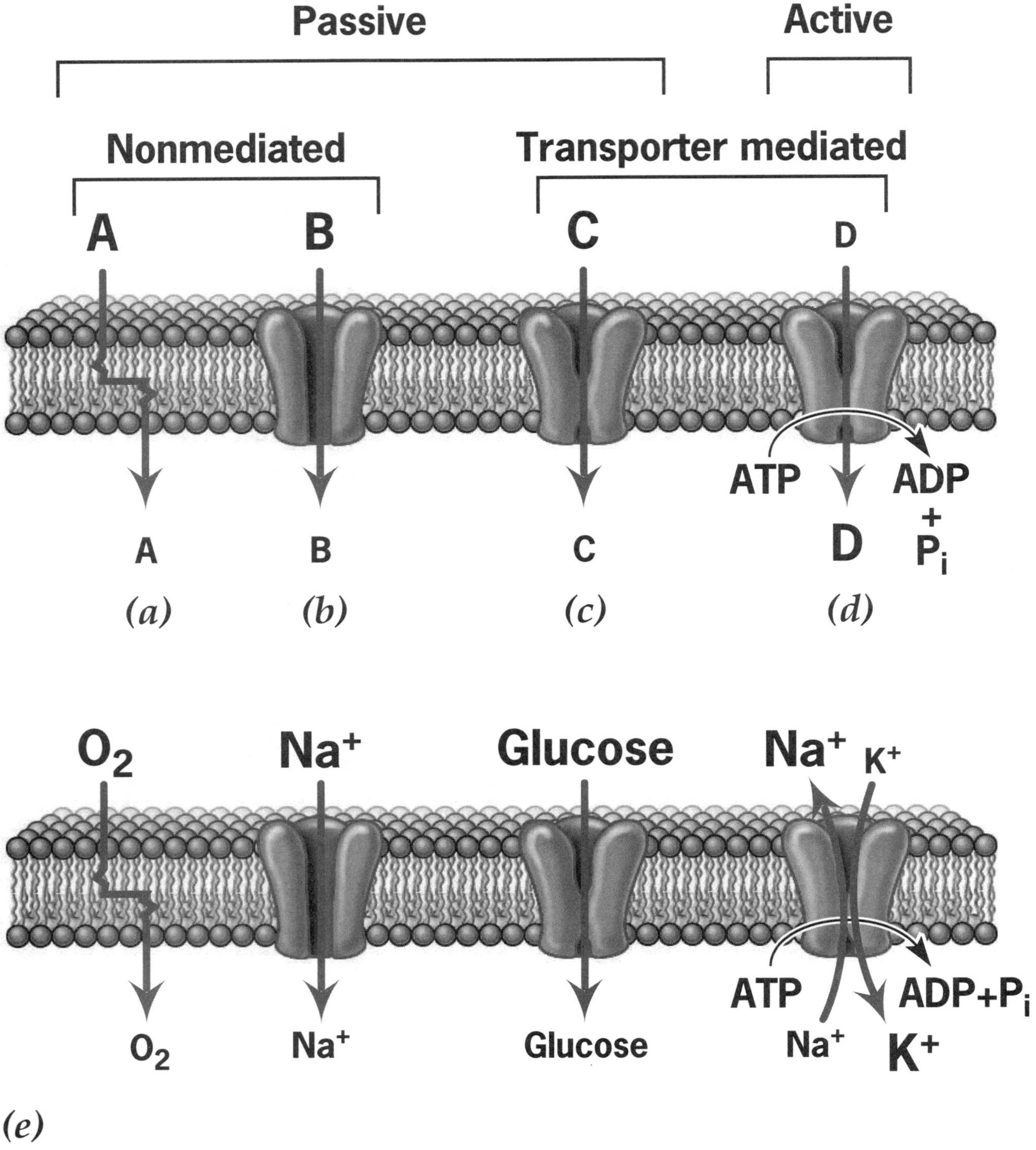

Figure 4.32 Four basic mechanisms by which solute molecules move across membranes.

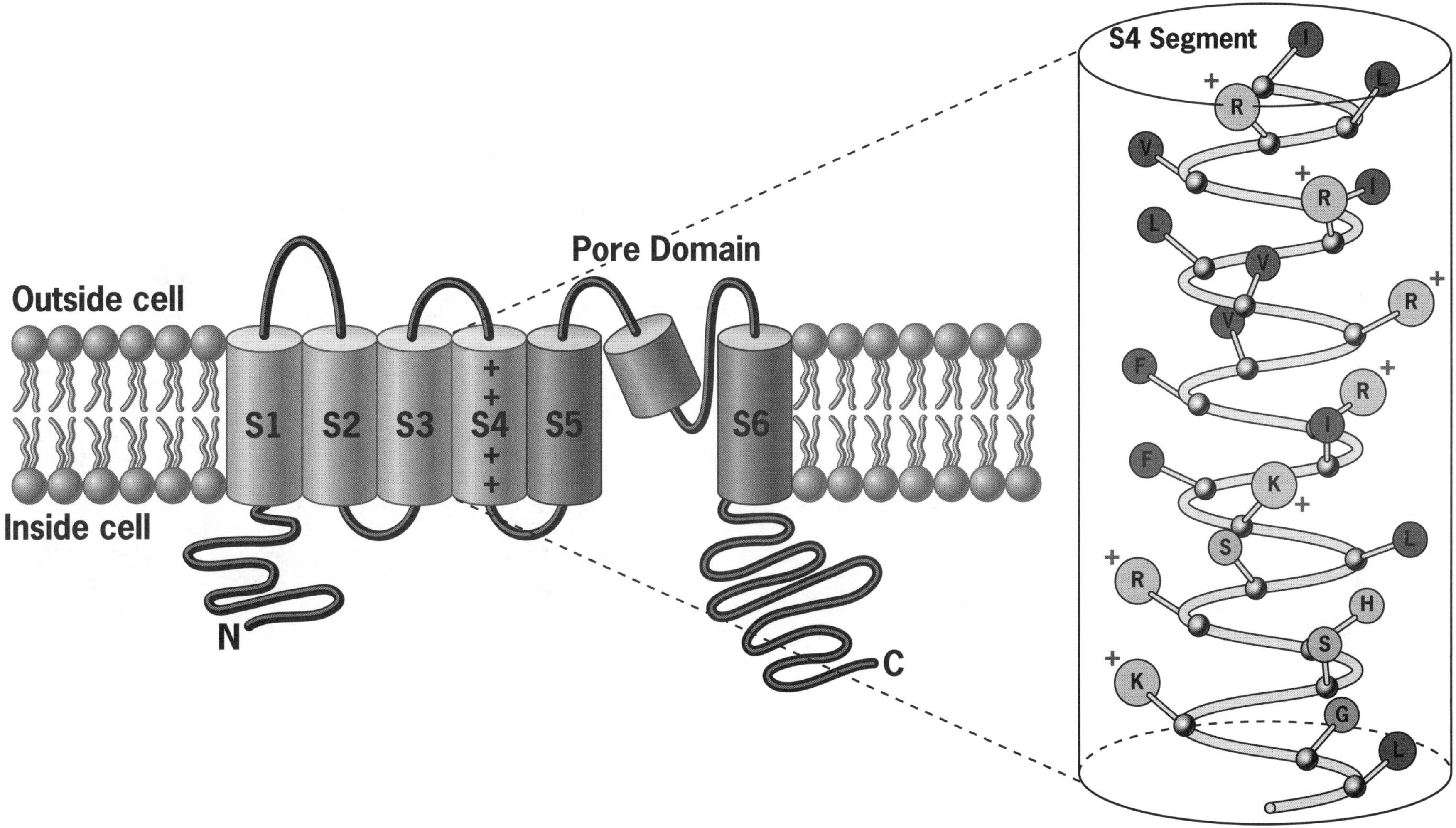

Figure 4.39 The structure of a eukaryotic, voltage-gated K^+ channel.

Copyright © 2005 John Wiley & Sons, Inc.

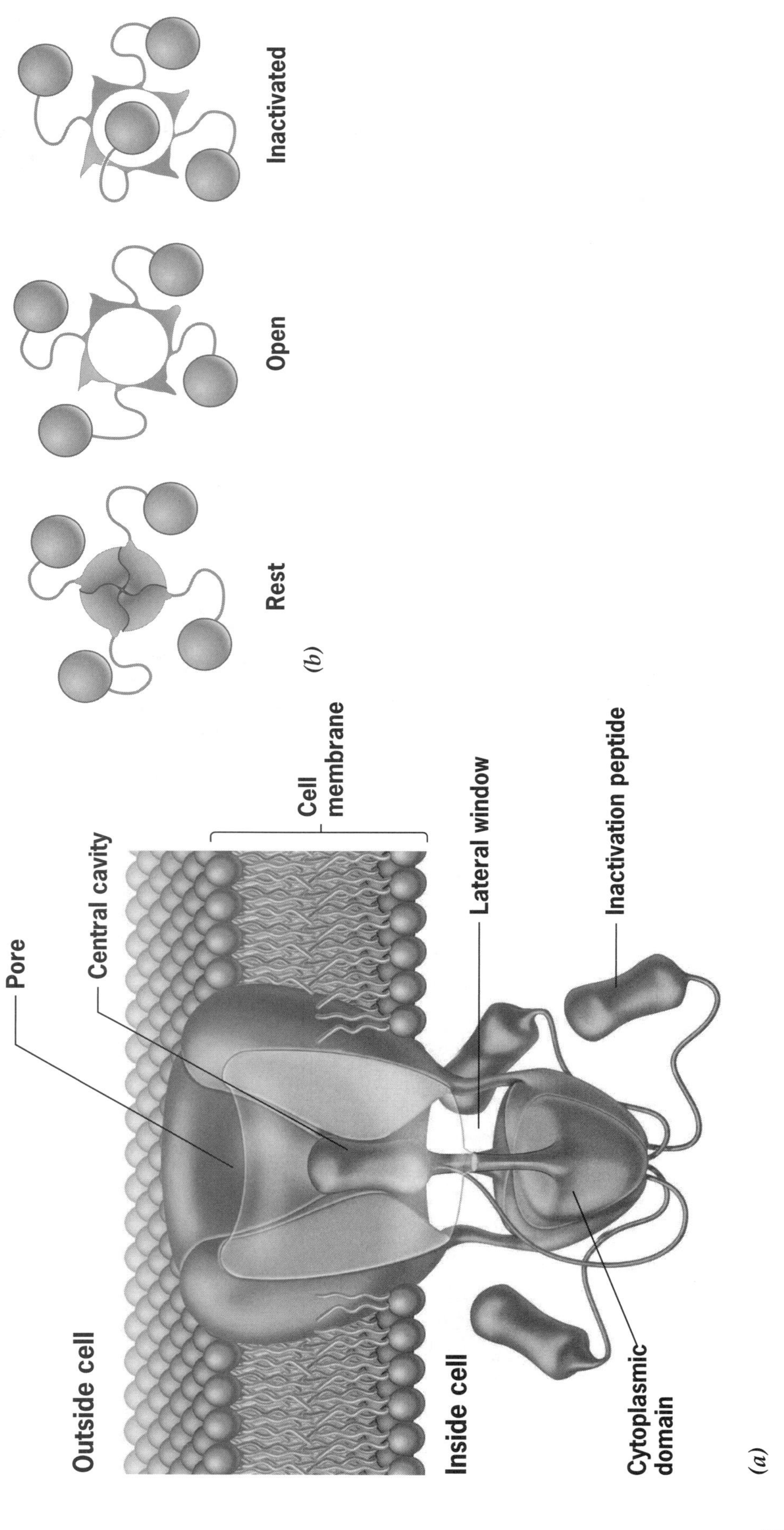

Figure 4.42 Conformational states of a voltage-gated K^+ ion channel.

Copyright © 2005 John Wiley & Sons, Inc.

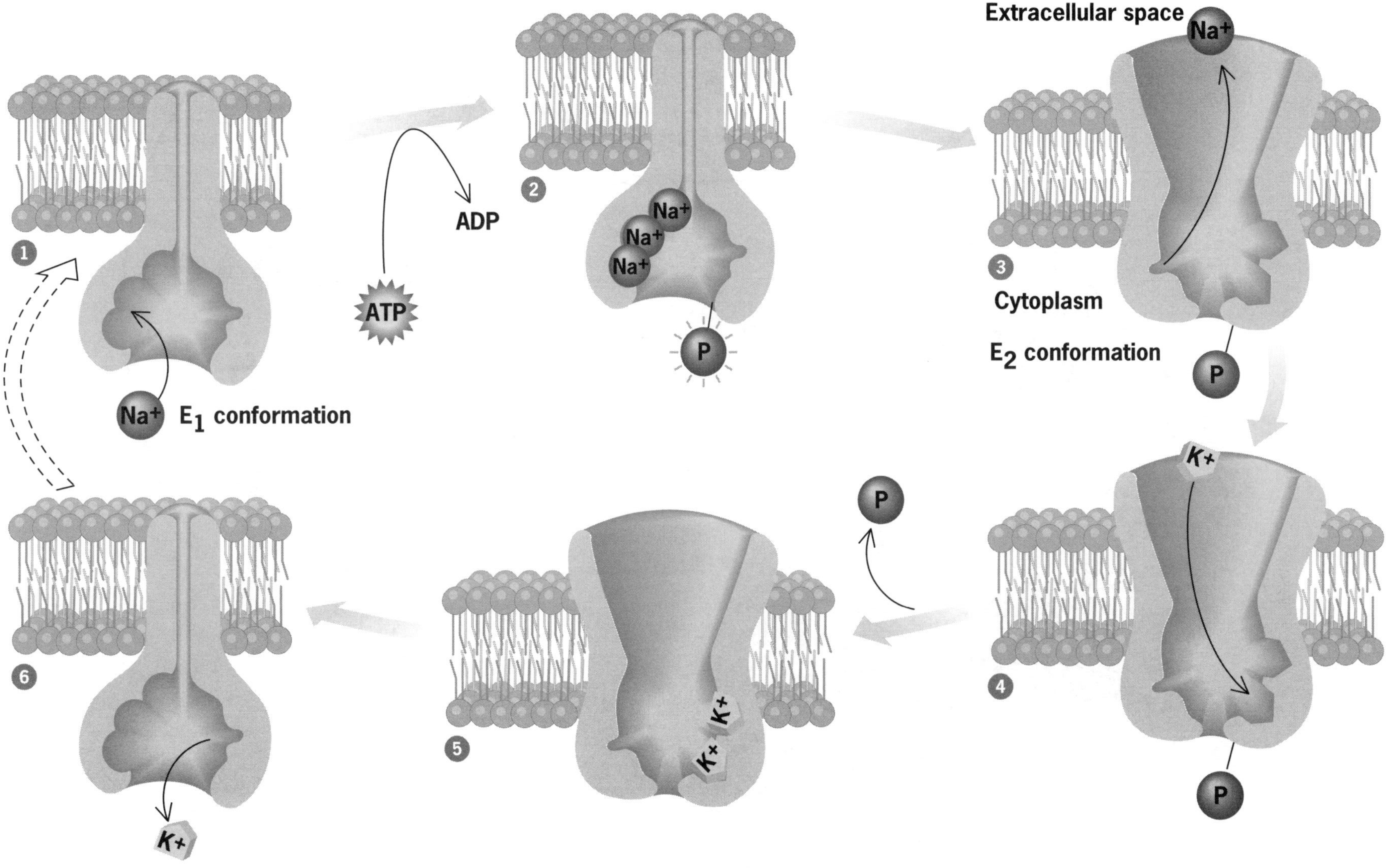

Figure 4.45 Simplified schematic model of the Na^+/K^+-ATPase transport cycle.

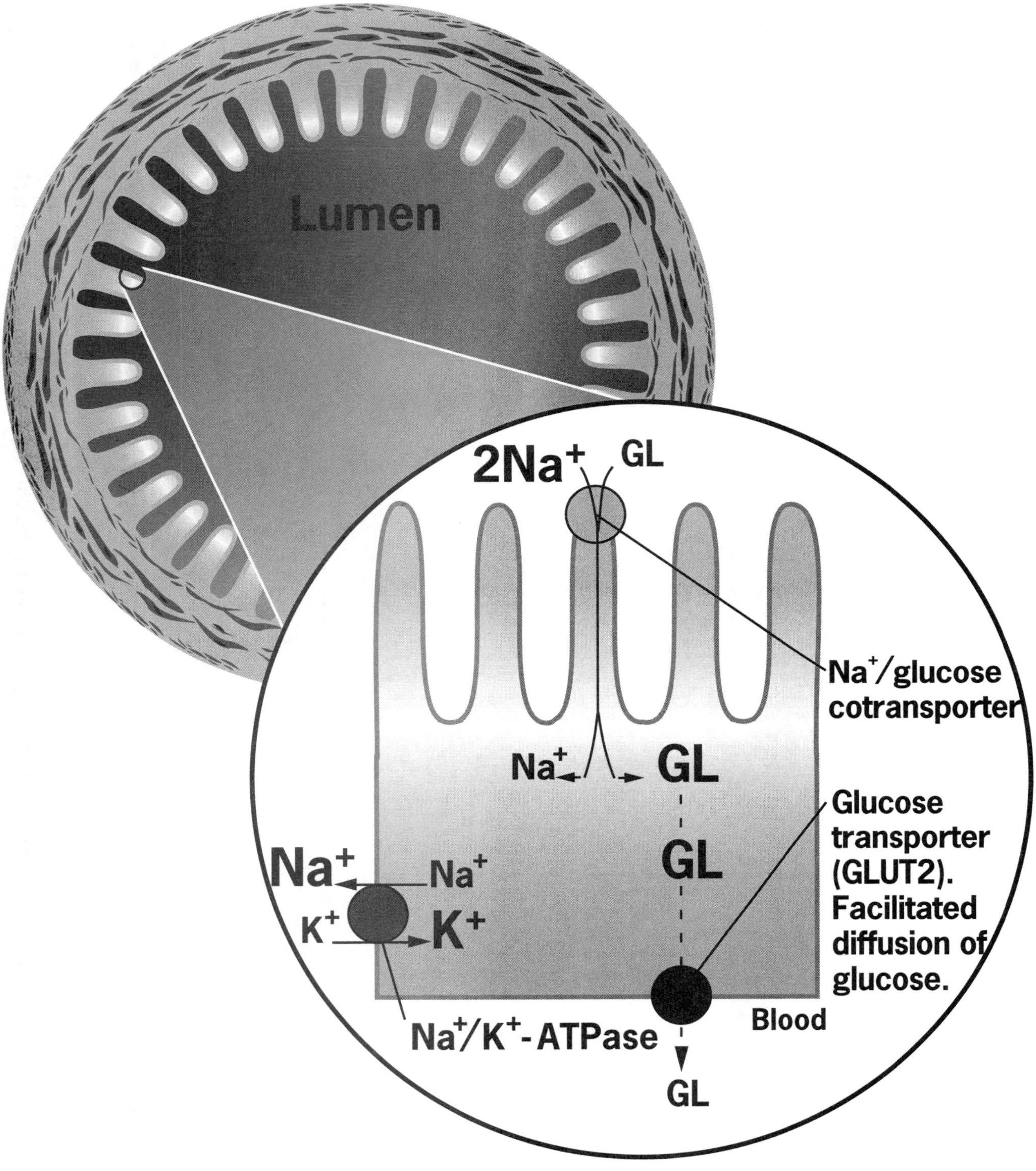

Figure 4.48 Secondary transport: the use of energy stored in an ionic gradient.

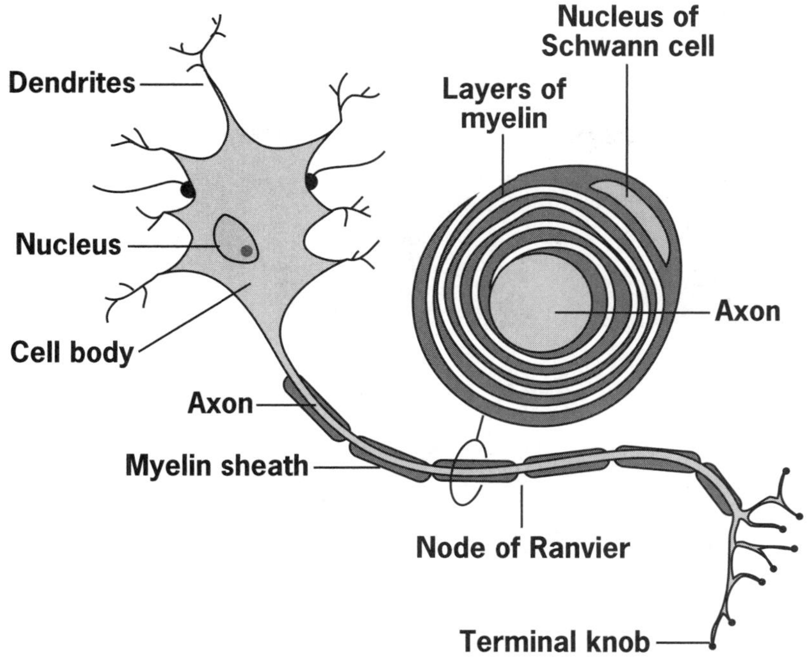

Figure 4.49 The structure of a nerve cell.

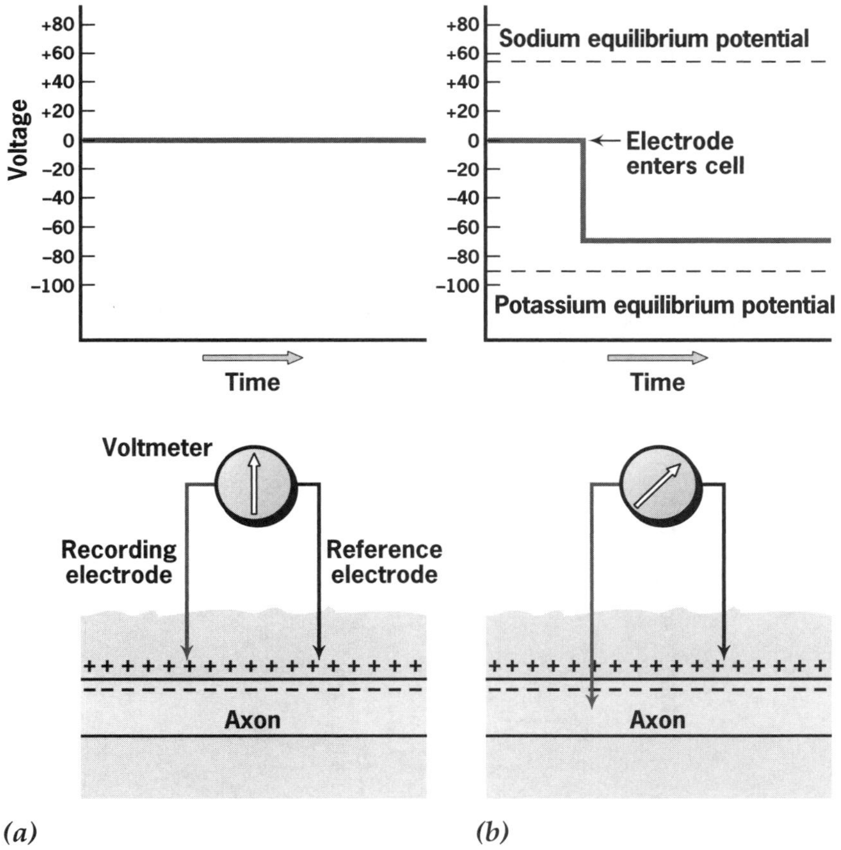

Figure 4.50 Measuring a membrane's resting potential.

Copyright © 2005 John Wiley & Sons, Inc.

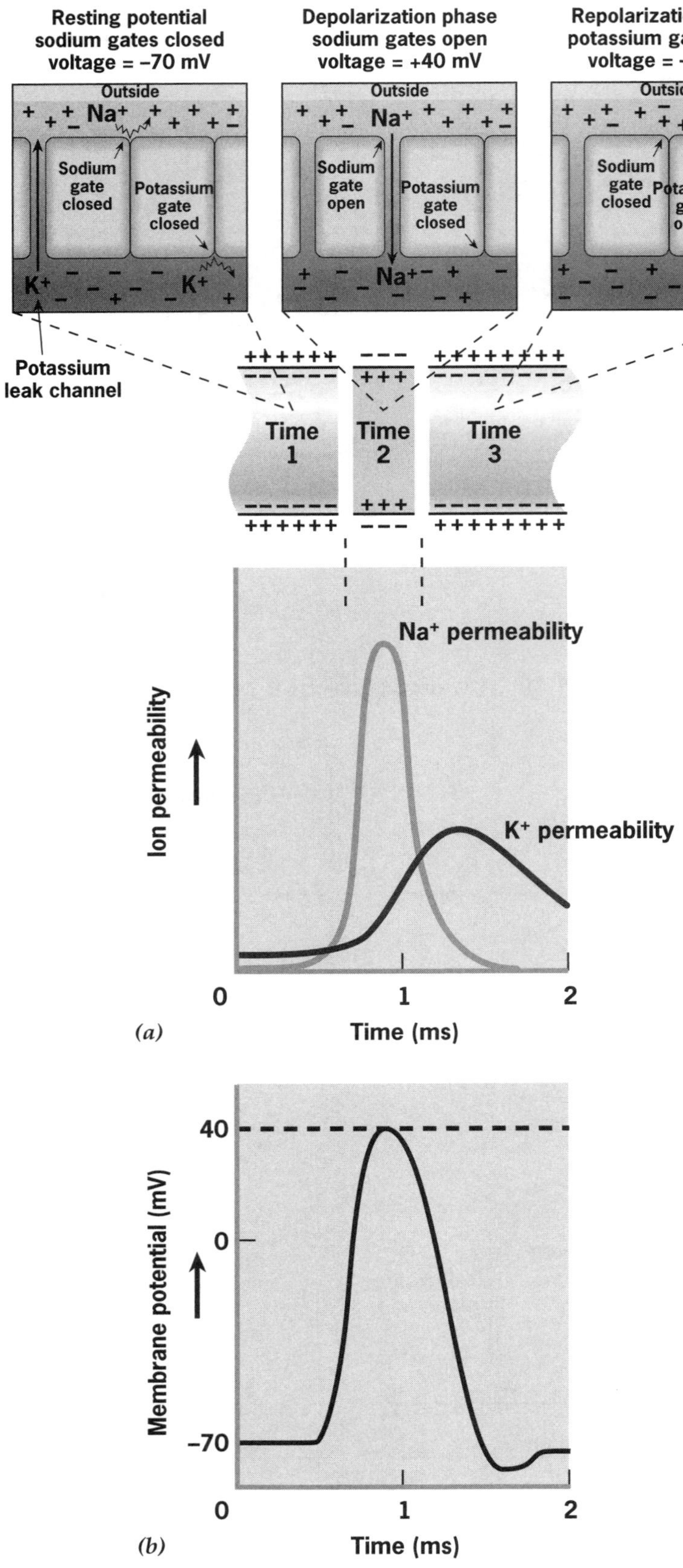

Figure 4.51 Formation of an action potential.

Copyright © 2005 John Wiley & Sons, Inc.

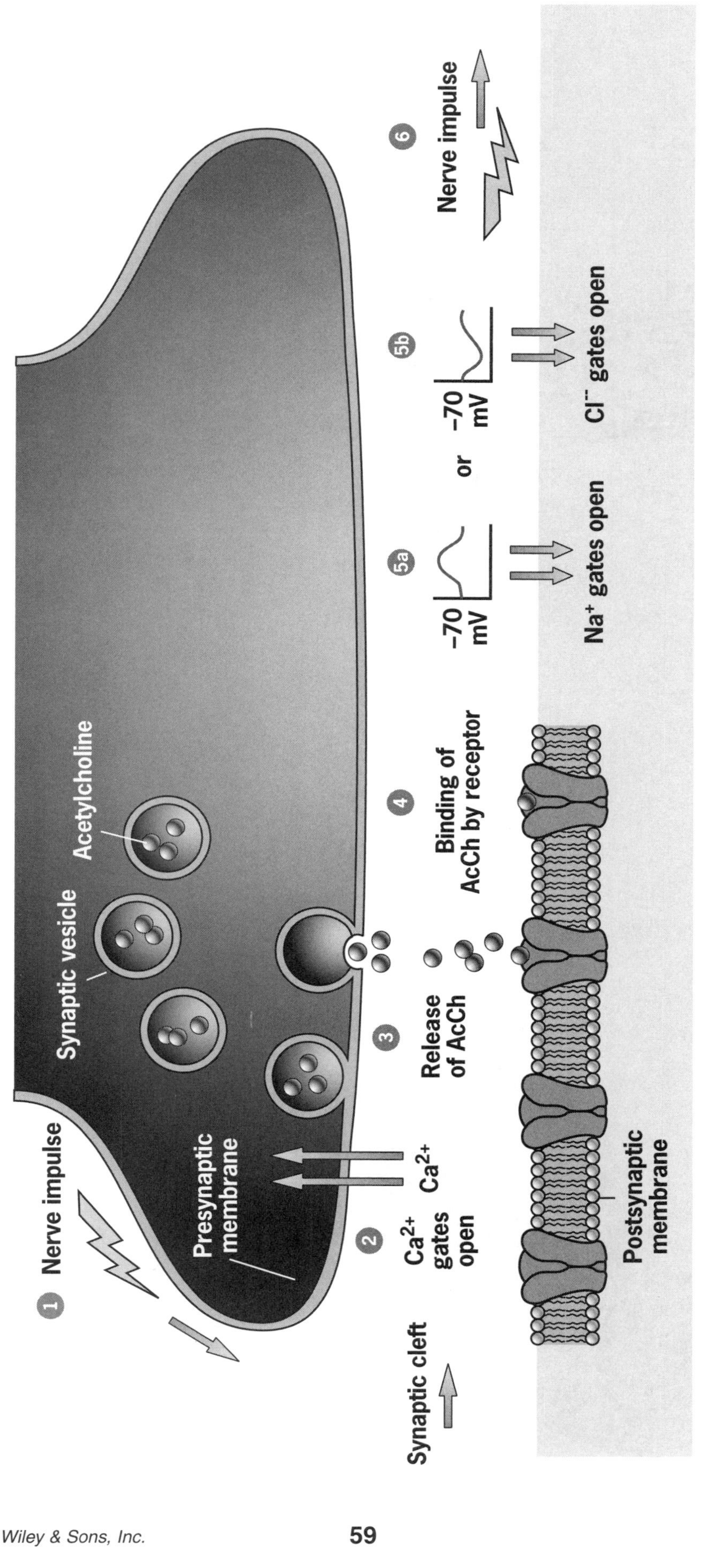

Figure 4.55 The sequence of events during synaptic transmission with acetylcholine as the neurotransmitter.

Copyright © 2005 John Wiley & Sons, Inc.

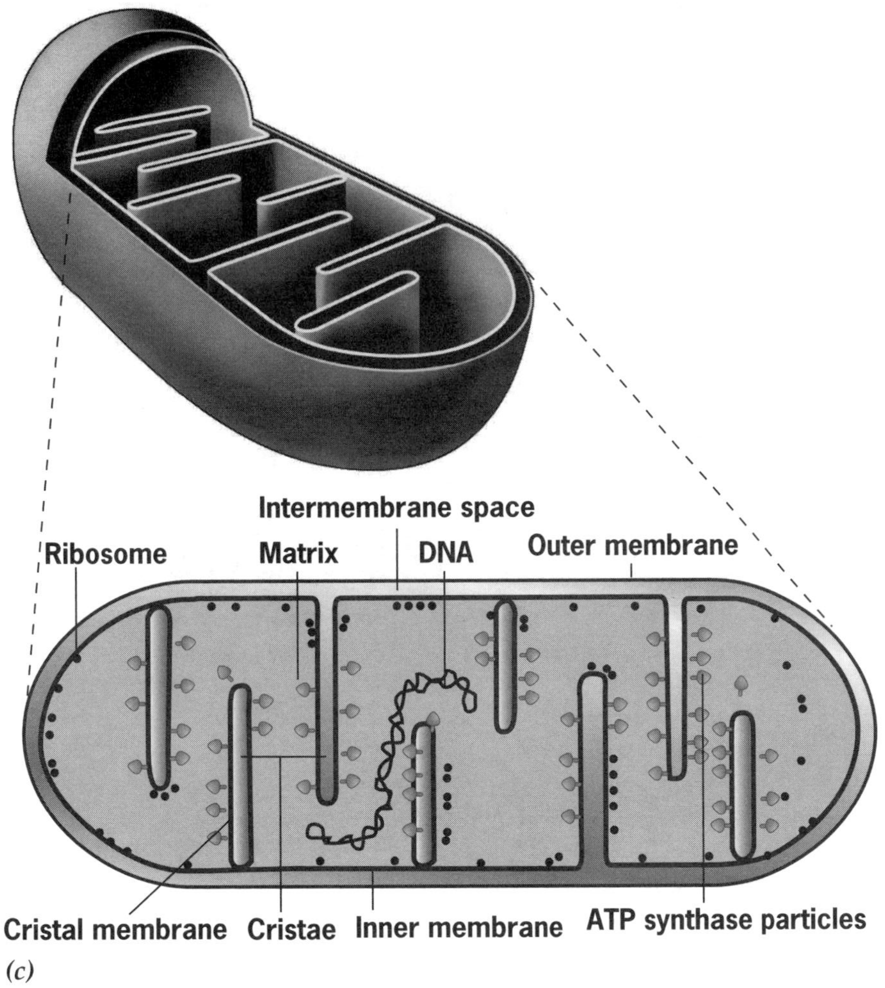

Figure 5.3c The structure of a mitochondrion.

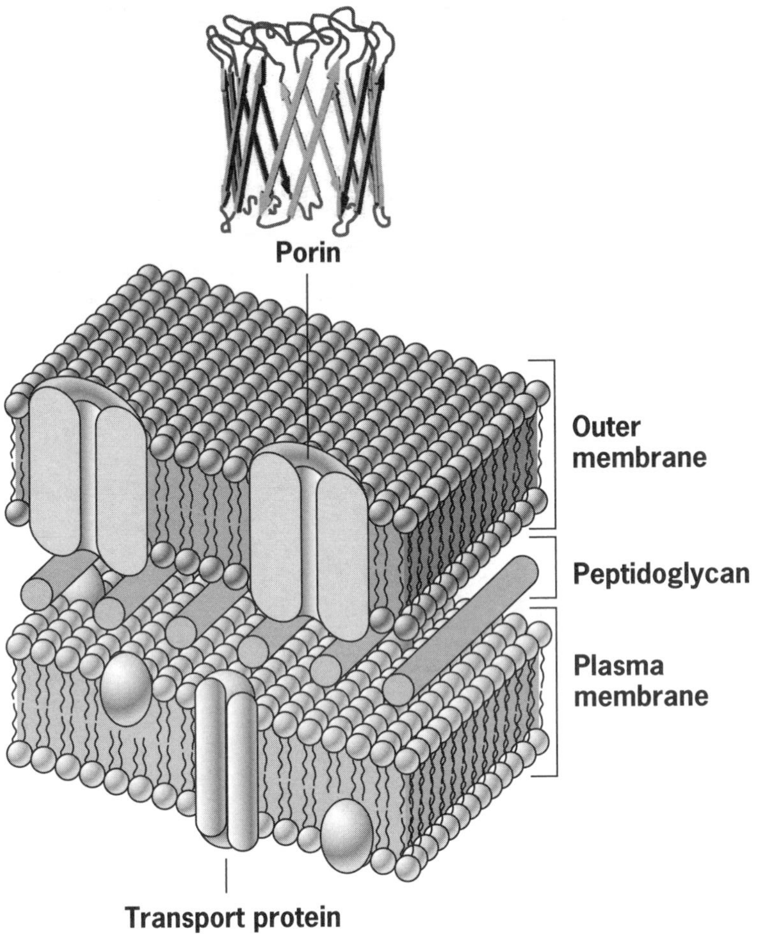

Figure 5.4 Porins.

Copyright © 2005 John Wiley & Sons, Inc.

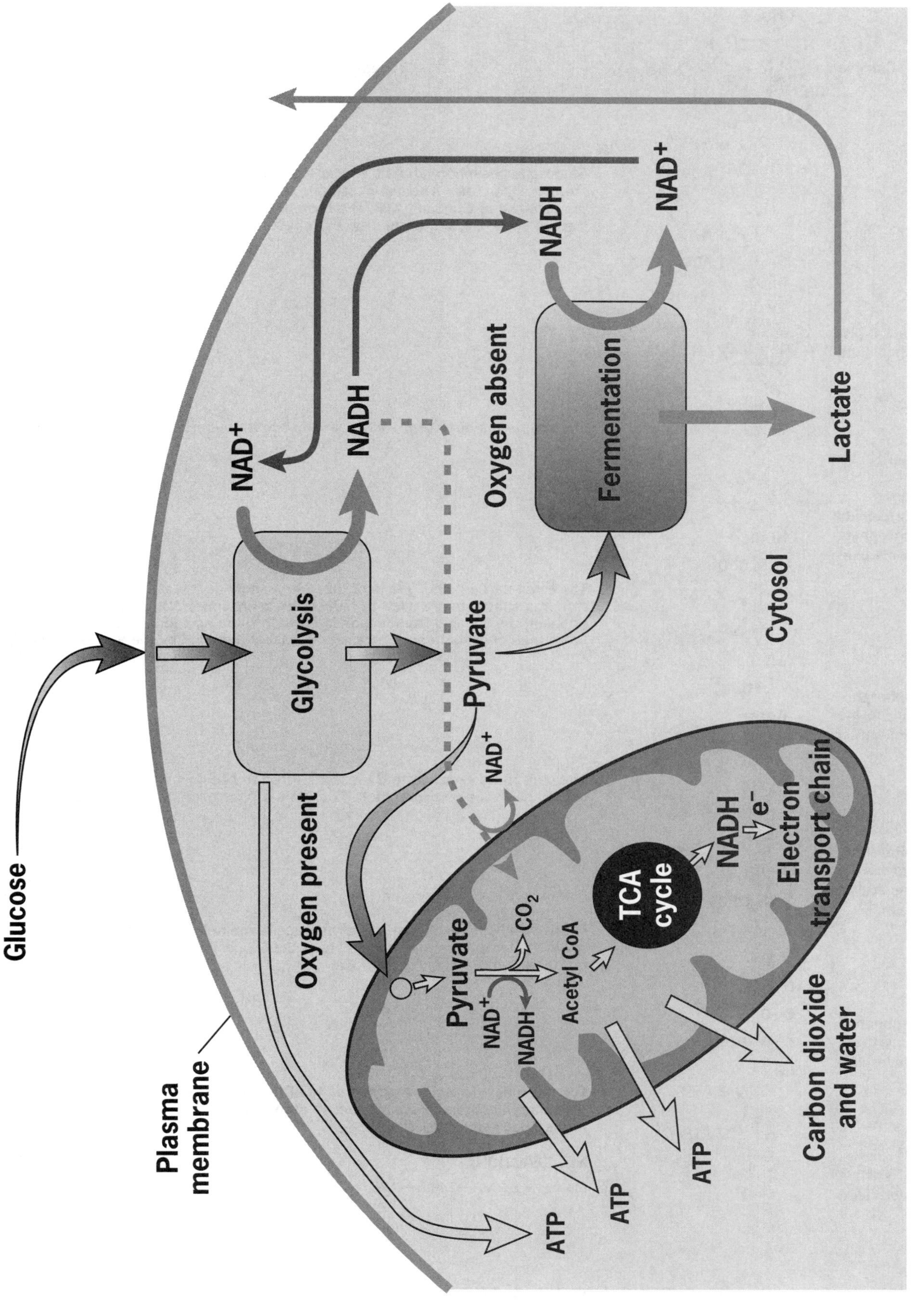

Figure 5.5 An overview of carbohydrate metabolism in eukaryotic cells.

Copyright © 2005 John Wiley & Sons, Inc.

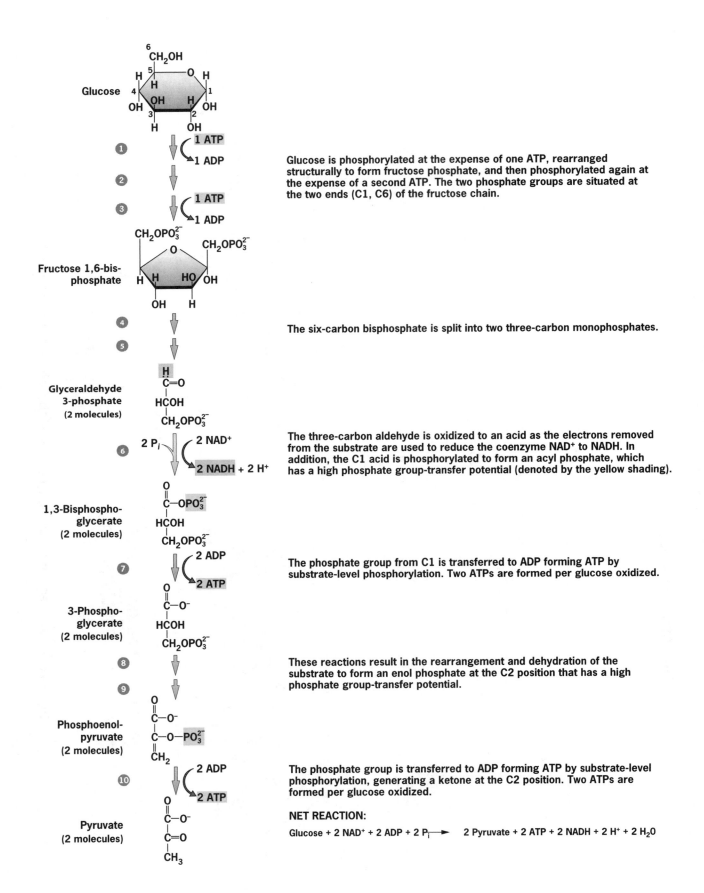

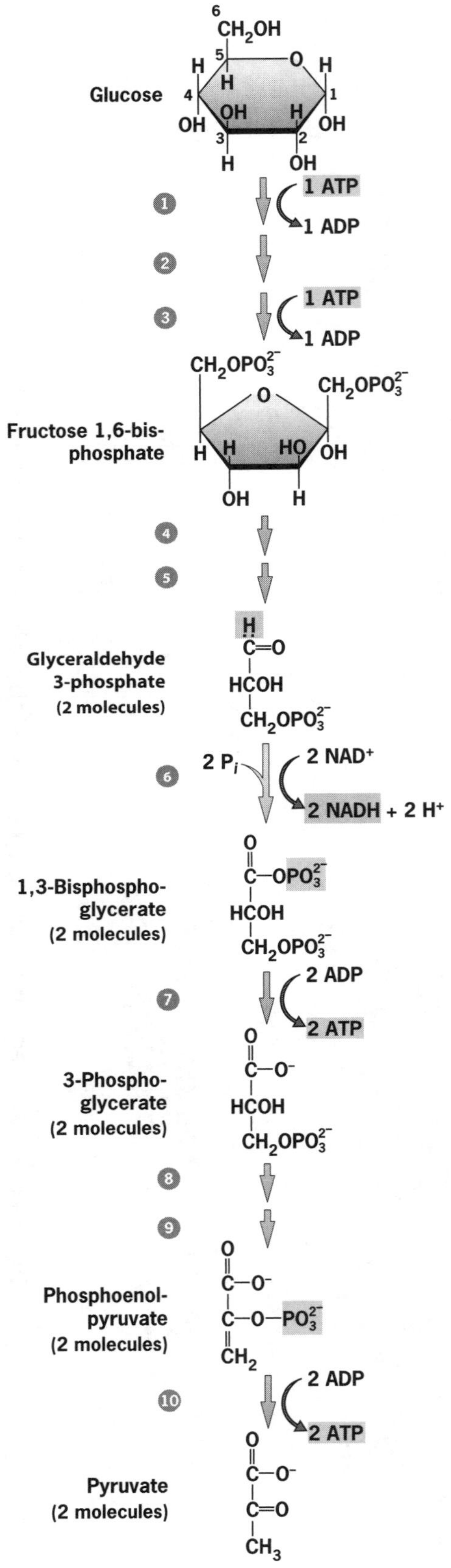

Glucose is phosphorylated at the expense of one ATP, rearranged structurally to form fructose phosphate, and then phosphorylated again at the expense of a second ATP. The two phosphate groups are situated at the two ends (C1, C6) of the fructose chain.

The six-carbon bisphosphate is split into two three-carbon monophosphates.

The three-carbon aldehyde is oxidized to an acid as the electrons removed from the substrate are used to reduce the coenzyme NAD^+ to NADH. In addition, the C1 acid is phosphorylated to form an acyl phosphate, which has a high phosphate group-transfer potential (denoted by the yellow shading).

The phosphate group from C1 is transferred to ADP forming ATP by substrate-level phosphorylation. Two ATPs are formed per glucose oxidized.

These reactions result in the rearrangement and dehydration of the substrate to form an enol phosphate at the C2 position that has a high phosphate group-transfer potential.

The phosphate group is transferred to ADP forming ATP by substrate-level phosphorylation, generating a ketone at the C2 position. Two ATPs are formed per glucose oxidized.

NET REACTION:

Glucose + 2 NAD^+ + 2 ADP + 2 P_i $\rightarrow$ 2 Pyruvate + 2 ATP + 2 NADH + 2 H^+ + 2 H_2O

Figure 5.6 An overview of glycolysis showing some of the key steps.

Copyright © 2005 John Wiley & Sons, Inc.

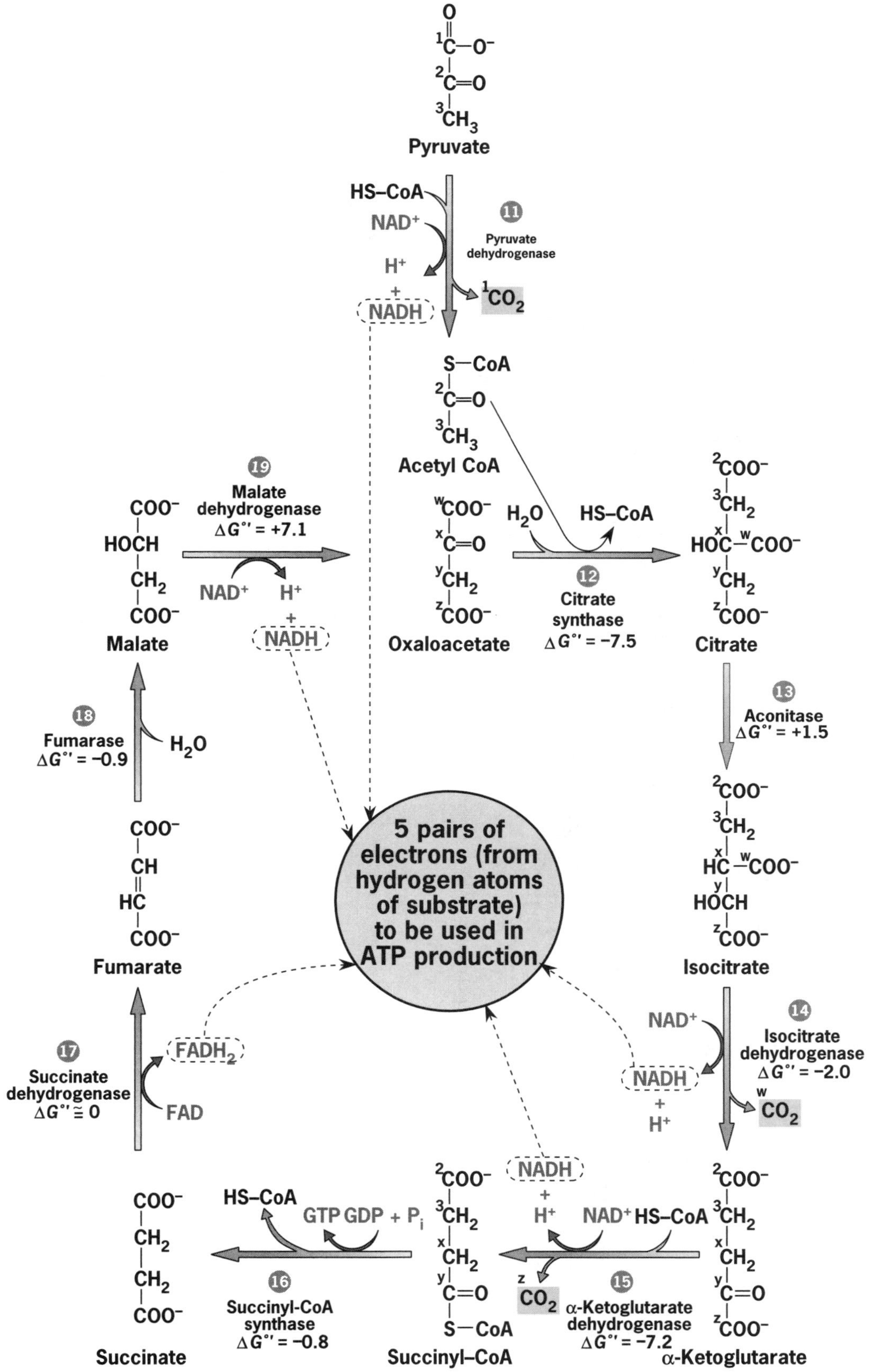

Figure 5.7 The tricarboxylic acid (TCA) cycle.

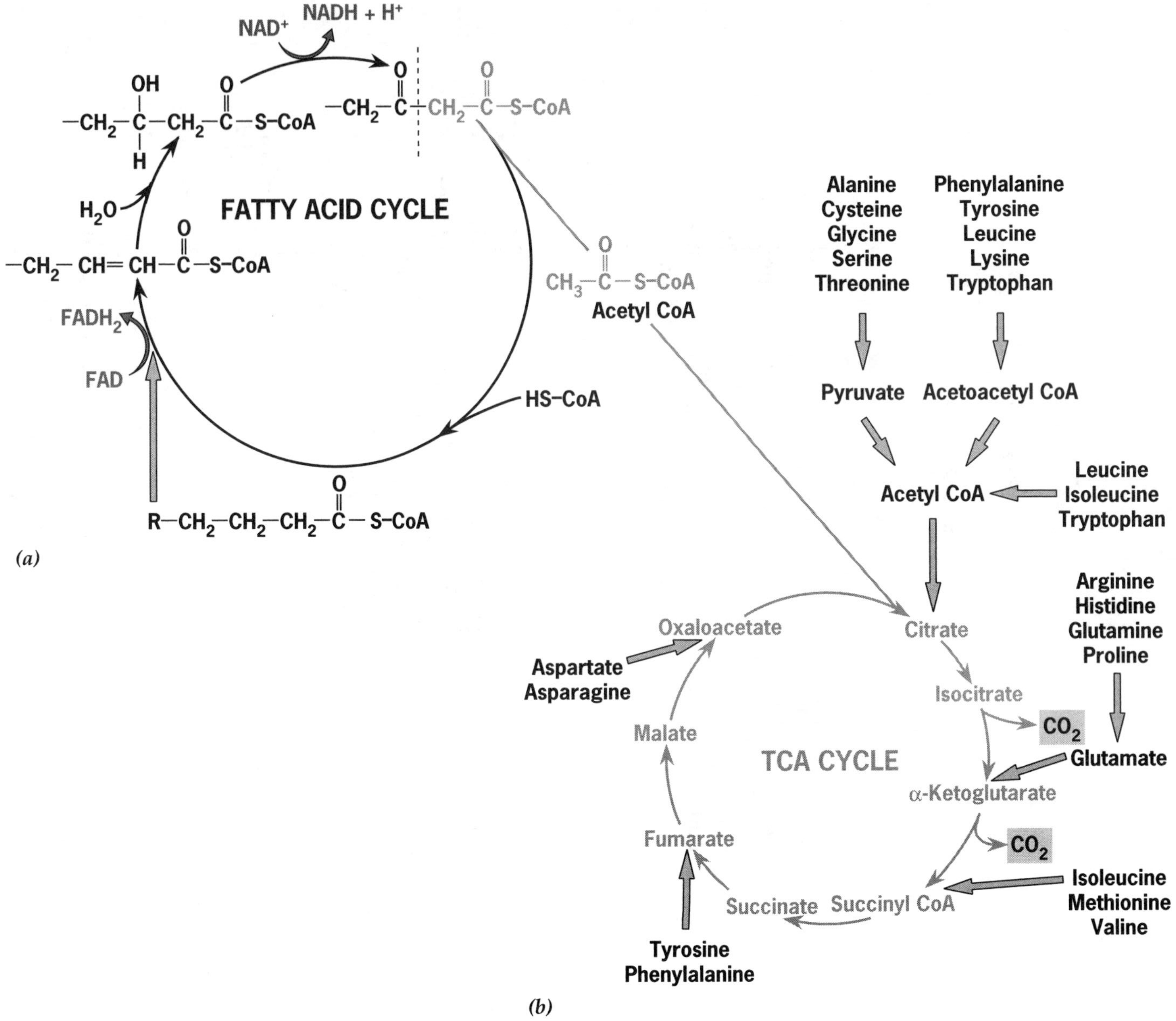

Figure 5.8 Catabolic pathways generate compounds that are fed into the TCA cycle.

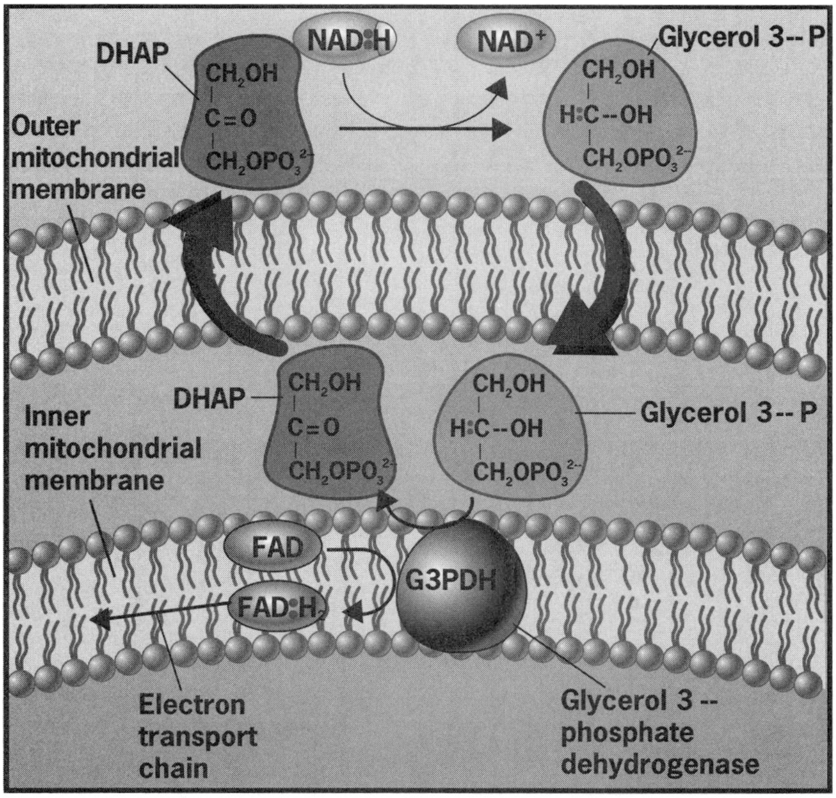

Figure 5.9 The glycerol phosphate shuttle.

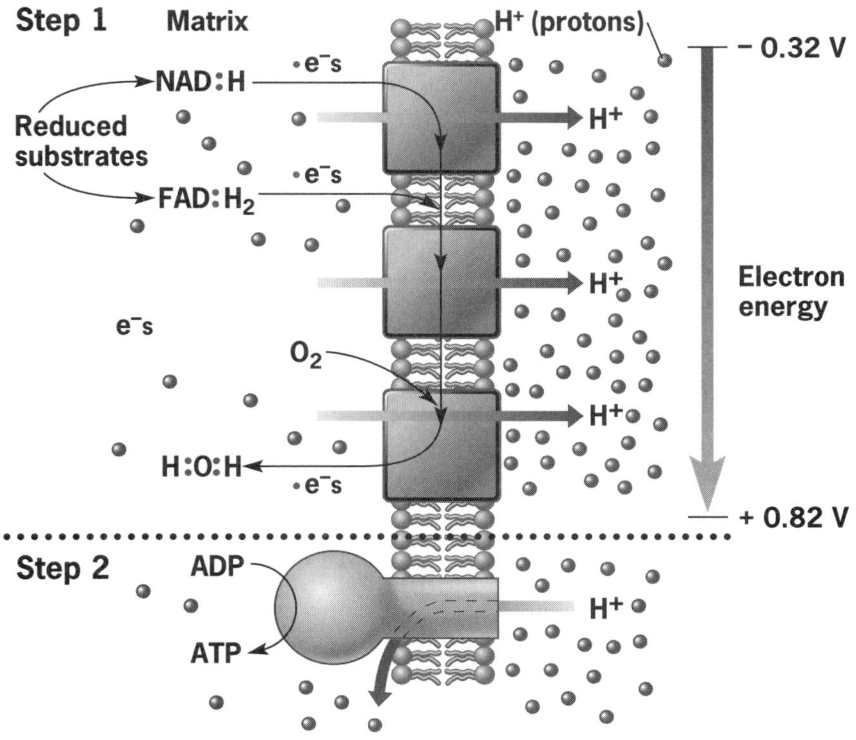

Figure 5.10 A summary of the process of oxidative phosphorylation.

Copyright © 2005 John Wiley & Sons, Inc.

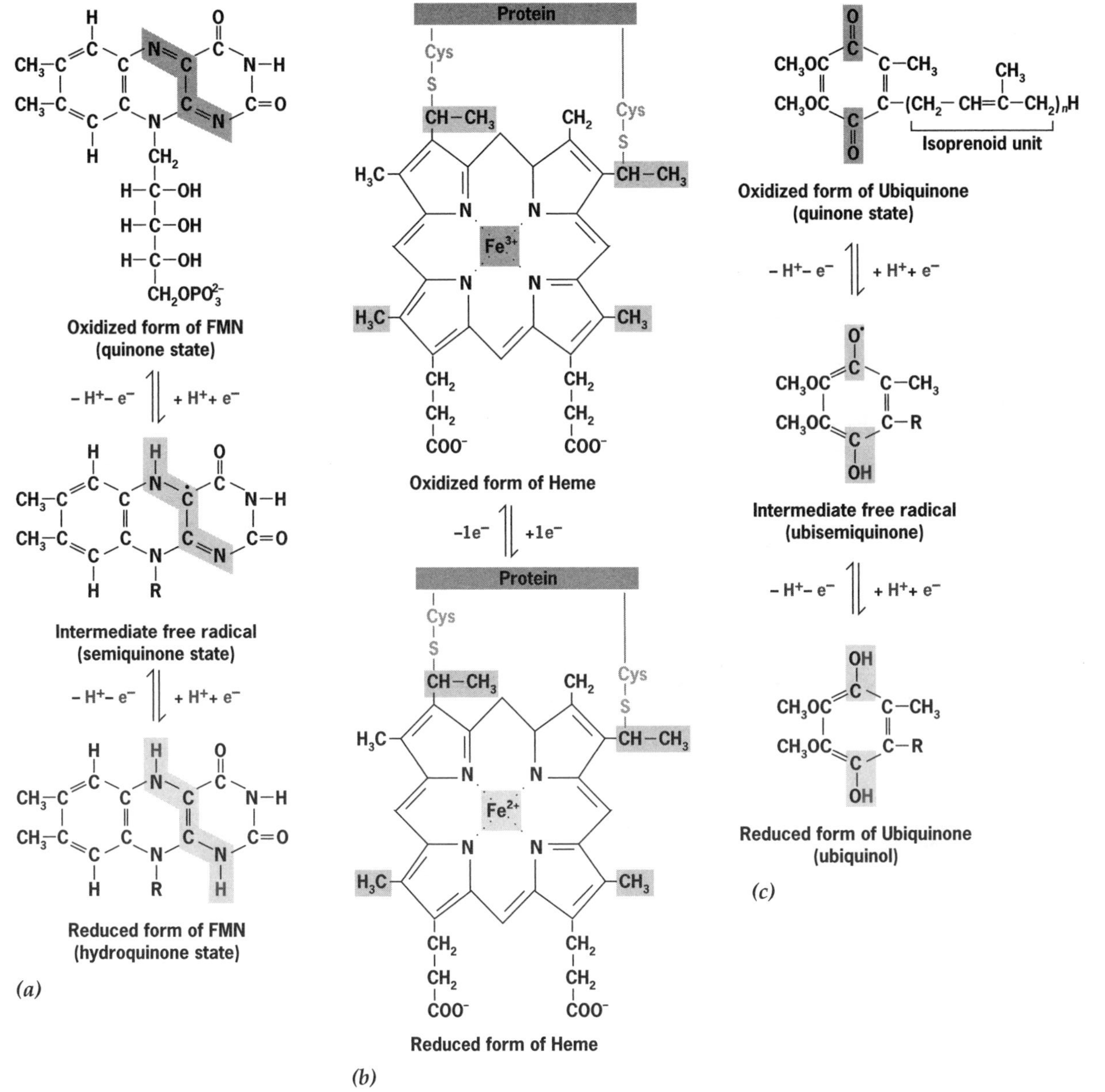

Figure 5.12 Structures of the oxidized and reduced forms of three types of electron carriers.

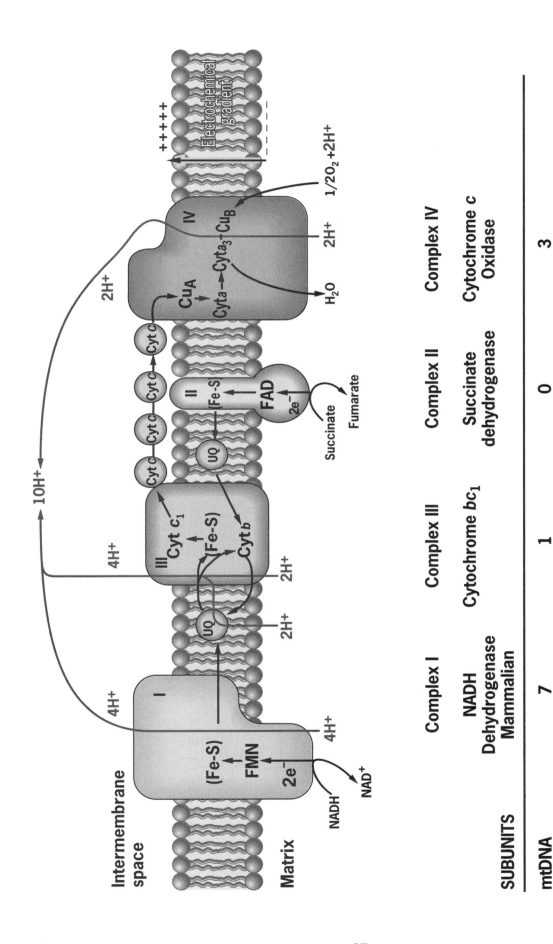

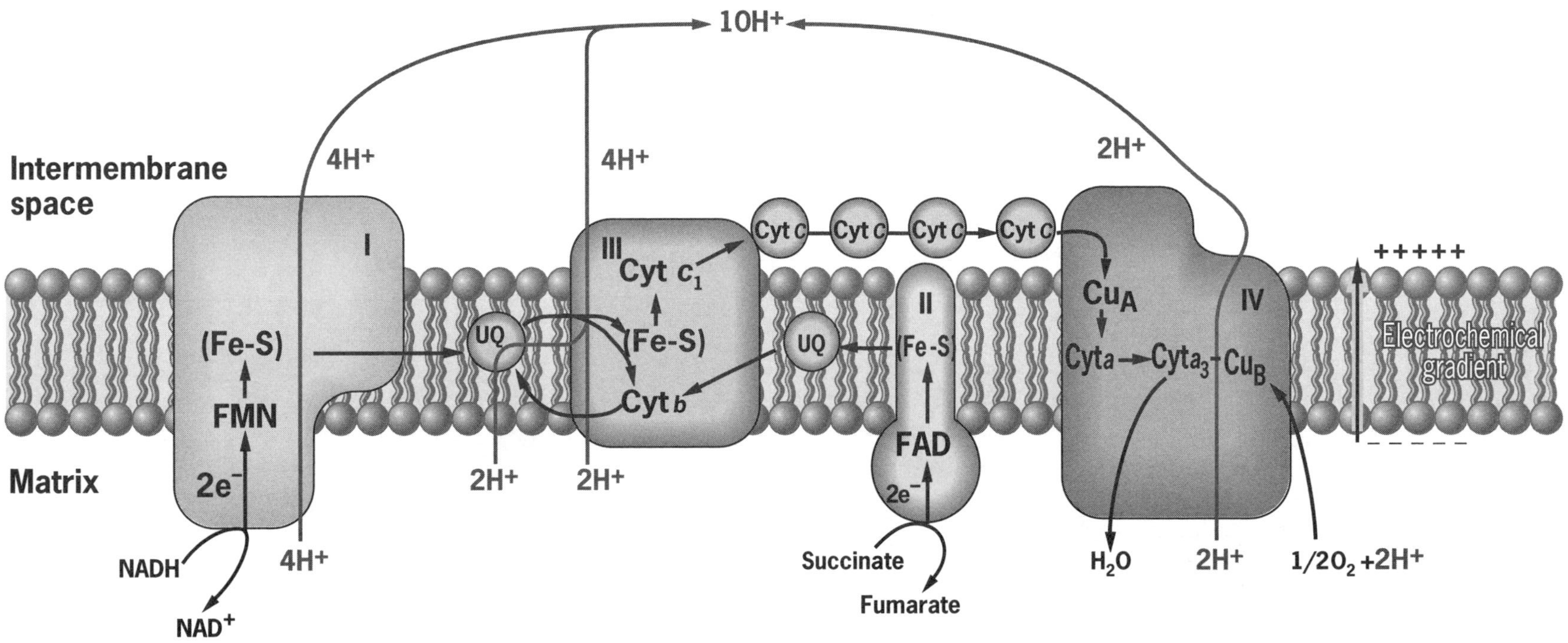

SUBUNITS	Complex I NADH Dehydrogenase Mammalian	Complex III Cytochrome bc_1	Complex II Succinate dehydrogenase	Complex IV Cytochrome *c* Oxidase
mtDNA	7	1	0	3
nDNA	38	10	4	10
TOTAL	45	11	4	13
Molecular mass (Da)	>900,000	~240,000	~125,000	~200,000

(a)

Figure 5.17a The electron-transport chain of the inner mitochondrial membrane.

Copyright © 2005 John Wiley & Sons, Inc.

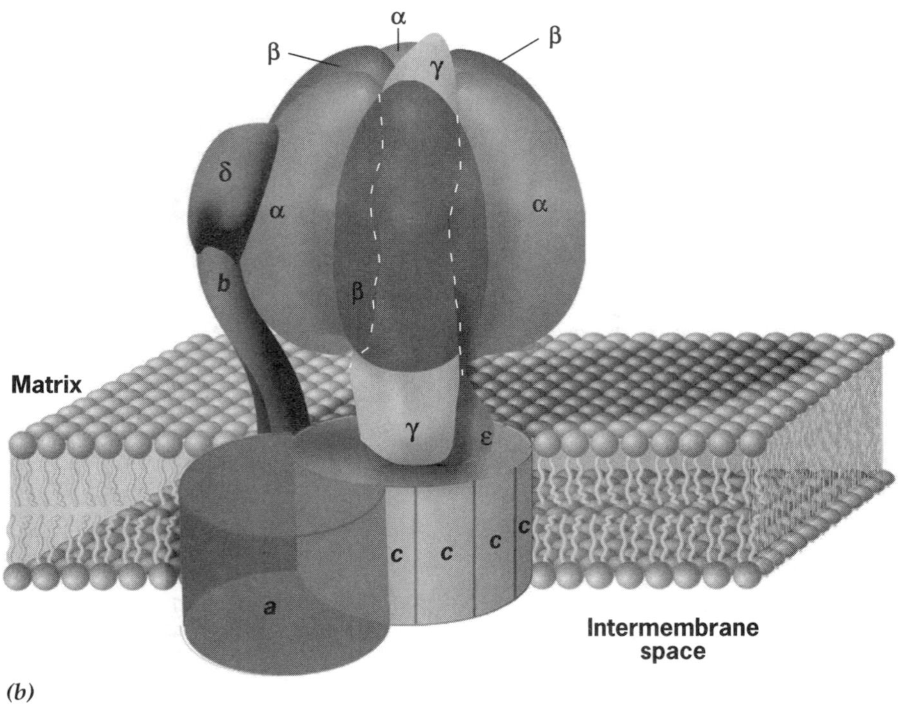

Figure 5.23b The structure of the ATP synthase.

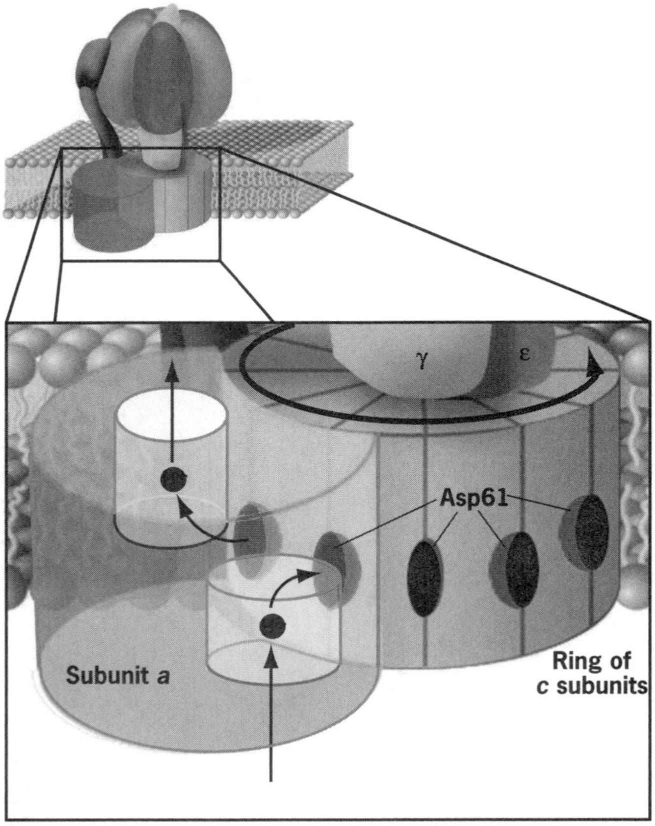

Figure 5.29 A model in which proton diffusion is coupled to the rotation of the *c* ring of the F_0 complex.

Copyright © 2005 John Wiley & Sons, Inc.

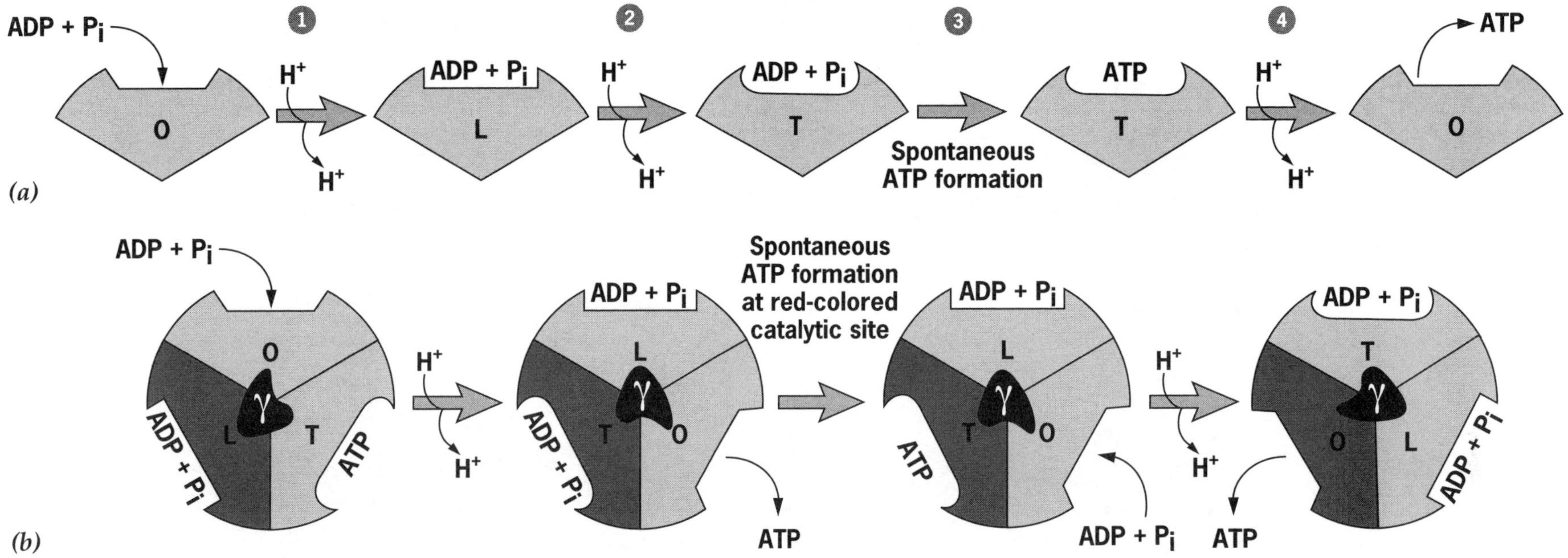

Figure 5.27 The binding change mechanism for ATP synthesis.

Copyright © 2005 John Wiley & Sons, Inc.

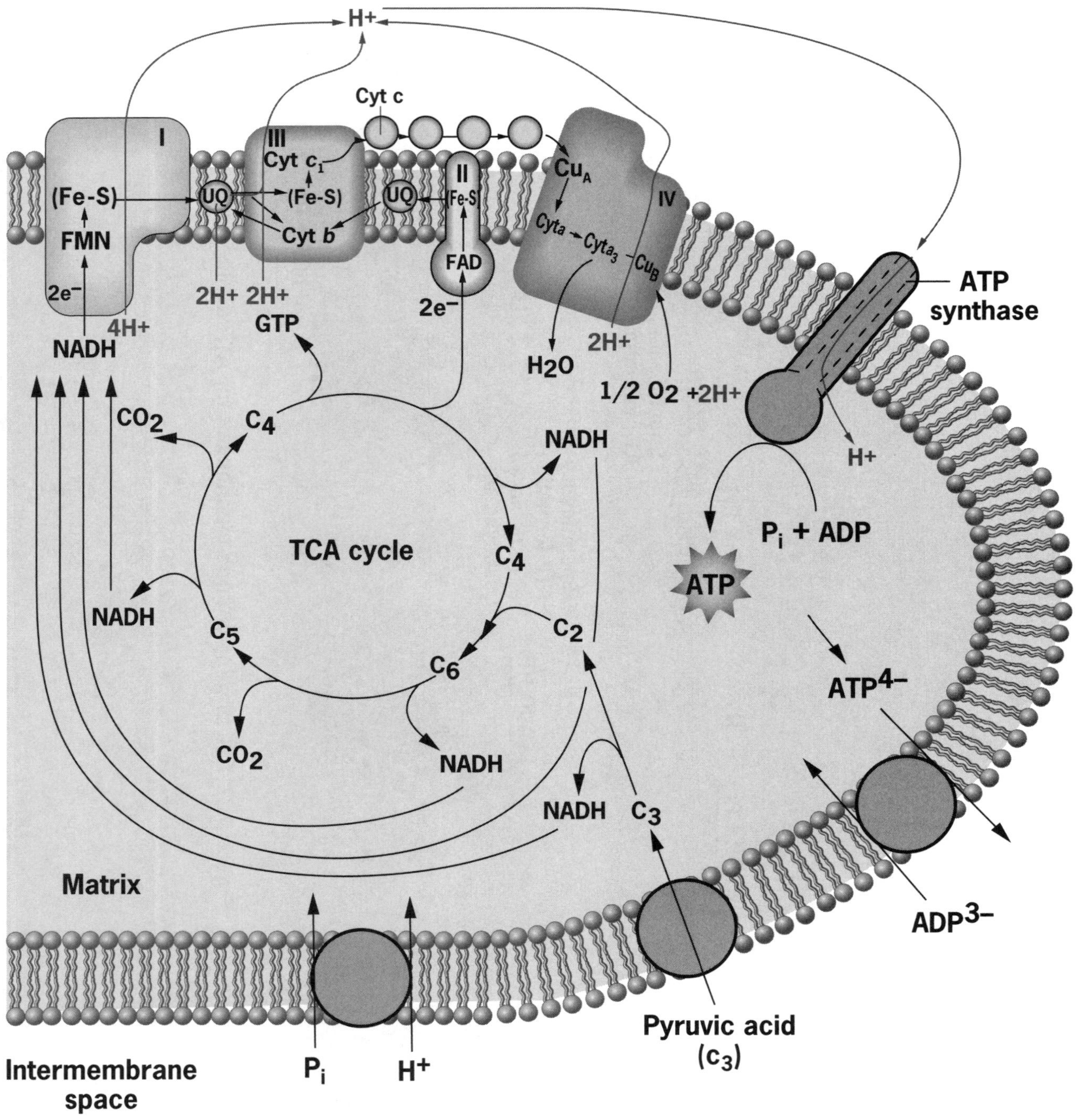

Figure 5.30 Summary of the major activities during aerobic respiration in a mitochondrion.

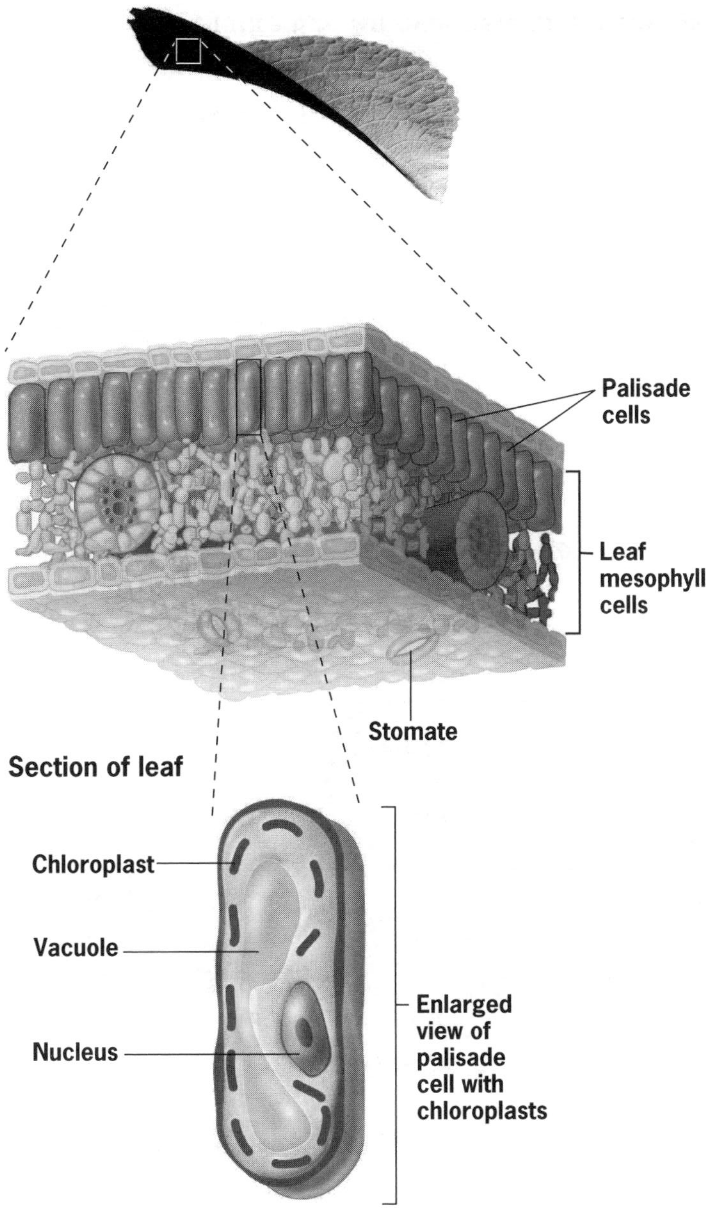

Figure 6.2 The functional organization of a leaf.

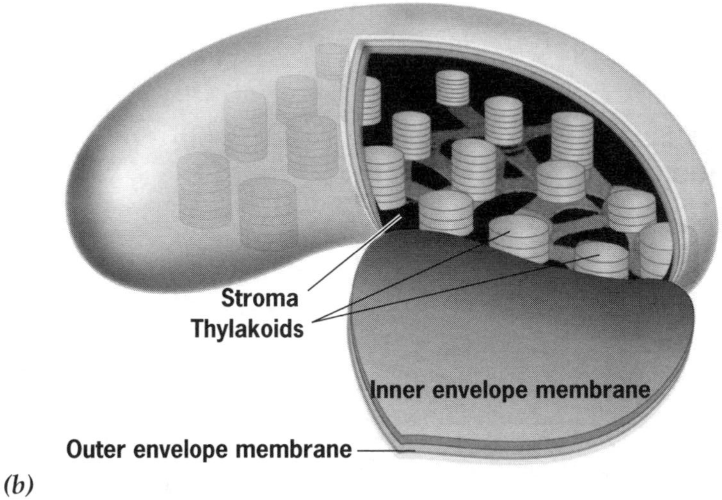

Figure 6.3b The internal structure of a chloroplast.

Copyright © 2005 John Wiley & Sons, Inc.

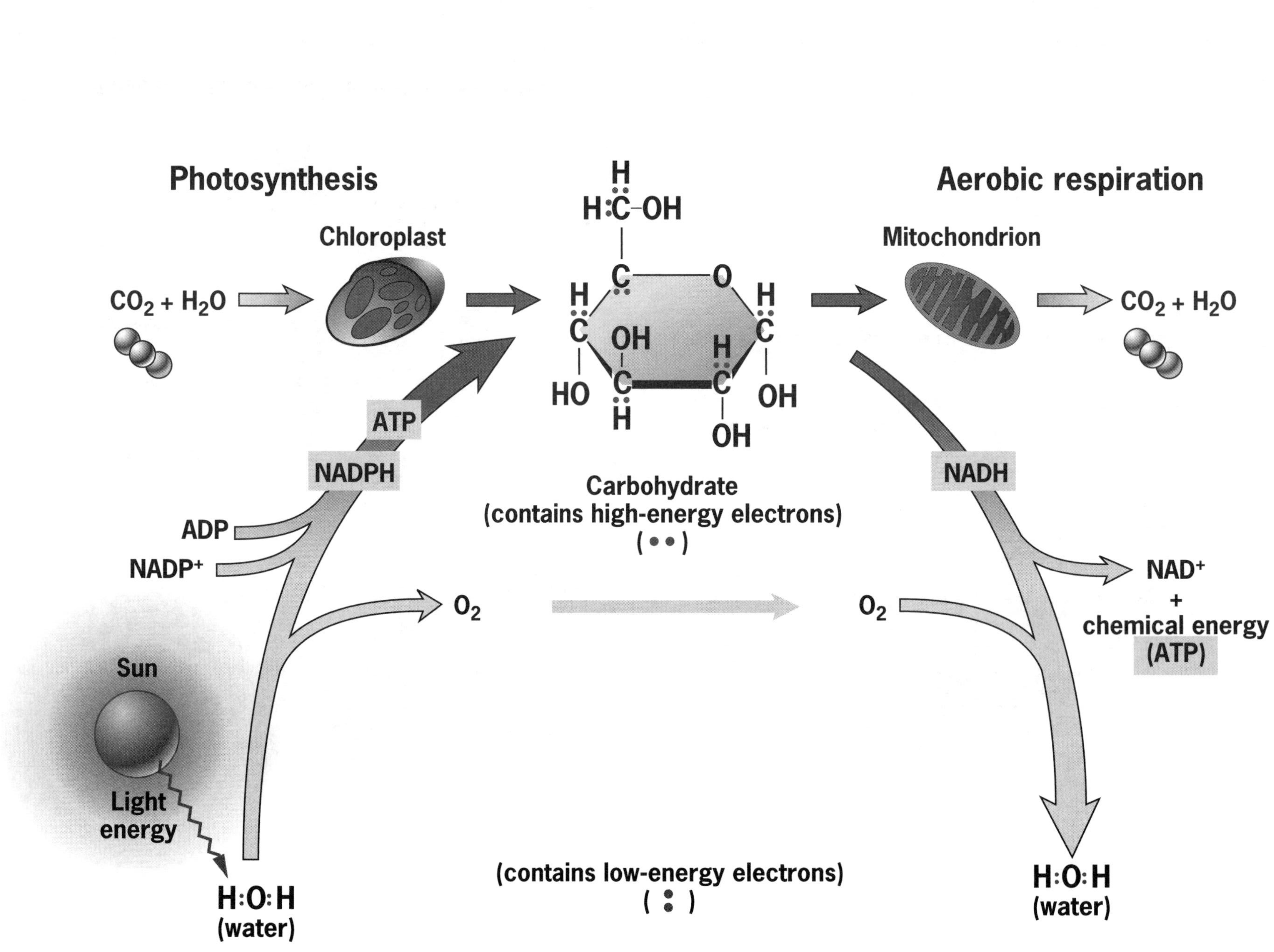

Figure 6.5 An overview of the energetics of photosynthesis and aerobic respiration.

Copyright © 2005 John Wiley & Sons, Inc.

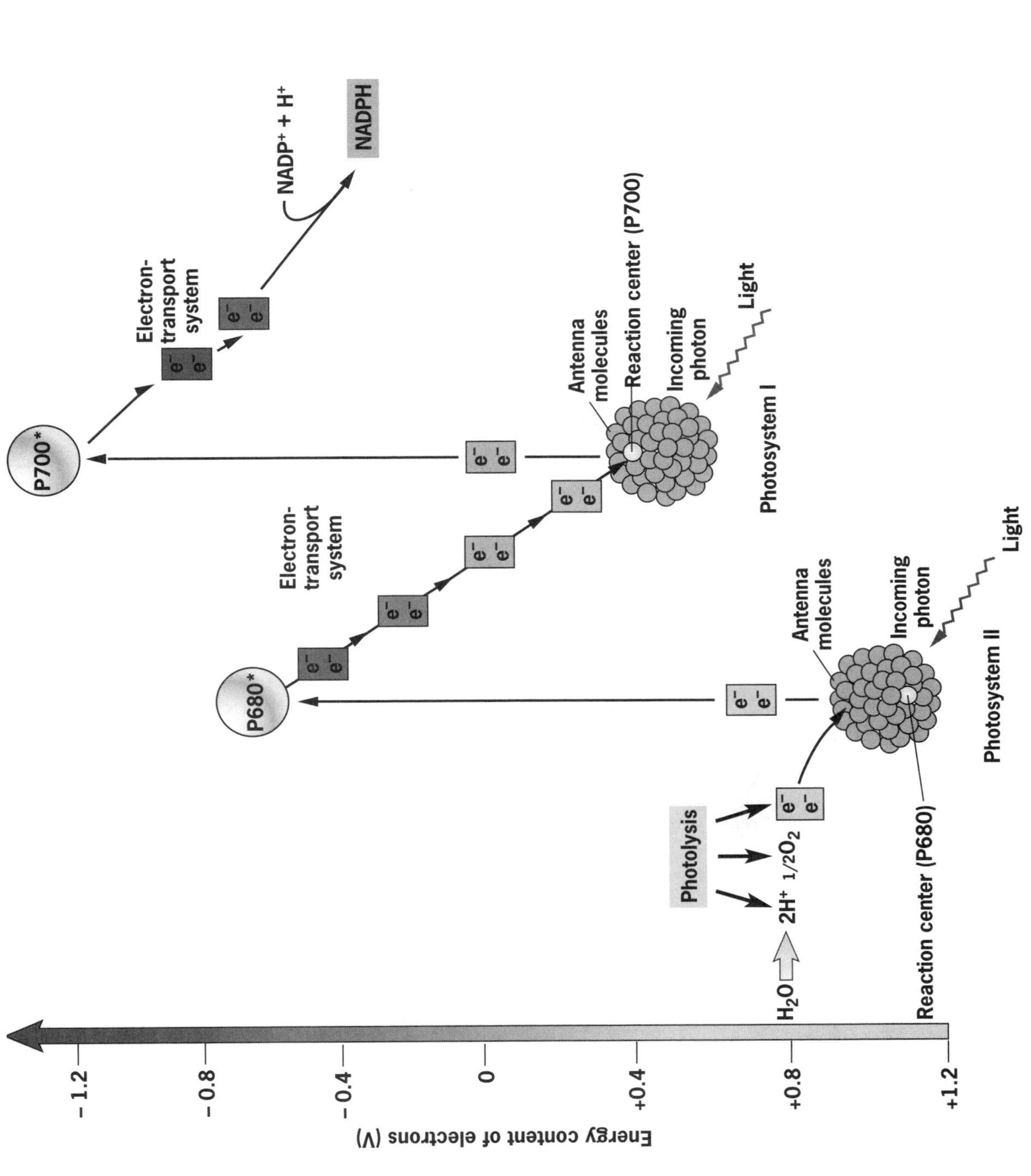

Figure 6.10 An overview of the flow of electrons during the light-dependent reactions of photosynthesis.

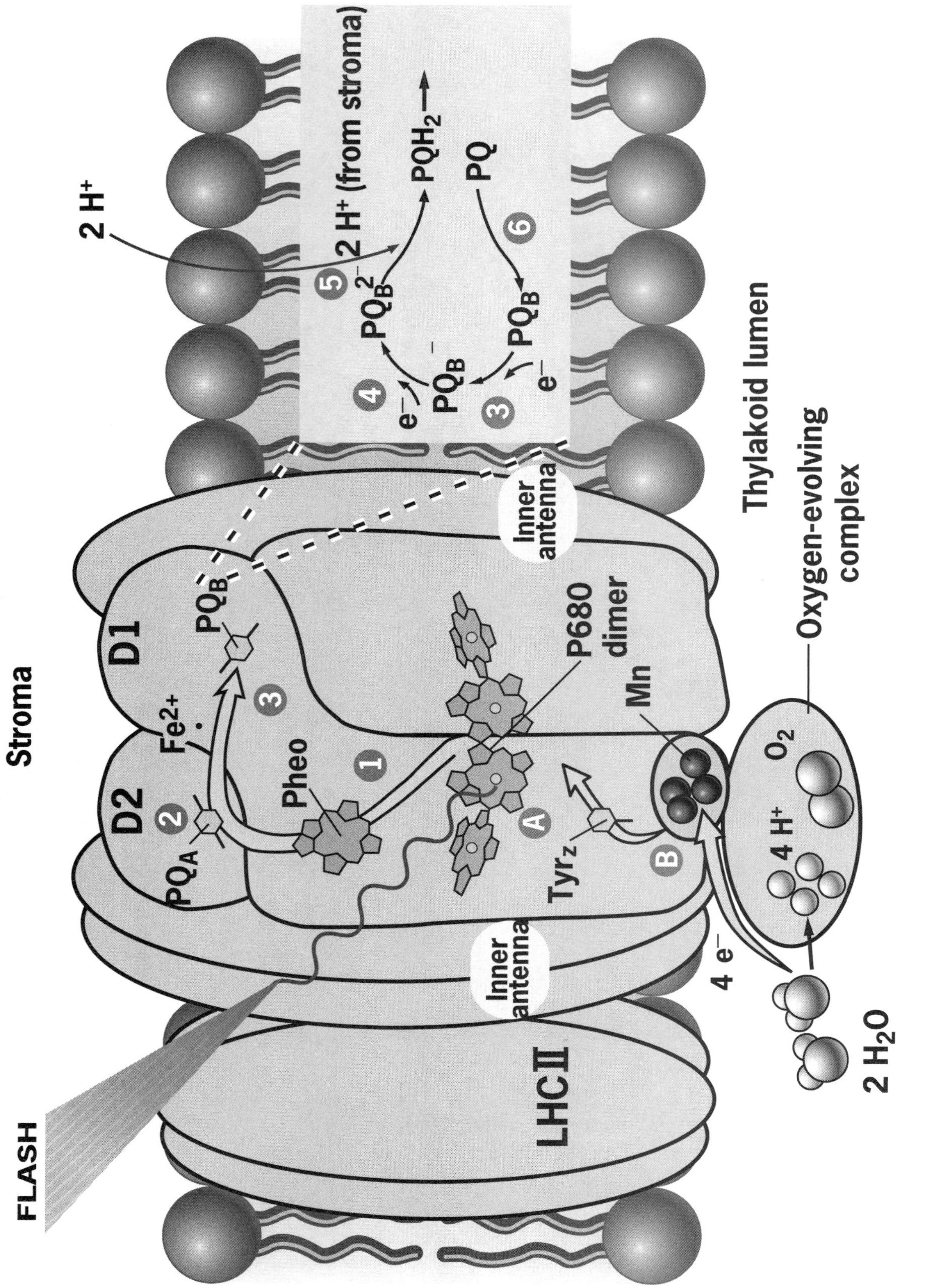

Figure 6.11 The functional organization of photosystem II.

Copyright © 2005 John Wiley & Sons, Inc.

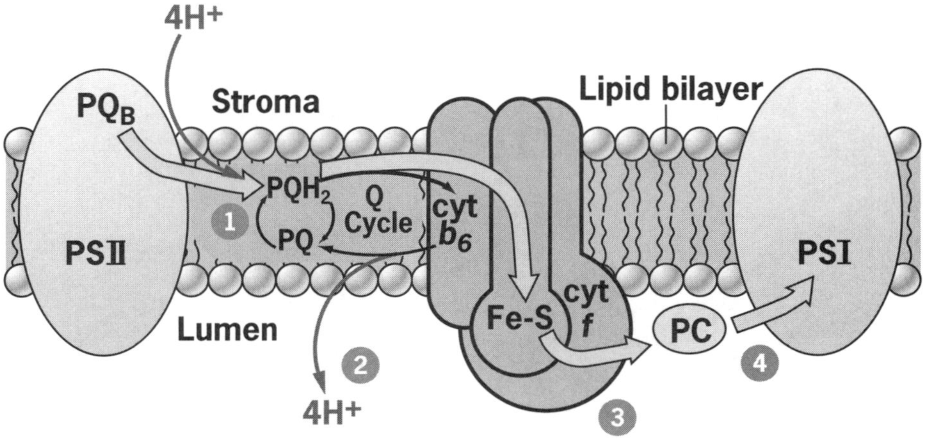

Figure 6.14 Electron transport between PSII and PSI.

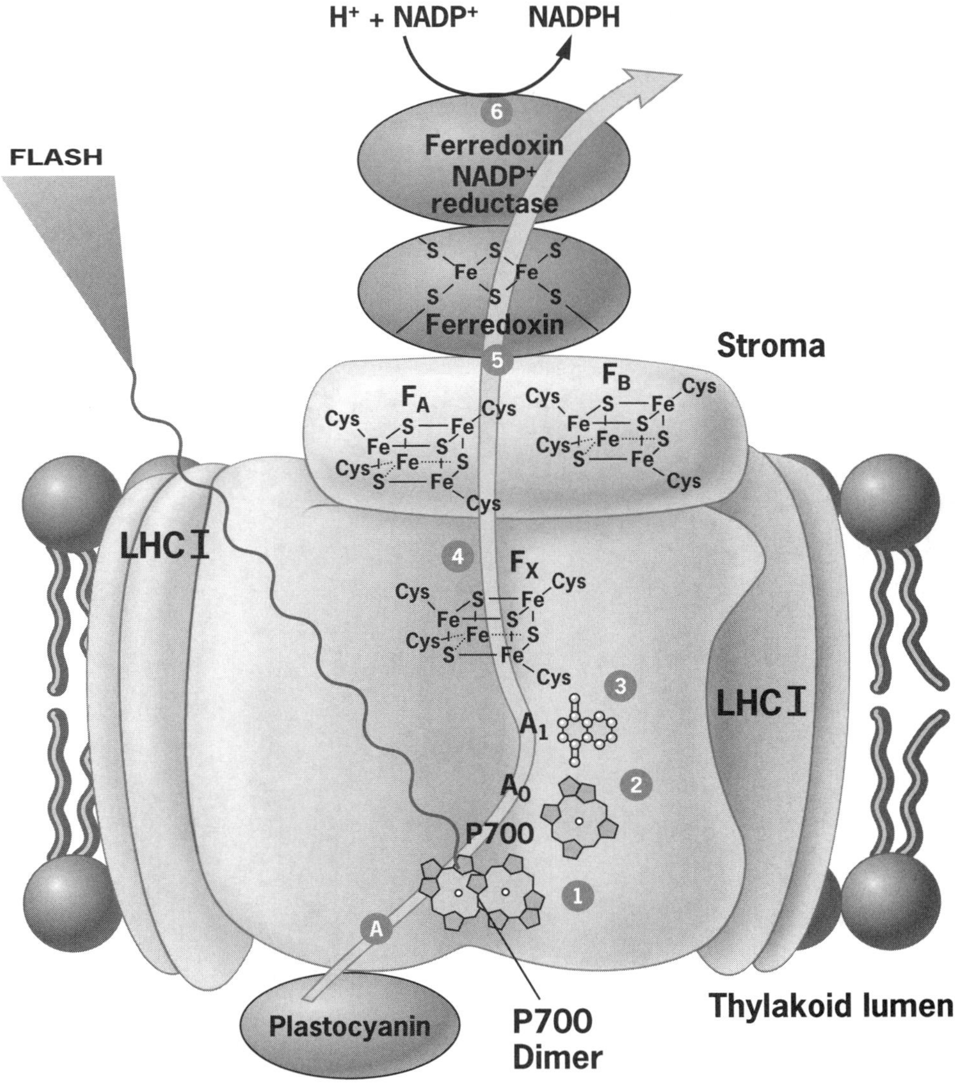

Figure 6.15 The functional organization of a plant photosystem I.

Copyright © 2005 John Wiley & Sons, Inc.

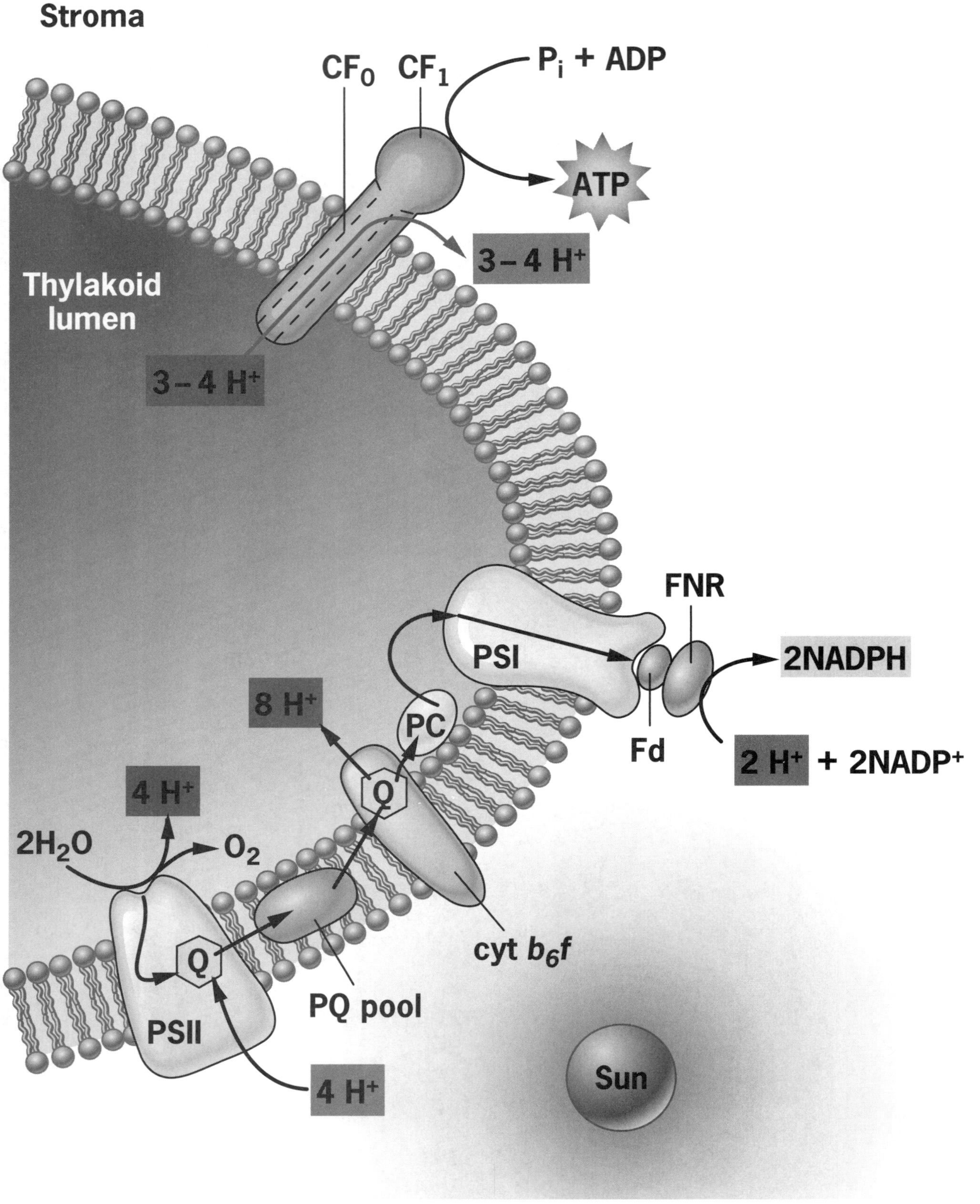

Figure 6.16 Summary of the light-dependent reactions.

Copyright © 2005 John Wiley & Sons, Inc.

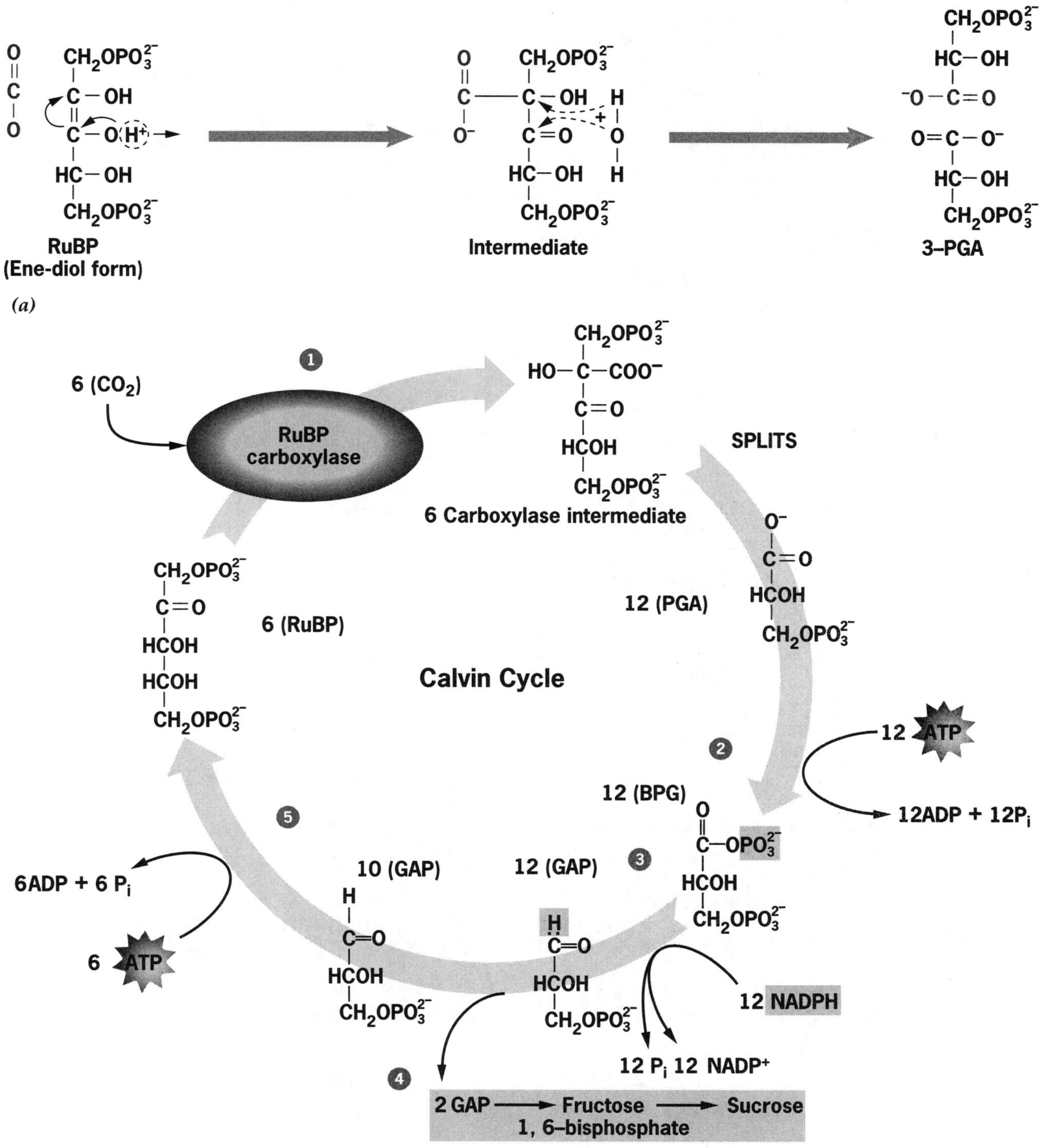

Figure 6.19 Converting CO2 into carbohydrate.

Copyright © 2005 John Wiley & Sons, Inc.

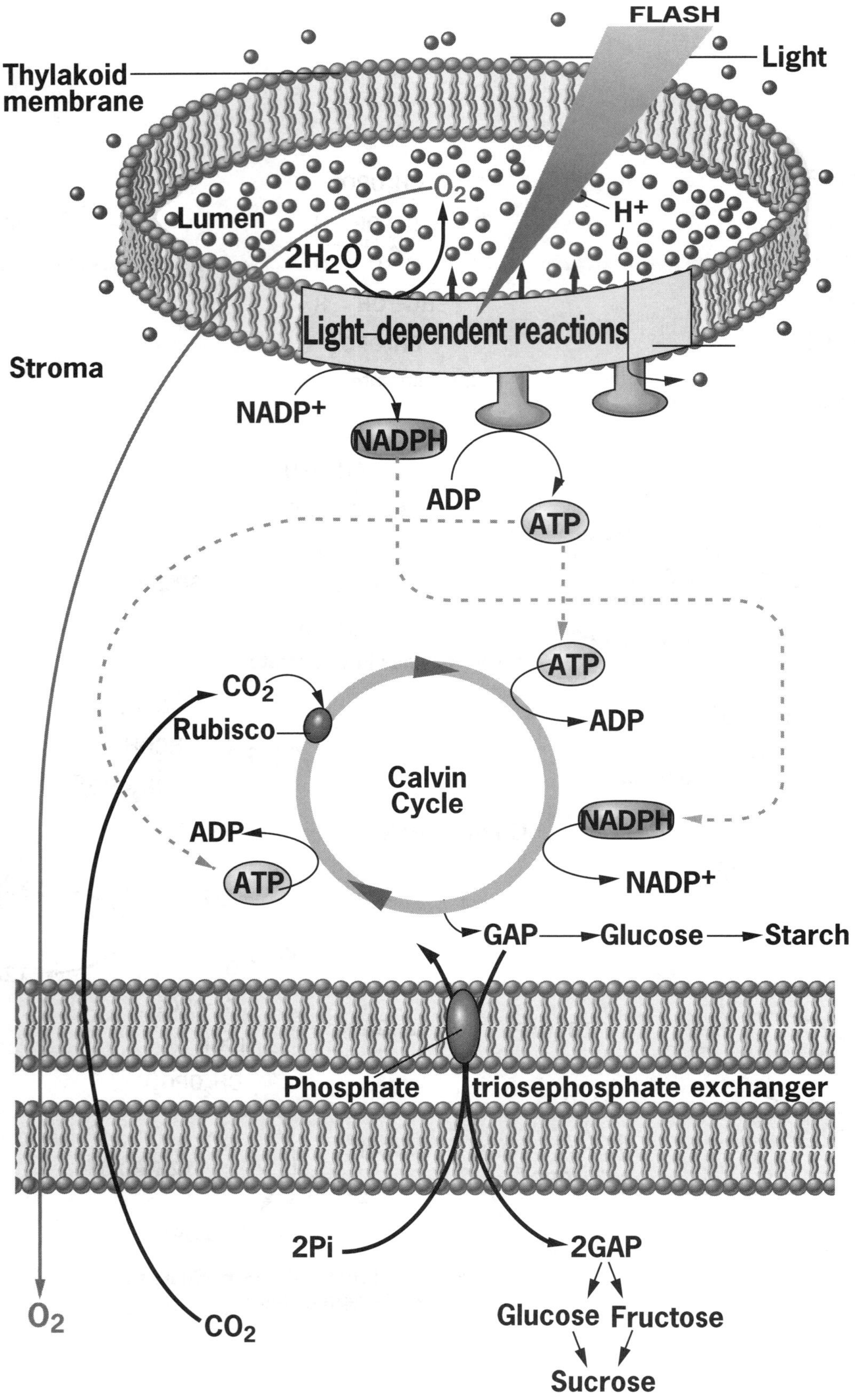

Figure 6.20 An overview of the various stages of photosynthesis.

Copyright © 2005 John Wiley & Sons, Inc.

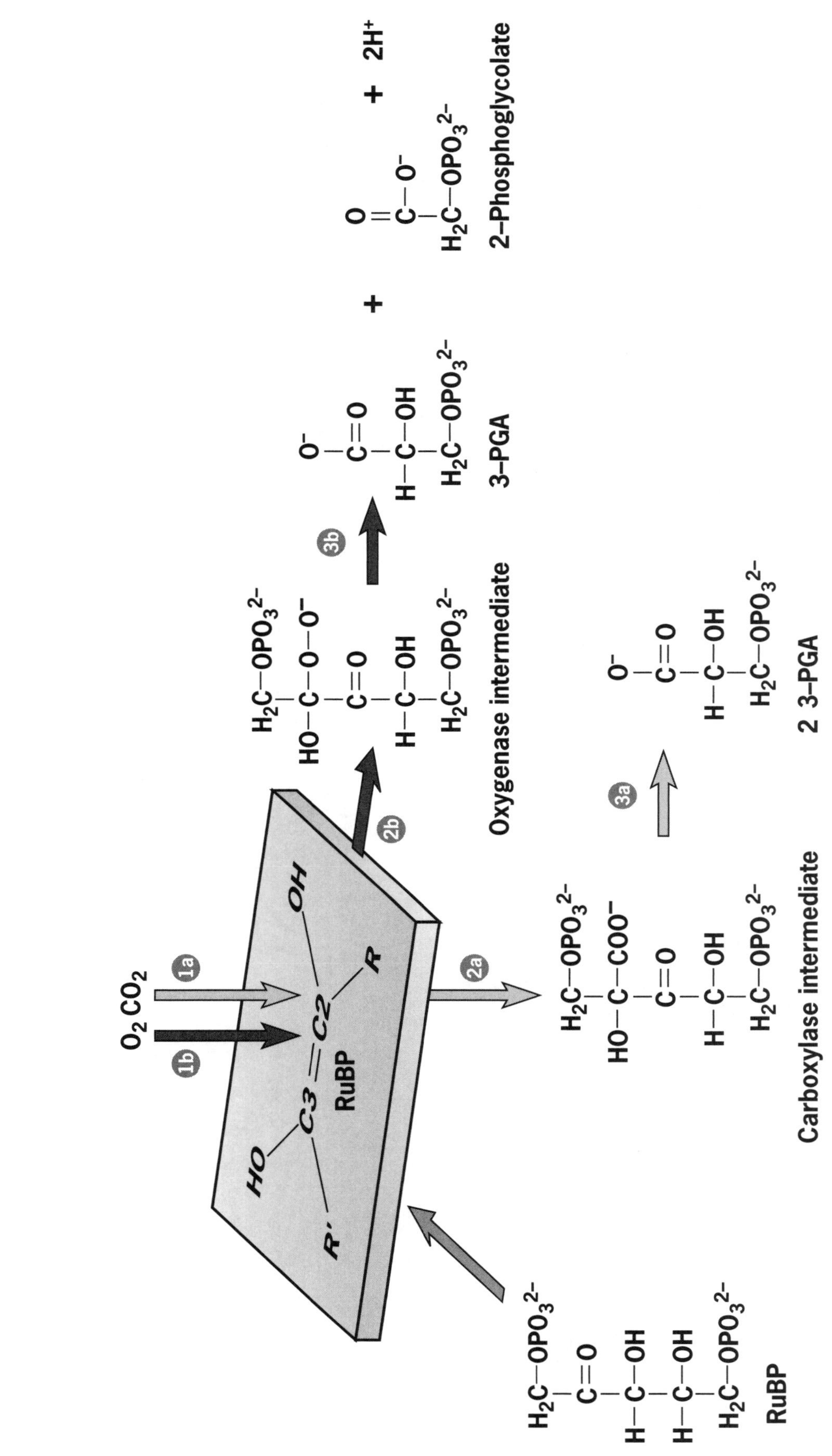
O_2 CO_2
1b
1a
HO
C3
C2
OH
R'
RuBP
R
2b
2a
3b
3a
Oxygenase intermediate
3-PGA
2-Phosphoglycolate
+
+ $2H^+$
RuBP
Carboxylase intermediate
2 3-PGA

Figure 6.22 The reactions of photorespiration.

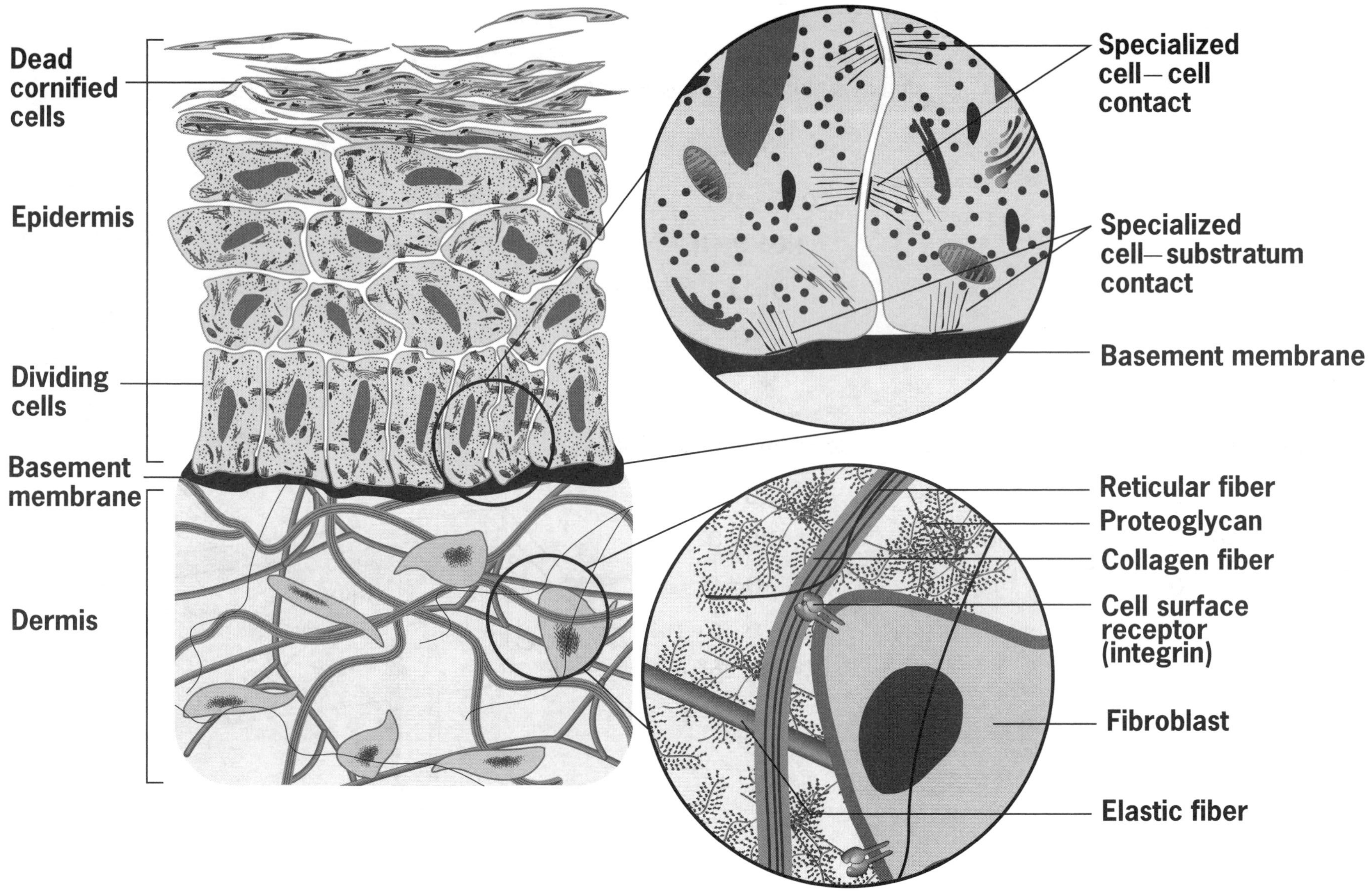

Figure 7.1 An overview of how cells are organized into tissues and how they interact with one another and with their extracellular environment.

Copyright © 2005 John Wiley & Sons, Inc.

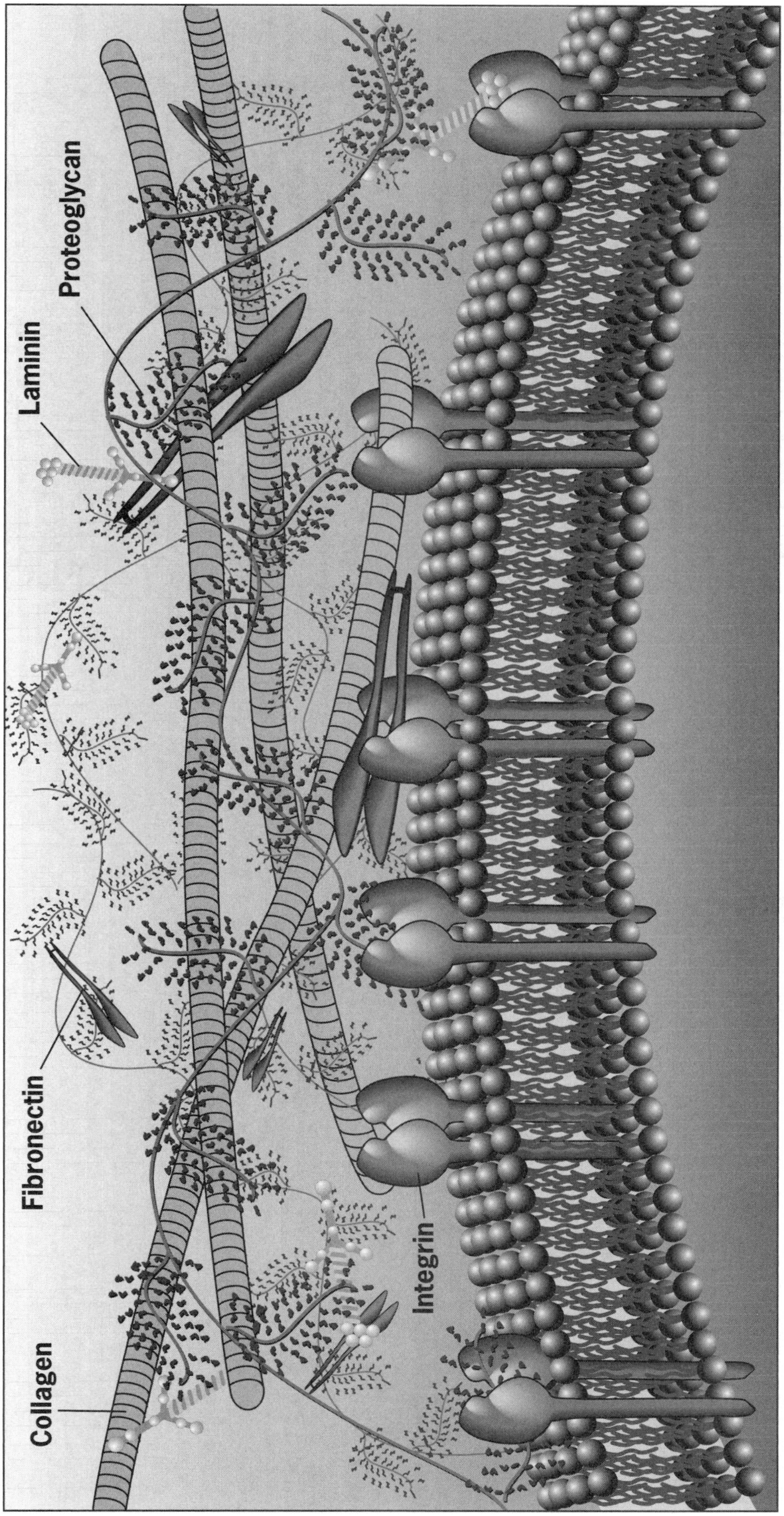

Figure 7.5 An overview of the macromolecular organization of the extracellular matrix.

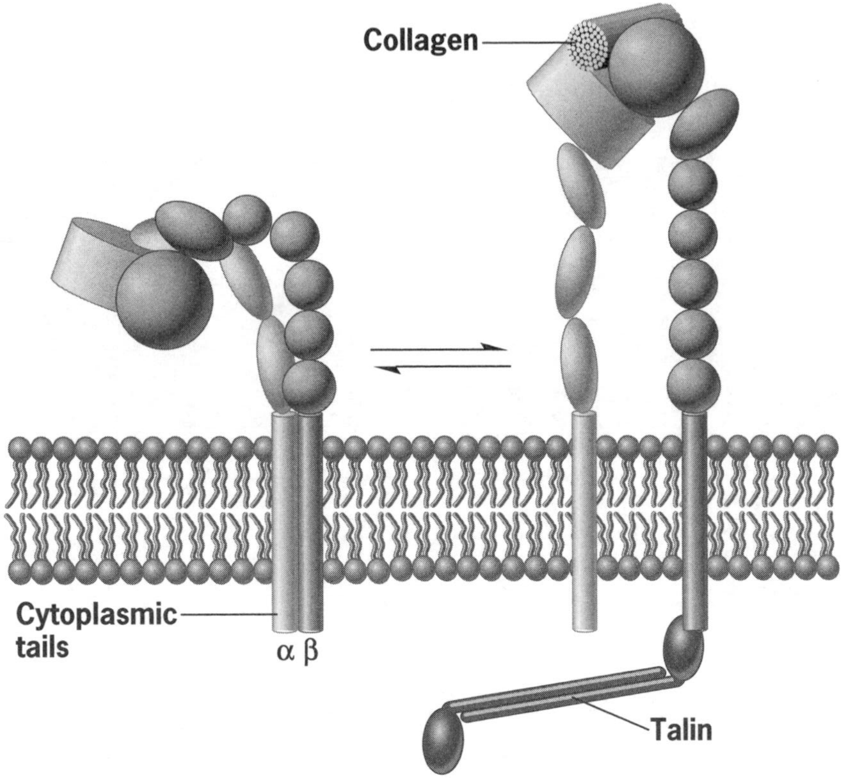

Figure 7.14 A model of integrin activation.

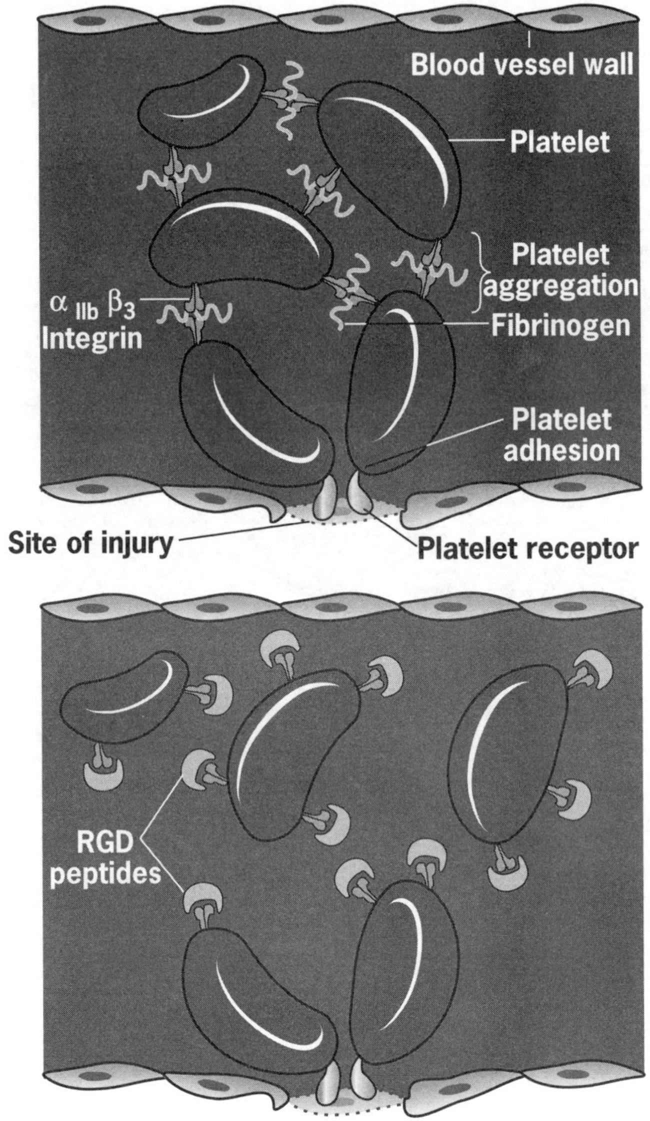

Figure 7.15 Blood clots form when platelets adhere to one another through fibrinogen bridges that bind to the platelet integrins.

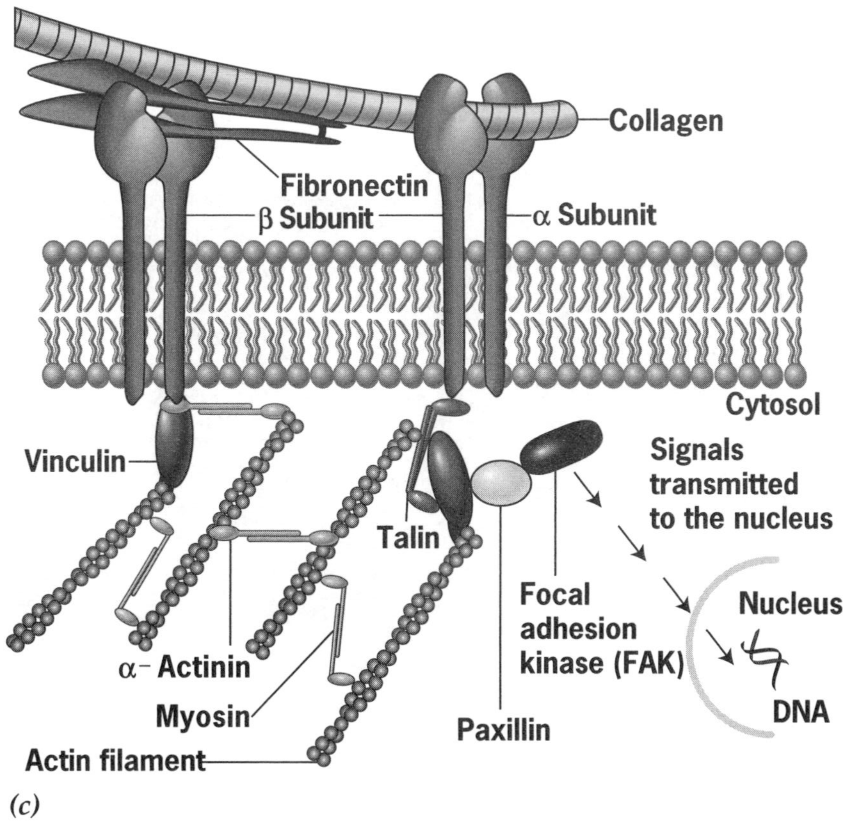

Figure 7.17c Focal adhesions are sites where cells adhere to their substratum.

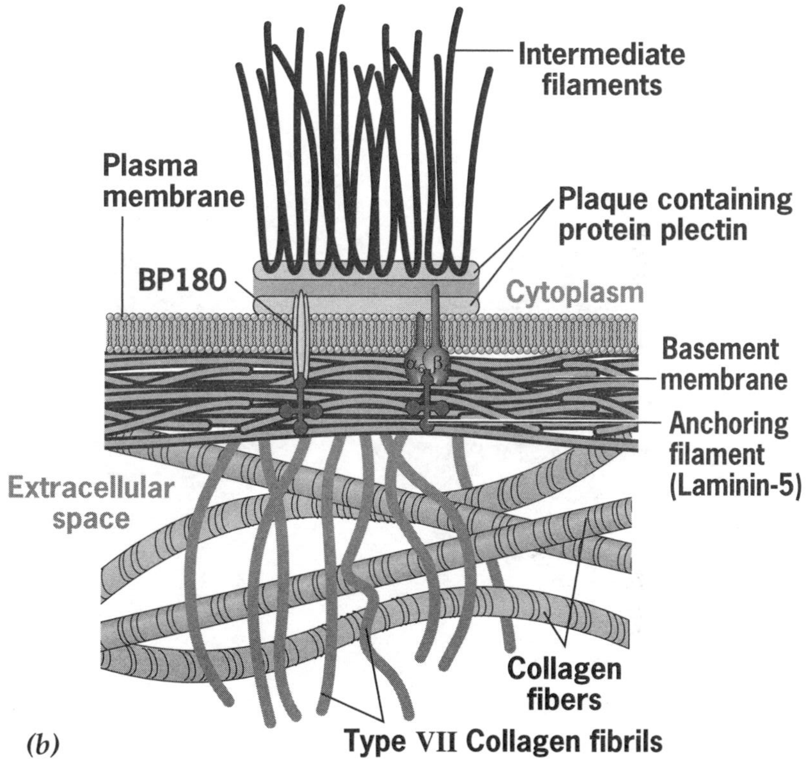

Figure 7.19b Hemidesmosomes.

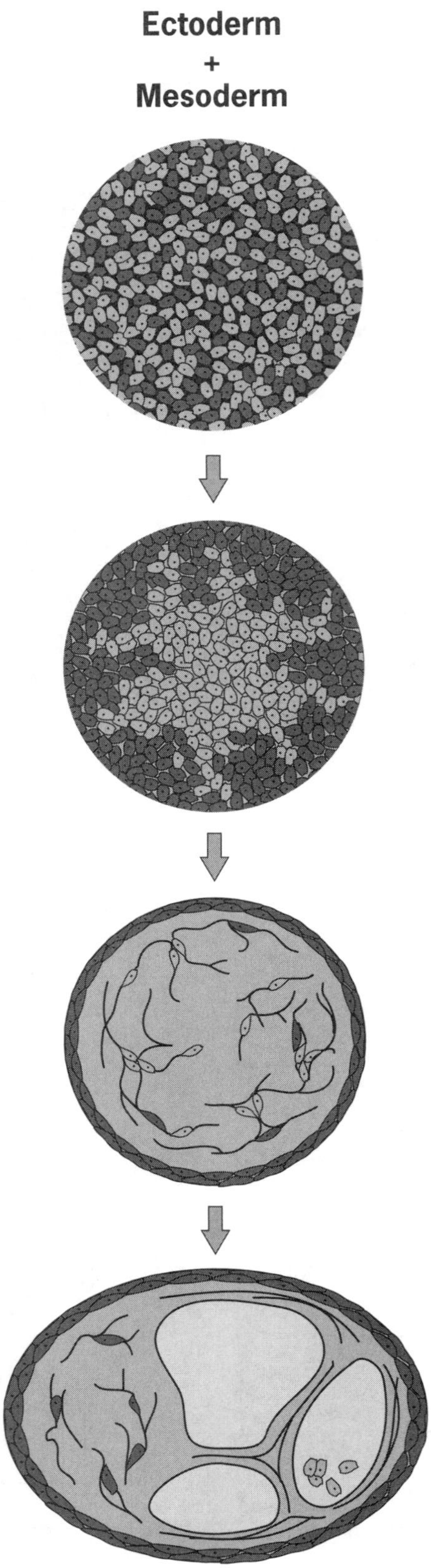

(a)

Figure 7.20a Experimental demonstration of cell-cell recognition.

L-Selectin
Lectin-like domain
EGF-like domain
Structural domain
E-Selectin
P-Selectin
(a)
Extracellular
Glycoprotein
(b)
Ligand
SiA
α2 → 3
Gal
β1 → 4
SO4
GlcNAc
Remainder of oligosaccharide
α1 → 3
Fuc
(c)

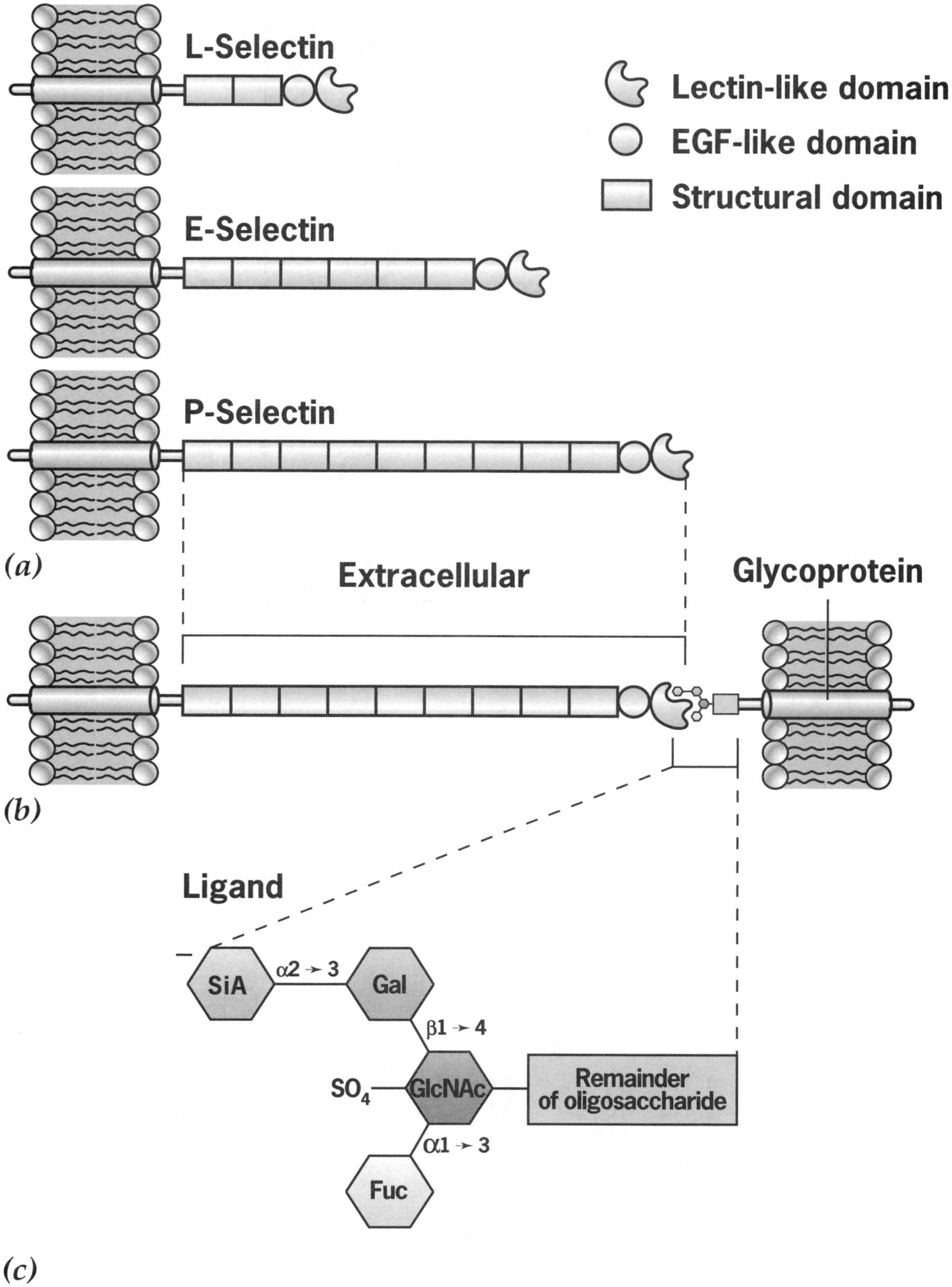

Figure 7.21a-c Selectins.

Copyright © 2005 John Wiley & Sons, Inc.

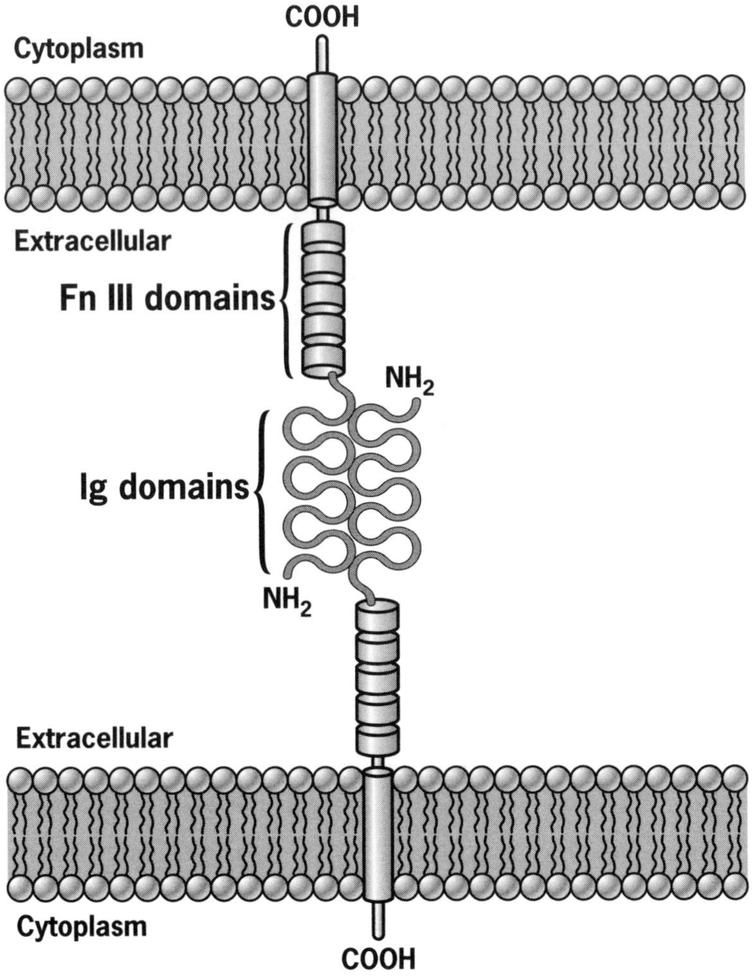

Figure 7.22 L1 is a cell-adhesion molecule of the immunoglobulin (Ig) superfamily.

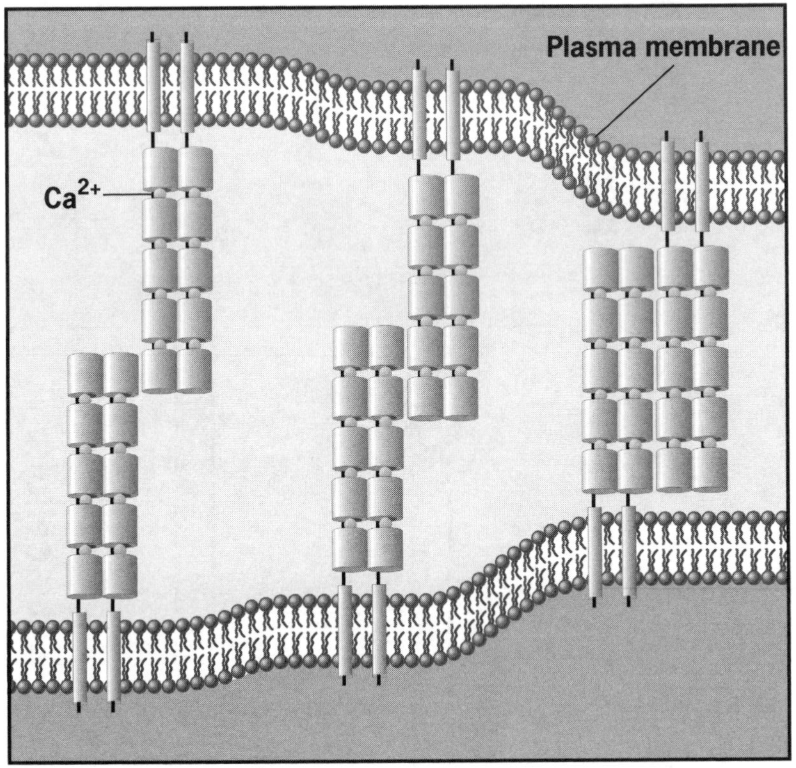

Figure 7.23 Cadherins and cell adhesion.

Copyright © 2005 John Wiley & Sons, Inc.

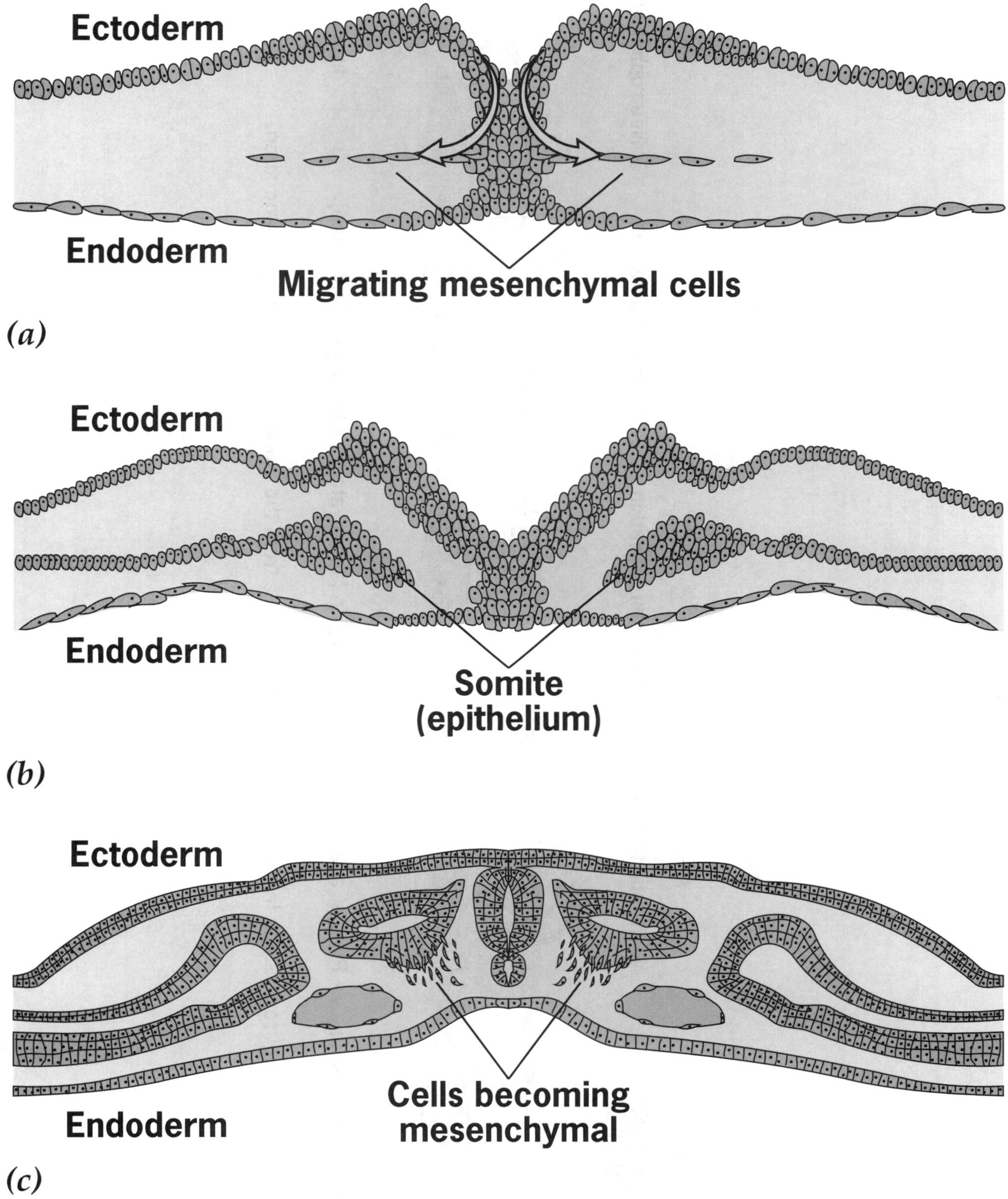

Figure 7.24a-c Cadherins and the epithelial–mesenchymal transformation.

Copyright © 2005 John Wiley & Sons, Inc.

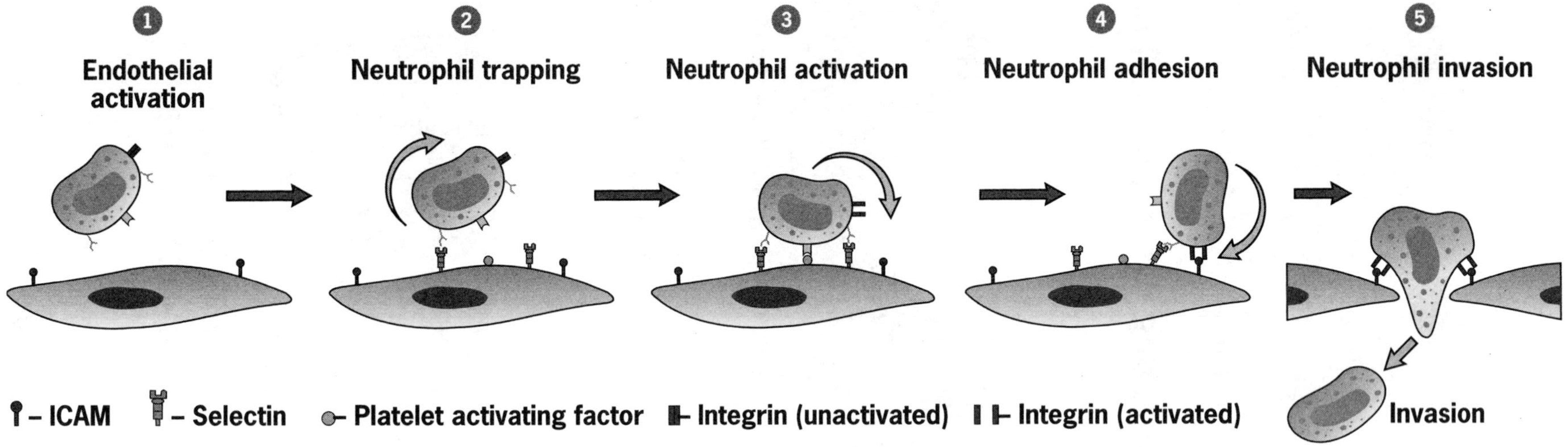

Figure 1 Steps in the movement of neutrophils from the bloodstream during inflammation.

Copyright © 2005 John Wiley & Sons, Inc.

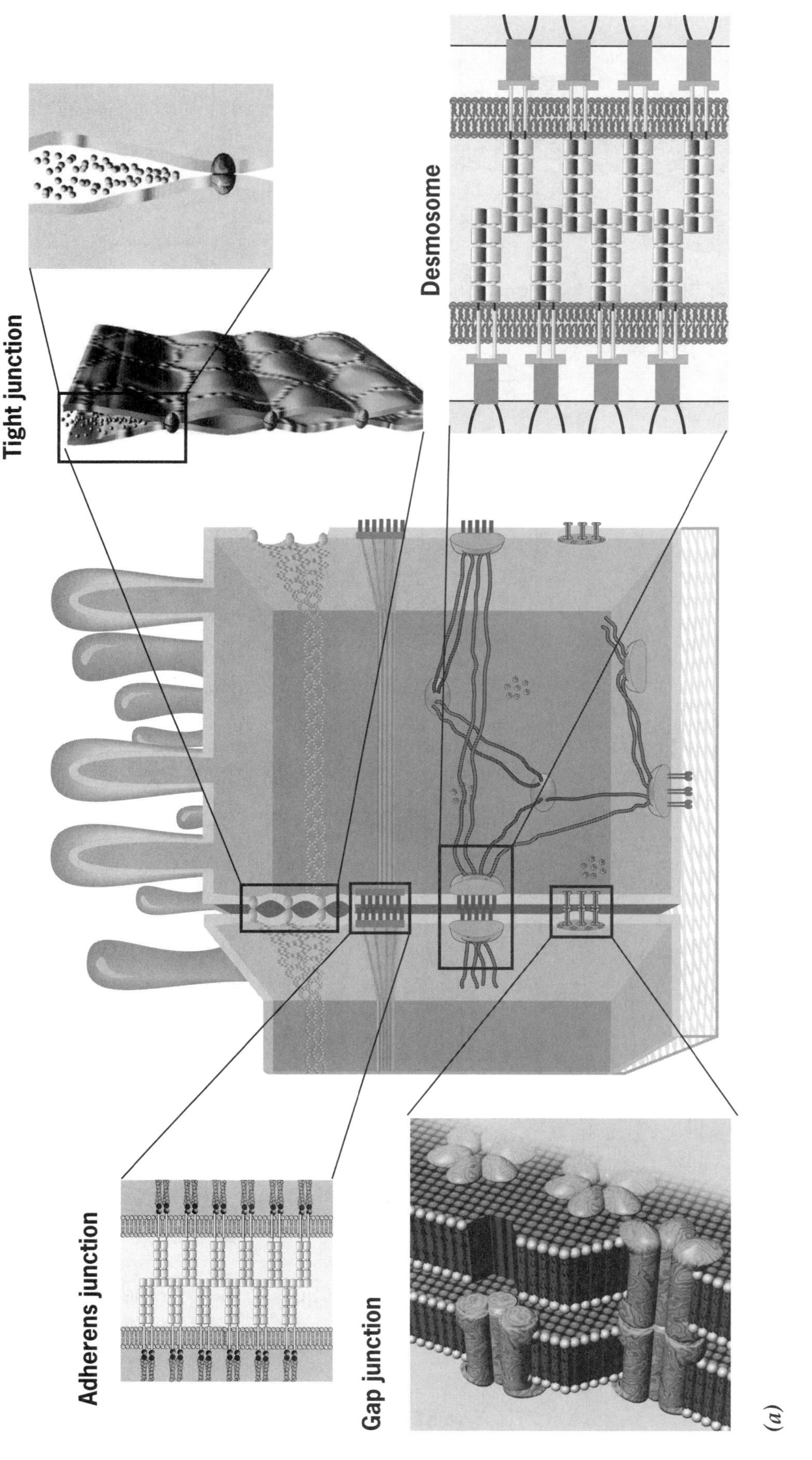

Figure 7.25a An intercellular junctional complex.

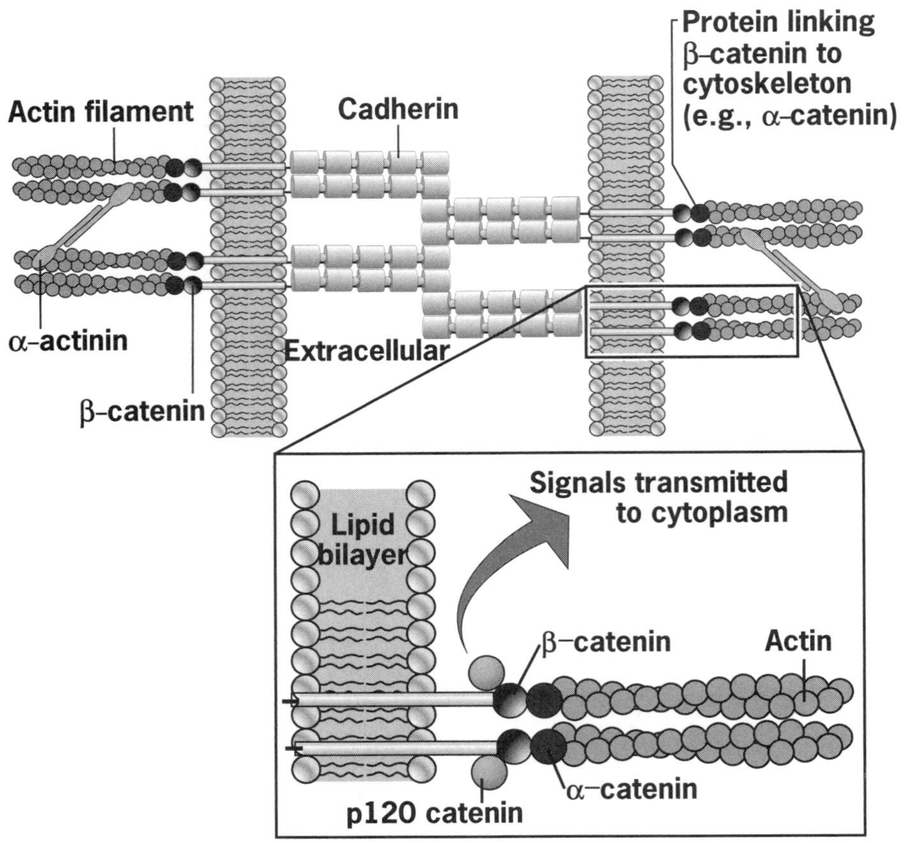

Figure 7.26 The structure of an adherens junction.

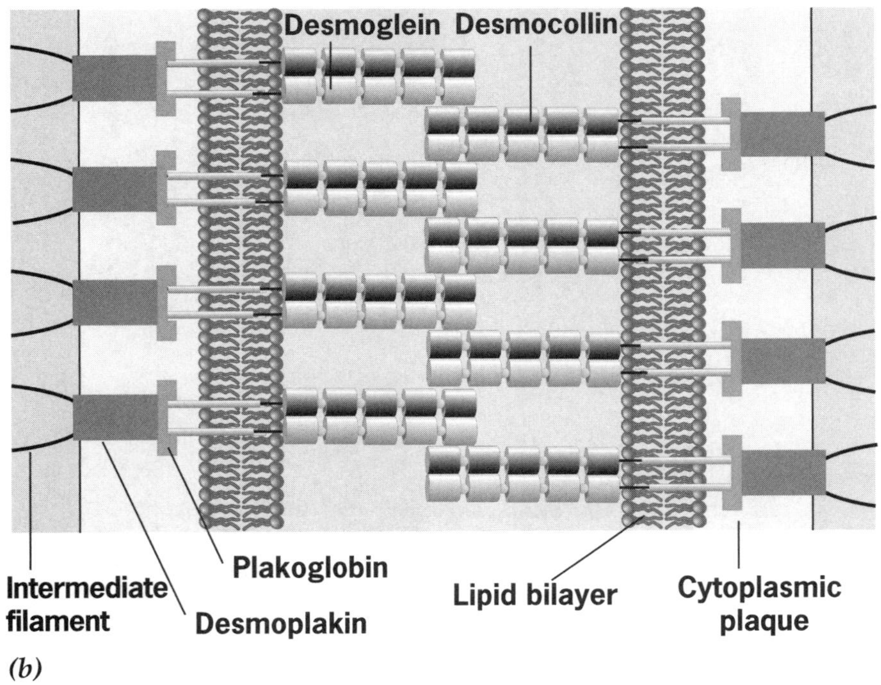

Figure 7.27b The structure of a desmosome.

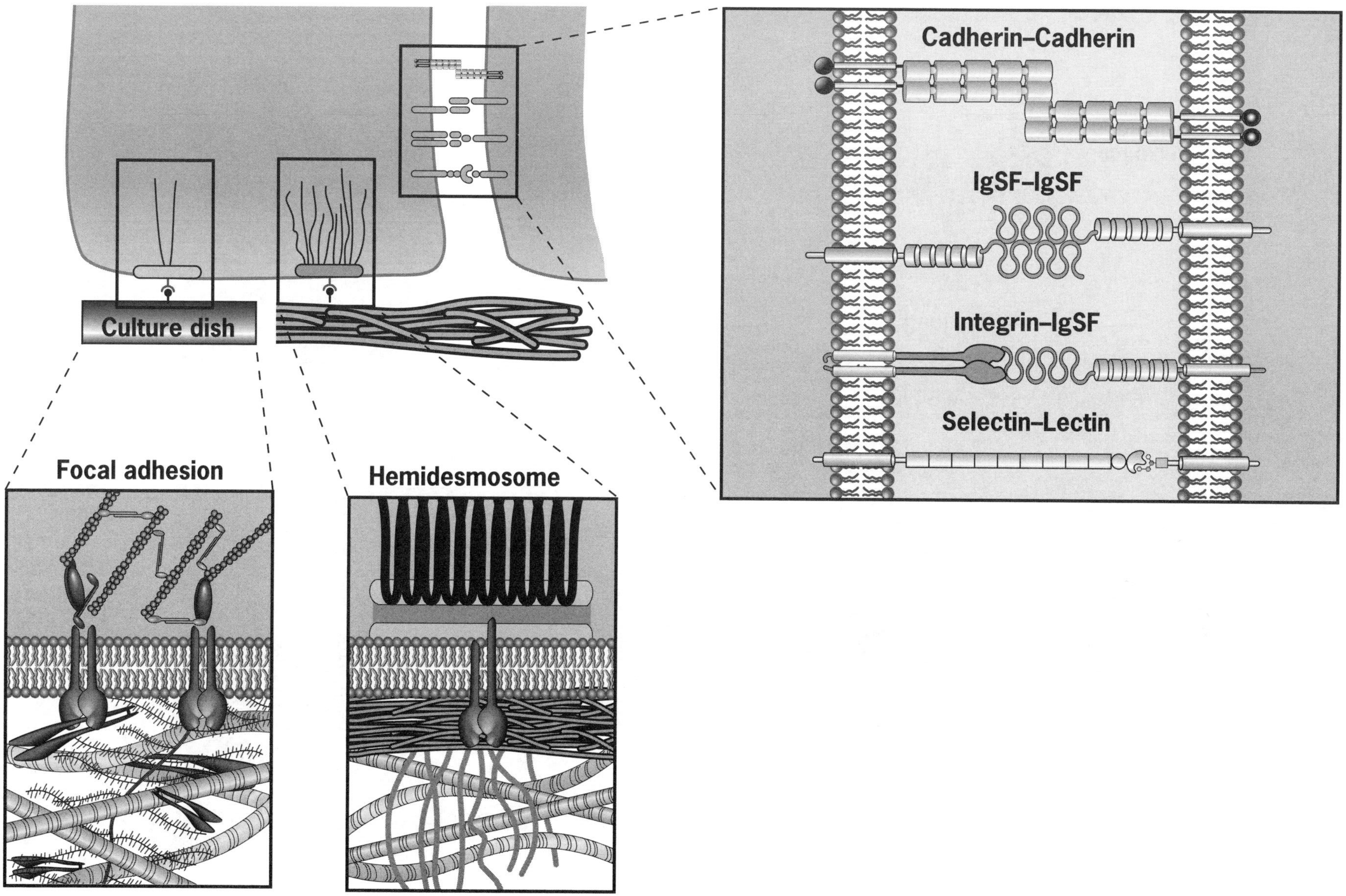

Figure 7.28 An overview of the types of interactions involving the cell surface.

Copyright © 2005 John Wiley & Sons, Inc.

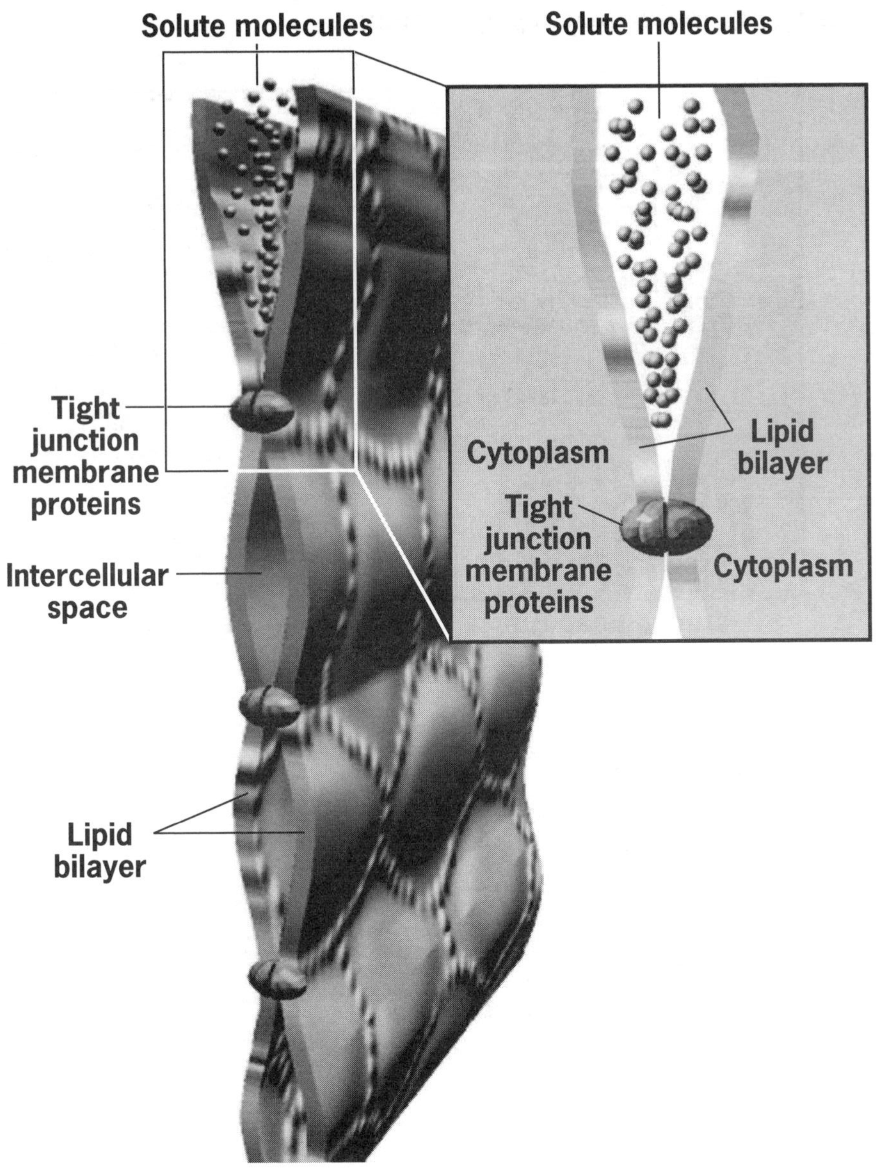

(b)

Figure 7.30b Tight junctions.

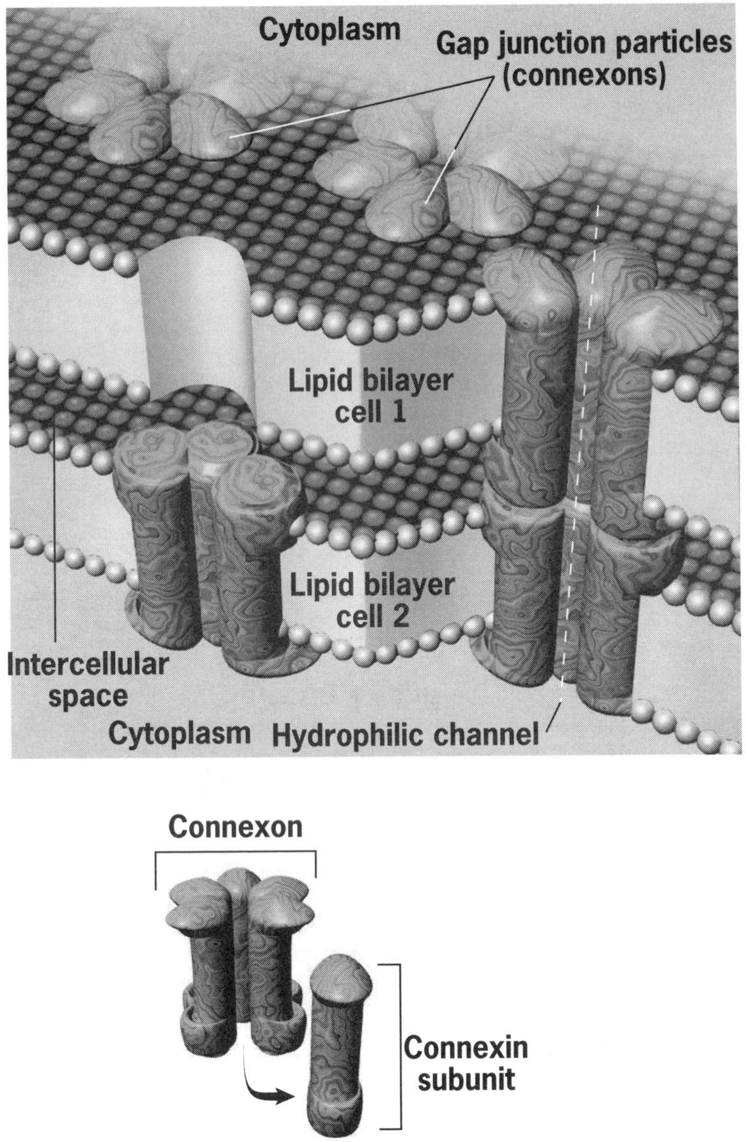

(b)

Figure 7.32b Gap junctions.

Copyright © 2005 John Wiley & Sons, Inc.

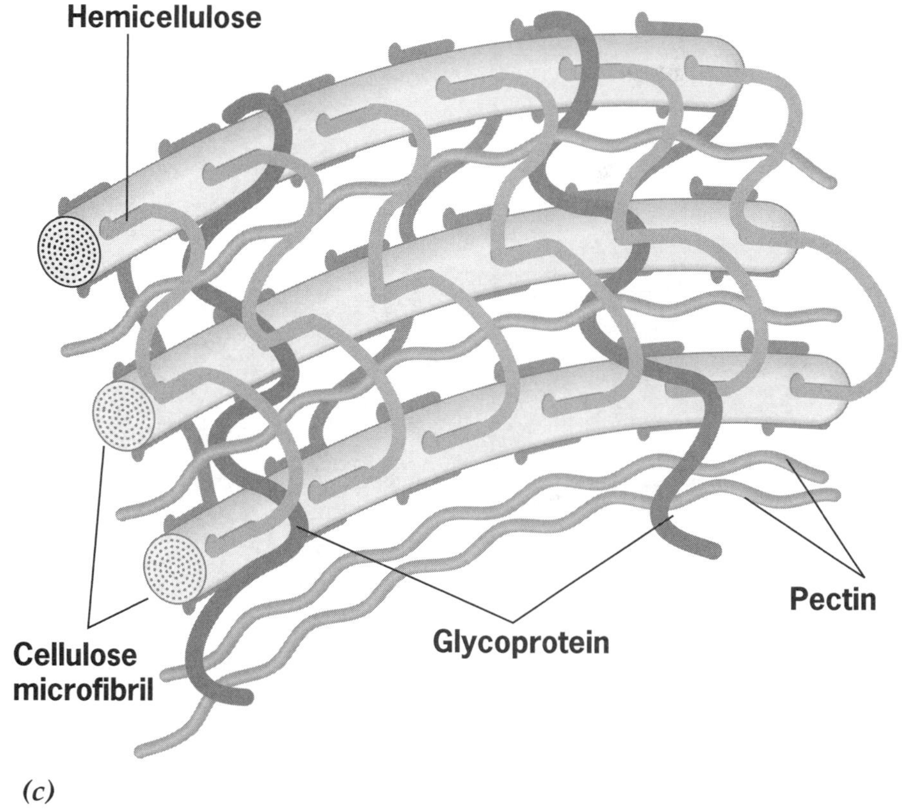

Figure 7.35c The plant cell wall.

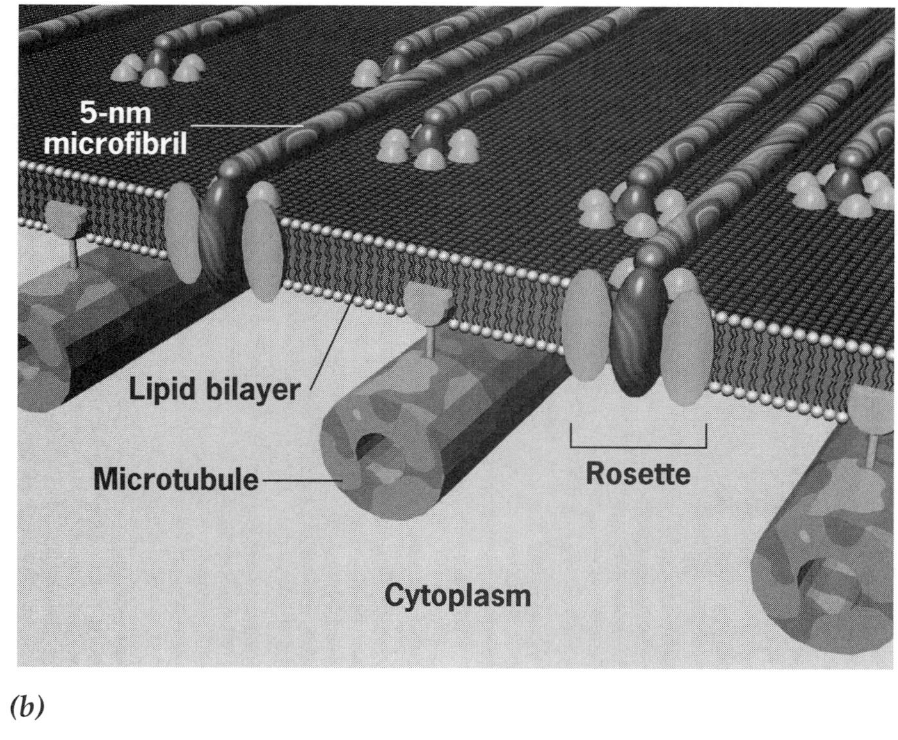

Figure 7.37b Synthesis of plant cell wall macromolecules.

Copyright © 2005 John Wiley & Sons, Inc.

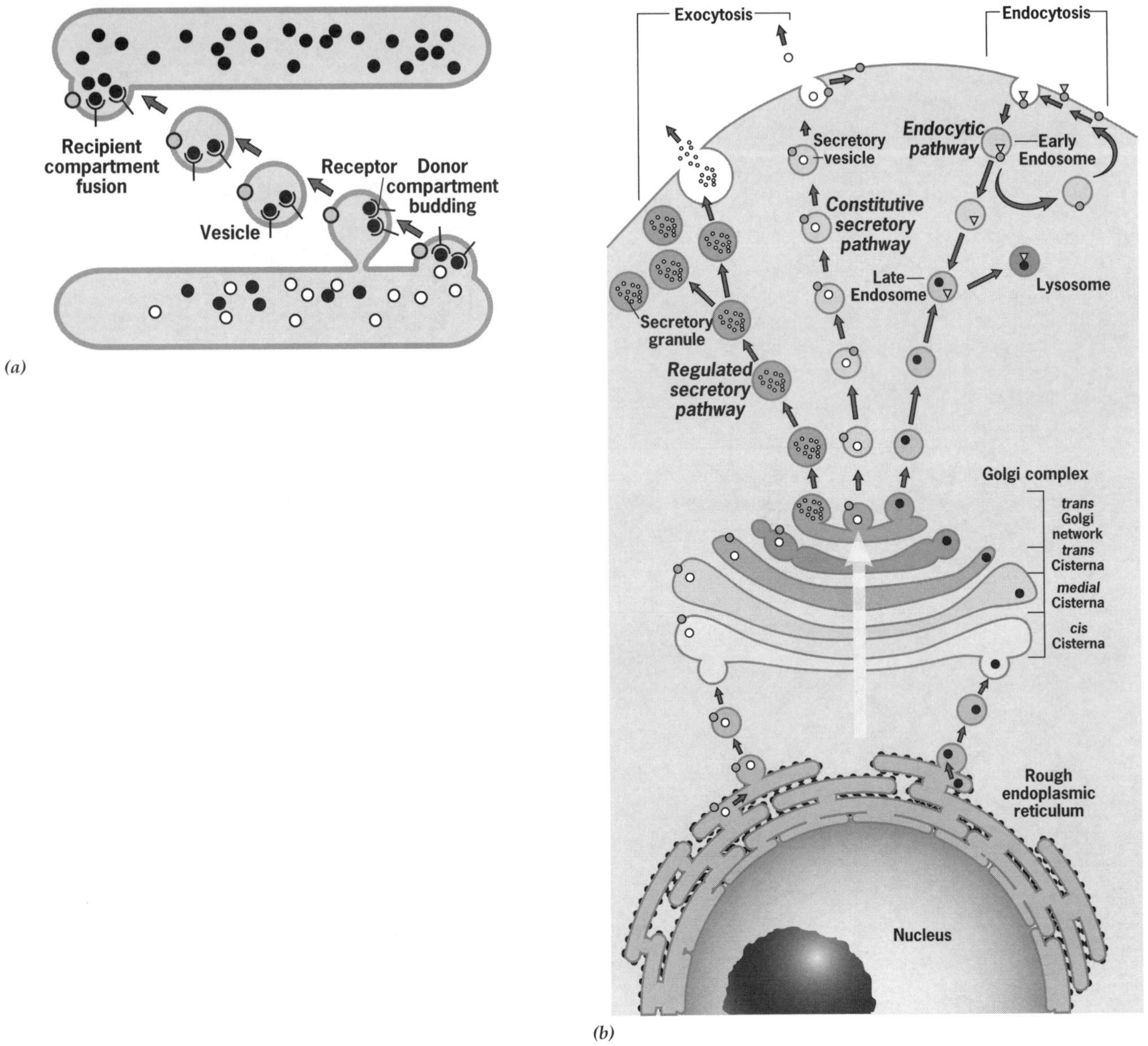

Figure 8.2 An overview of the biosynthetic/secretory and endocytic pathways that unite endomembranes into a dynamic, interconnected network.

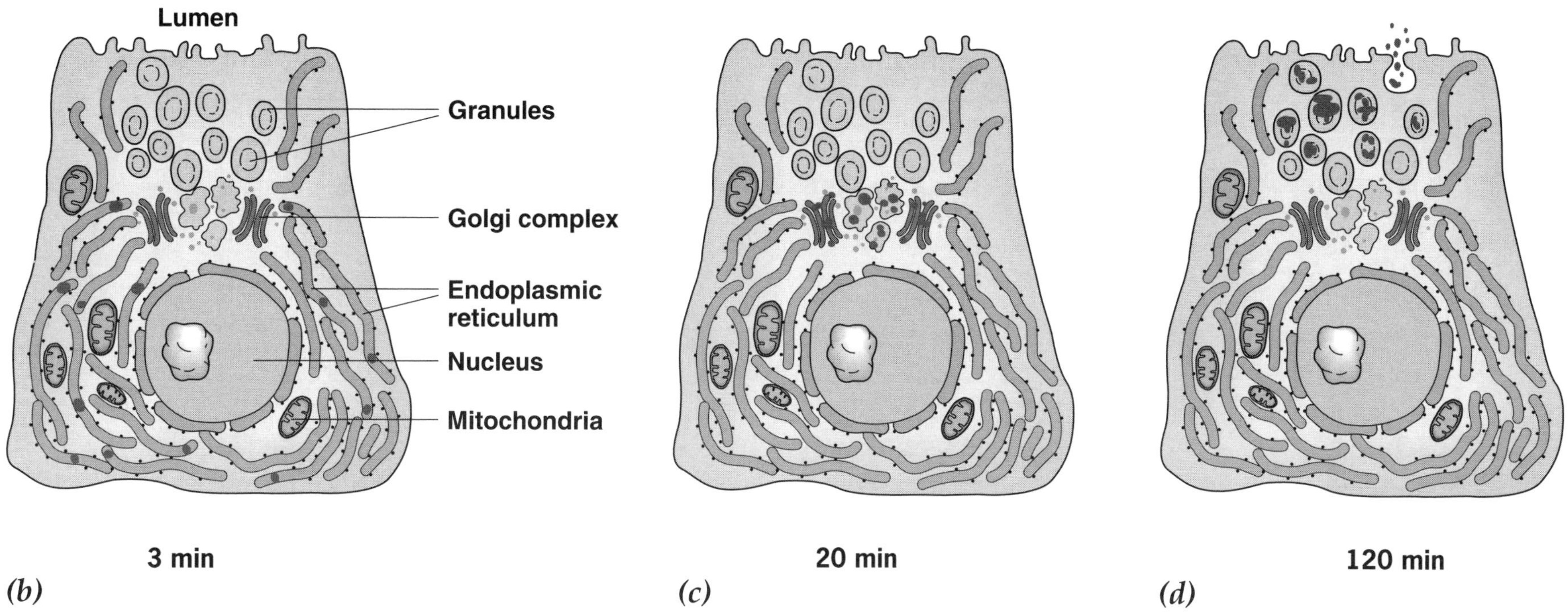

Figure 8.3b-d Autoradiography reveals the sites of synthesis and subsequent transport of secretory proteins.

Copyright © 2005 John Wiley & Sons, Inc.

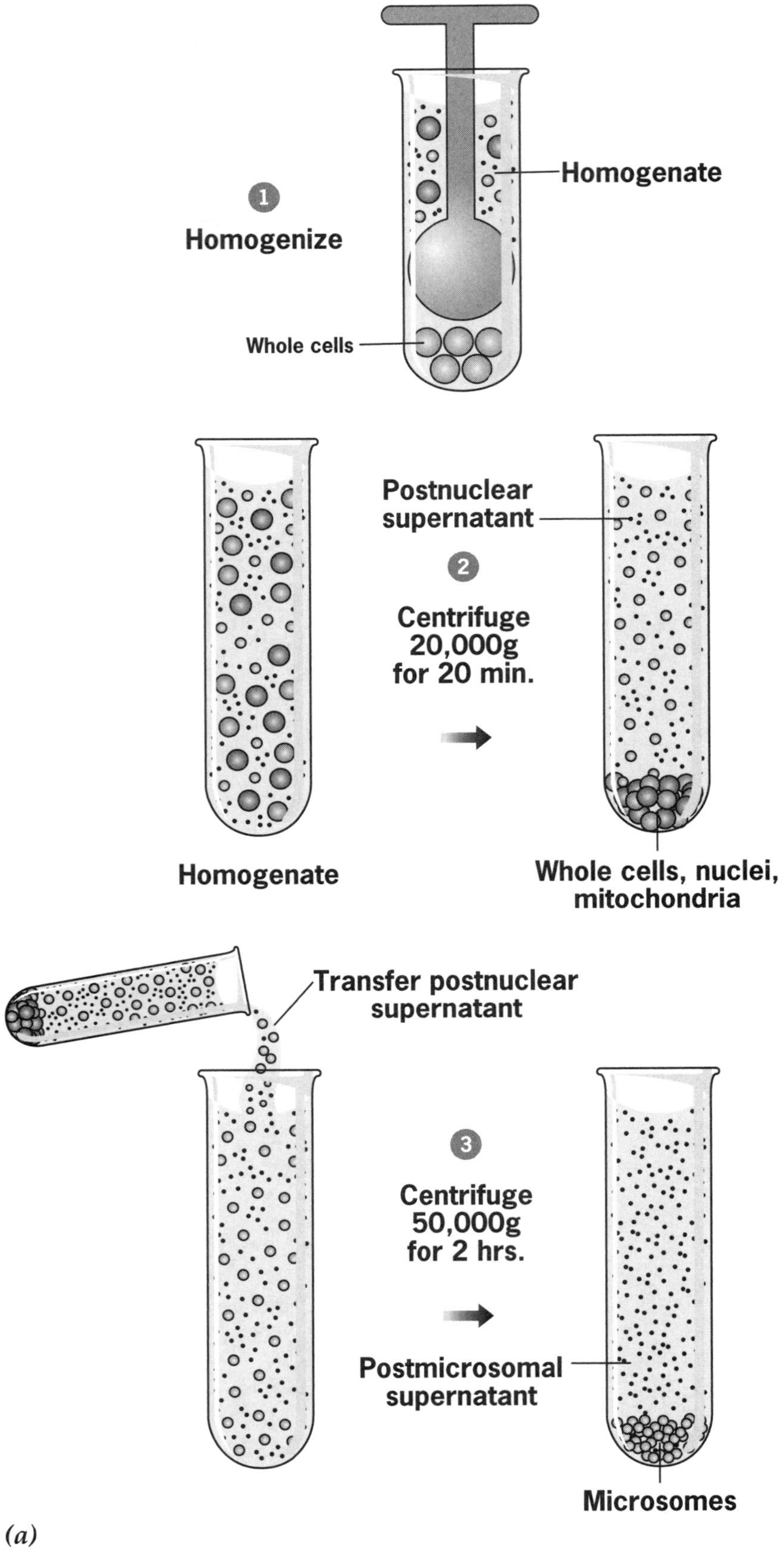

Figure 8.5a Isolation of a microsomal fraction by differential centrifugation.

Copyright © 2005 John Wiley & Sons, Inc.

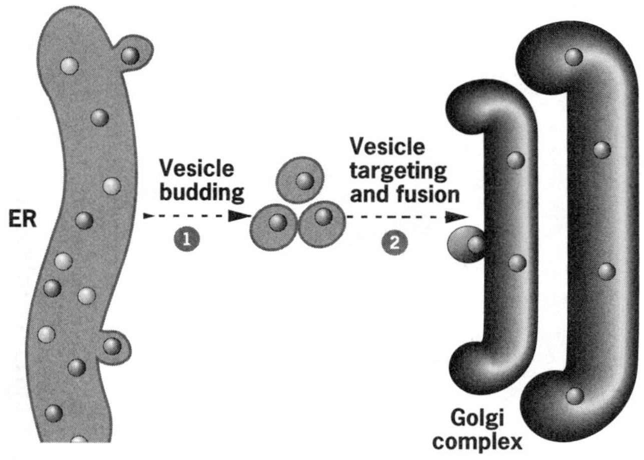

(a)

Figure 8.7a The use of genetic mutants in the study of secretion.

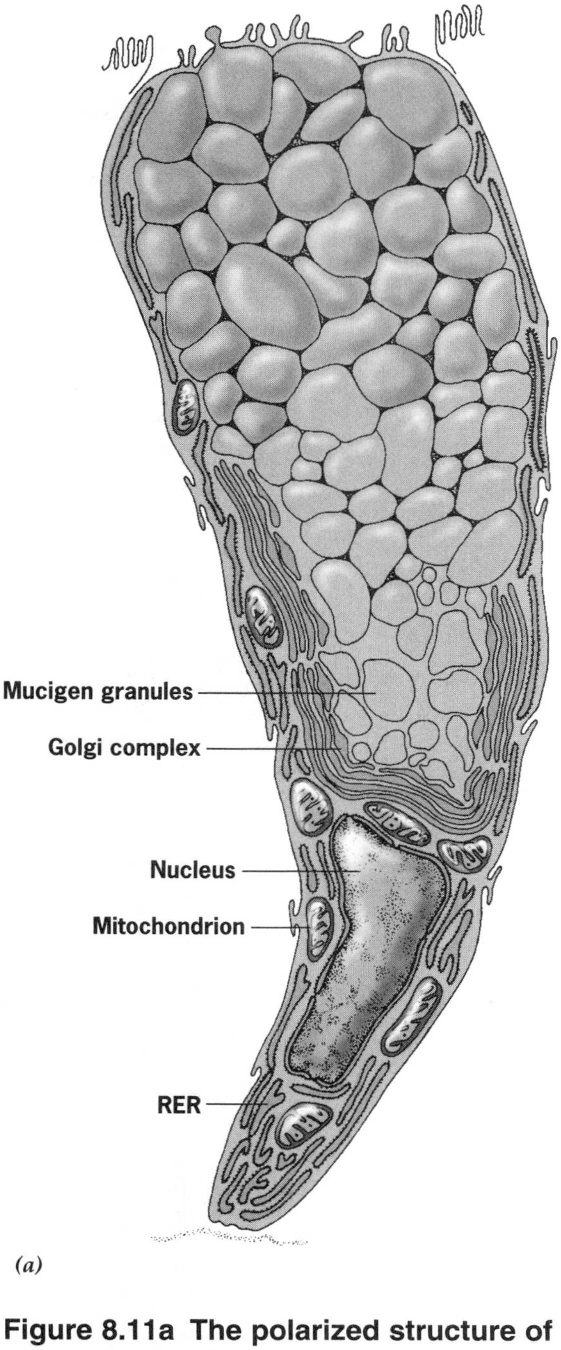

(a)

Figure 8.11a The polarized structure of a secretory cell.

Copyright © 2005 John Wiley & Sons, Inc.

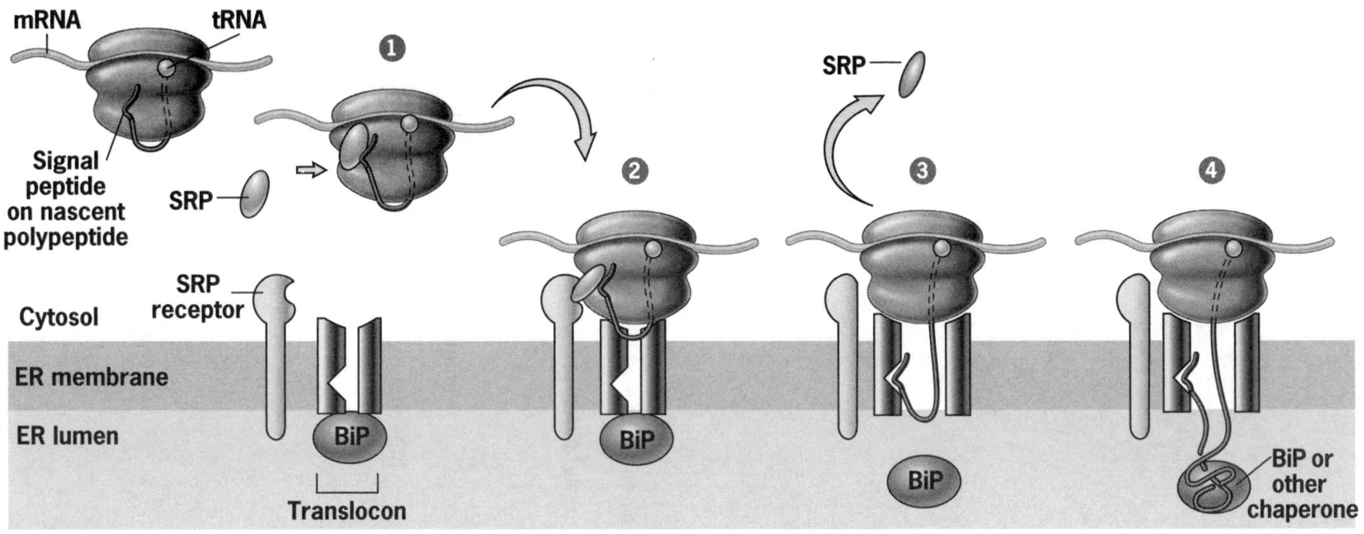

Figure 8.12 A schematic model for the synthesis of a secretory protein (or a lysosomal enzyme) on a membrane-bound ribosome of the RER.

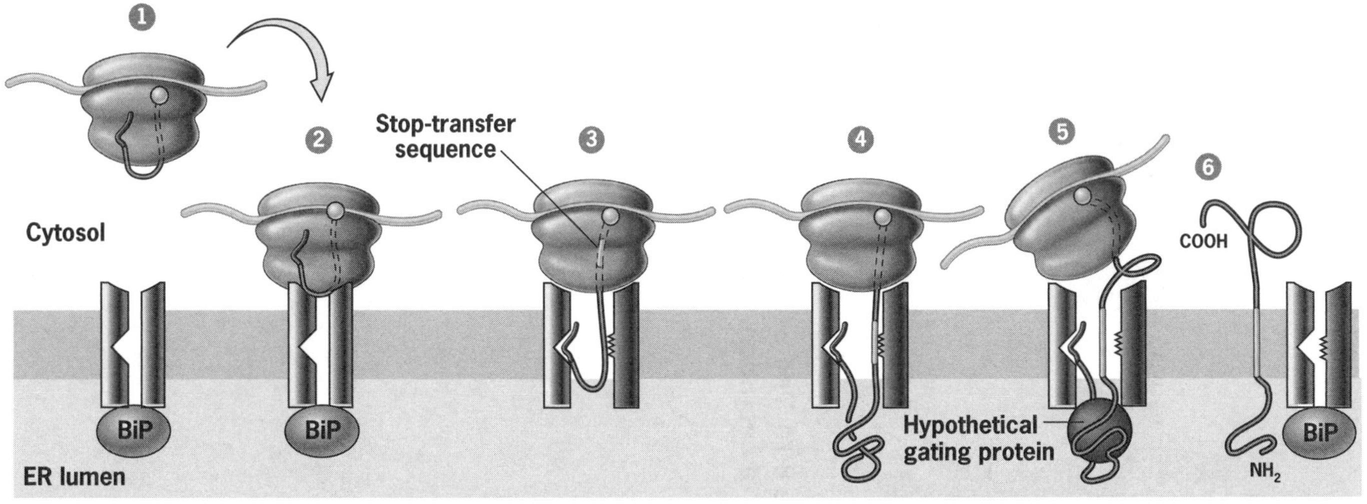

Figure 8.13 A schematic model for the synthesis of an integral membrane protein.

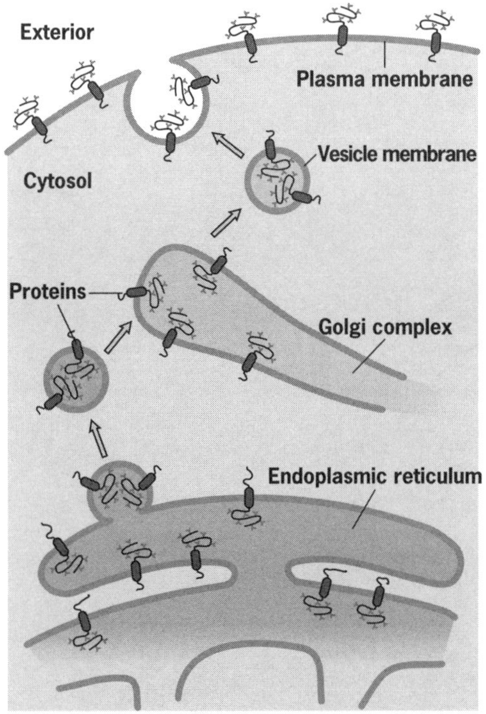

Figure 8.14 Maintenance of membrane asymmetry.

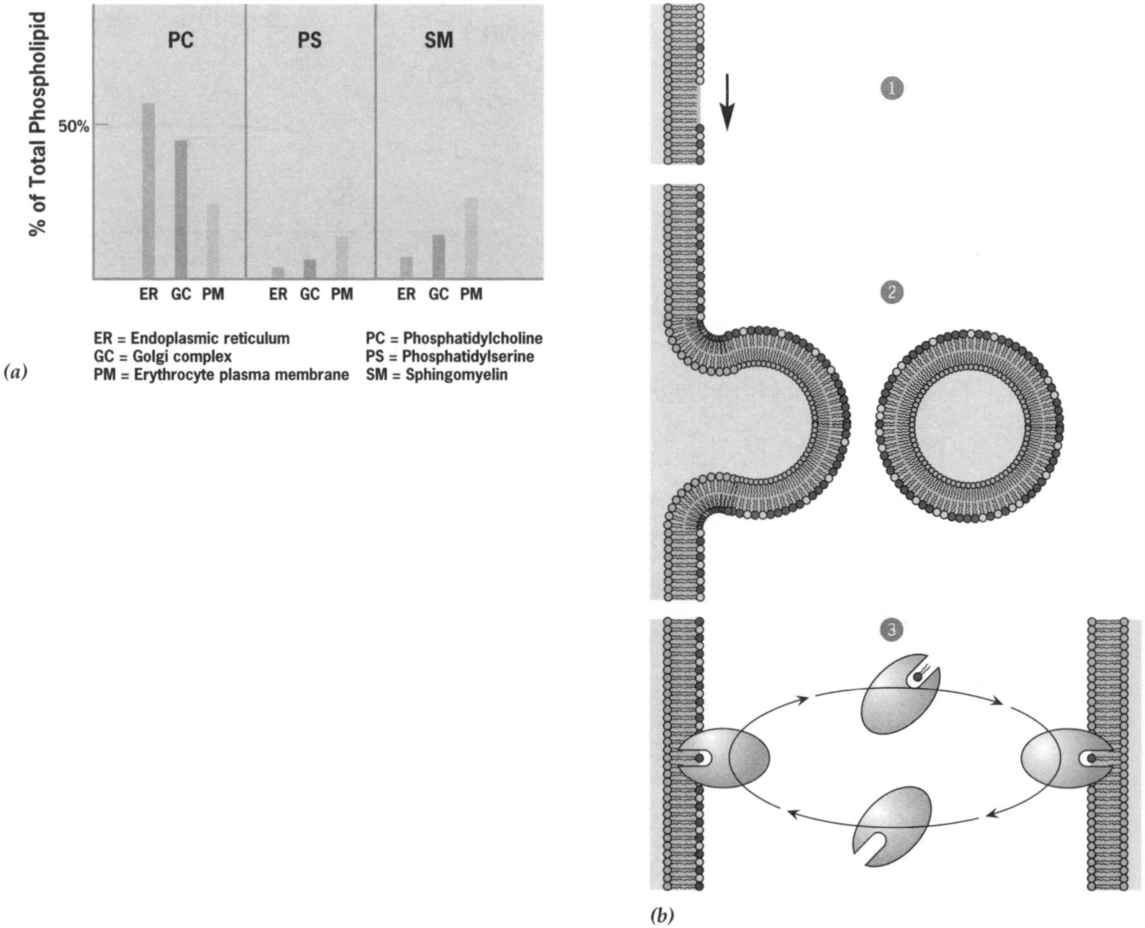

Figure 8.15 Modifying the lipid composition of membranes.

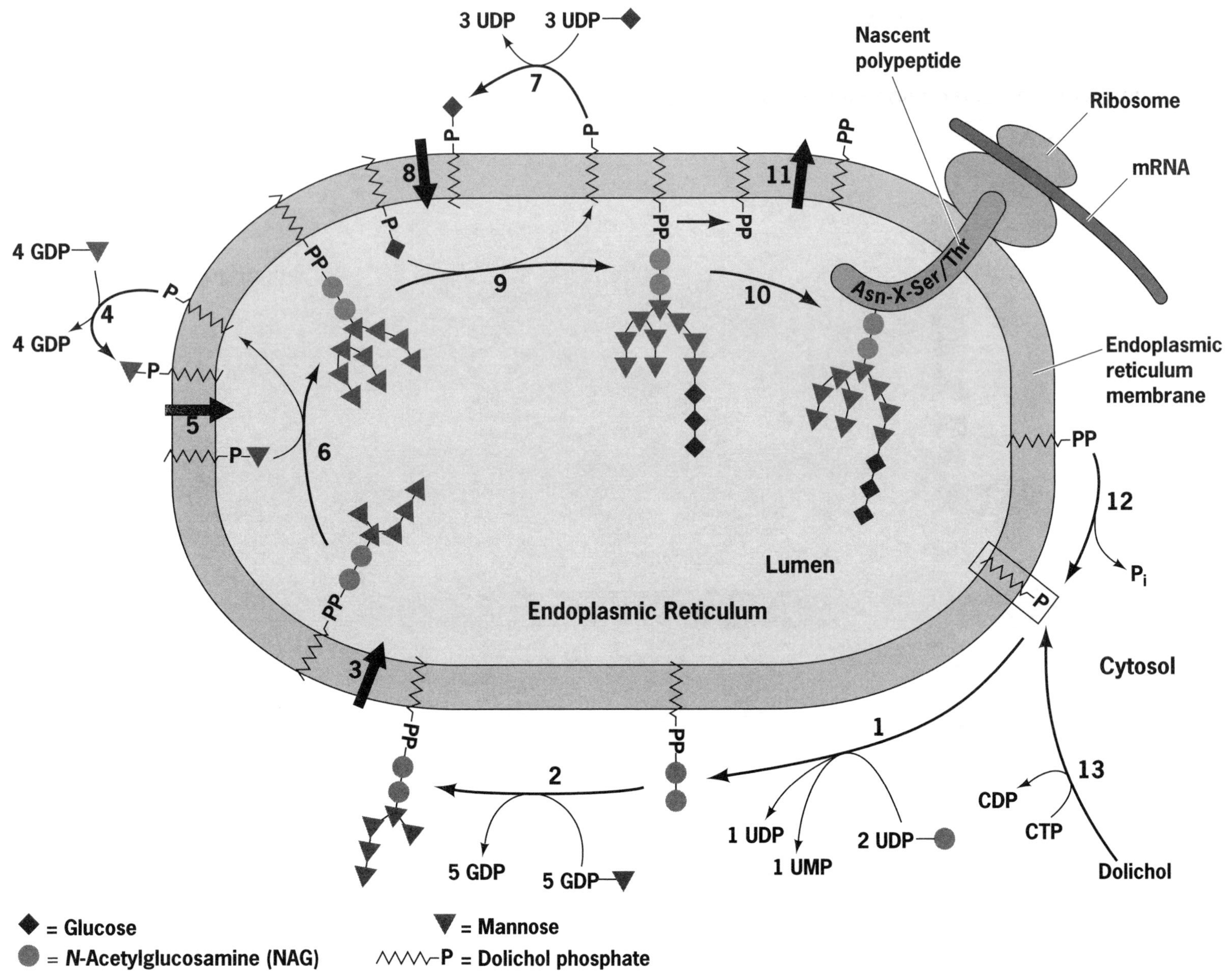

Figure 8.17 The steps in the synthesis of the core portion of an *N*-linked oligosaccharide in the rough ER.

Copyright © 2005 John Wiley & Sons, Inc.

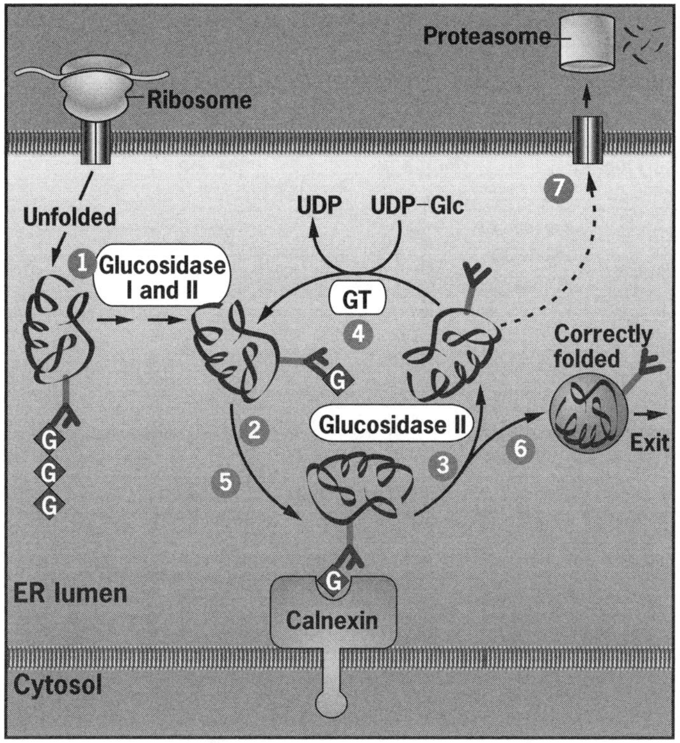

Figure 8.18 Quality control: ensuring that misfolded proteins do not leave the ER.

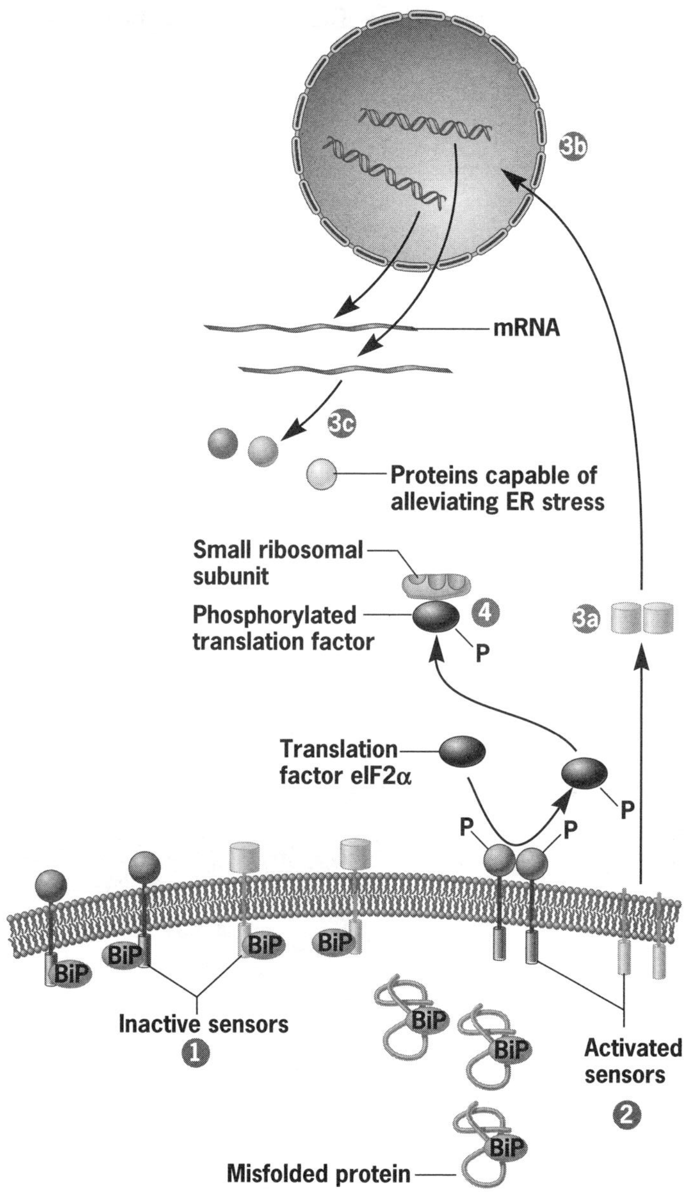

Figure 8.19 A model of the mammalian unfolded protein response (UPR).

Figure 8.21a The Golgi complex.

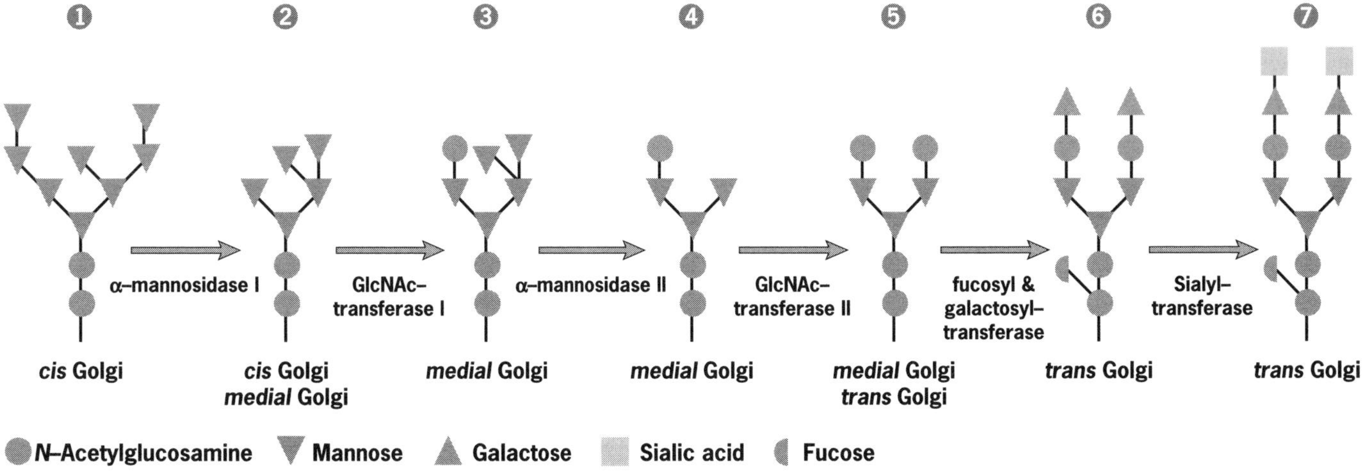

Figure 8.23 Steps in the glycosylation of a typical mammalian *N*-linked oligosaccharide in the Golgi complex.

Copyright © 2005 John Wiley & Sons, Inc.

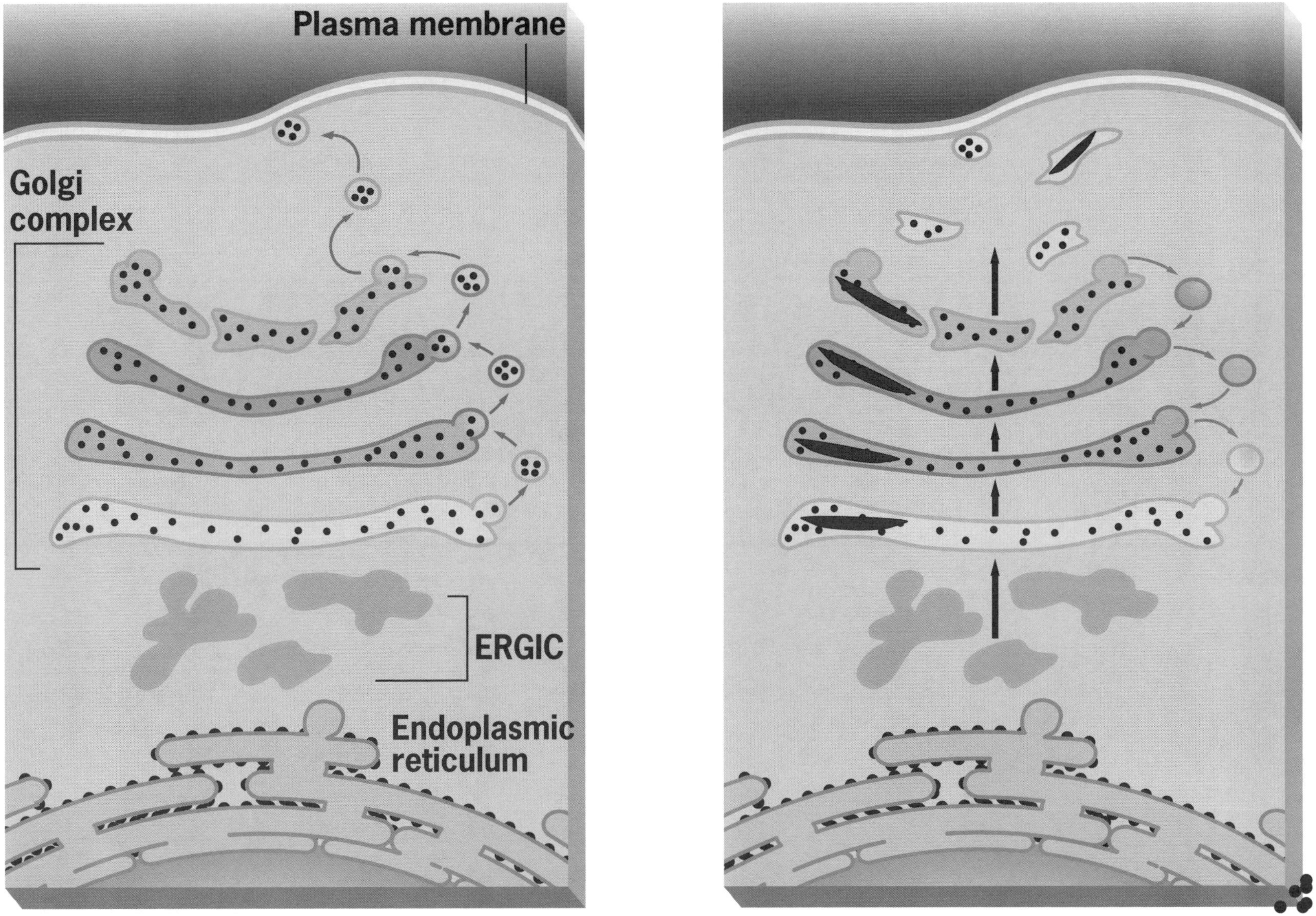

(*a*) **Vesicular transport model** (*b*) **Cisternal maturation model**

Figure 8.24 The dynamics of transport through the Golgi complex.

Copyright © 2005 John Wiley & Sons, Inc.

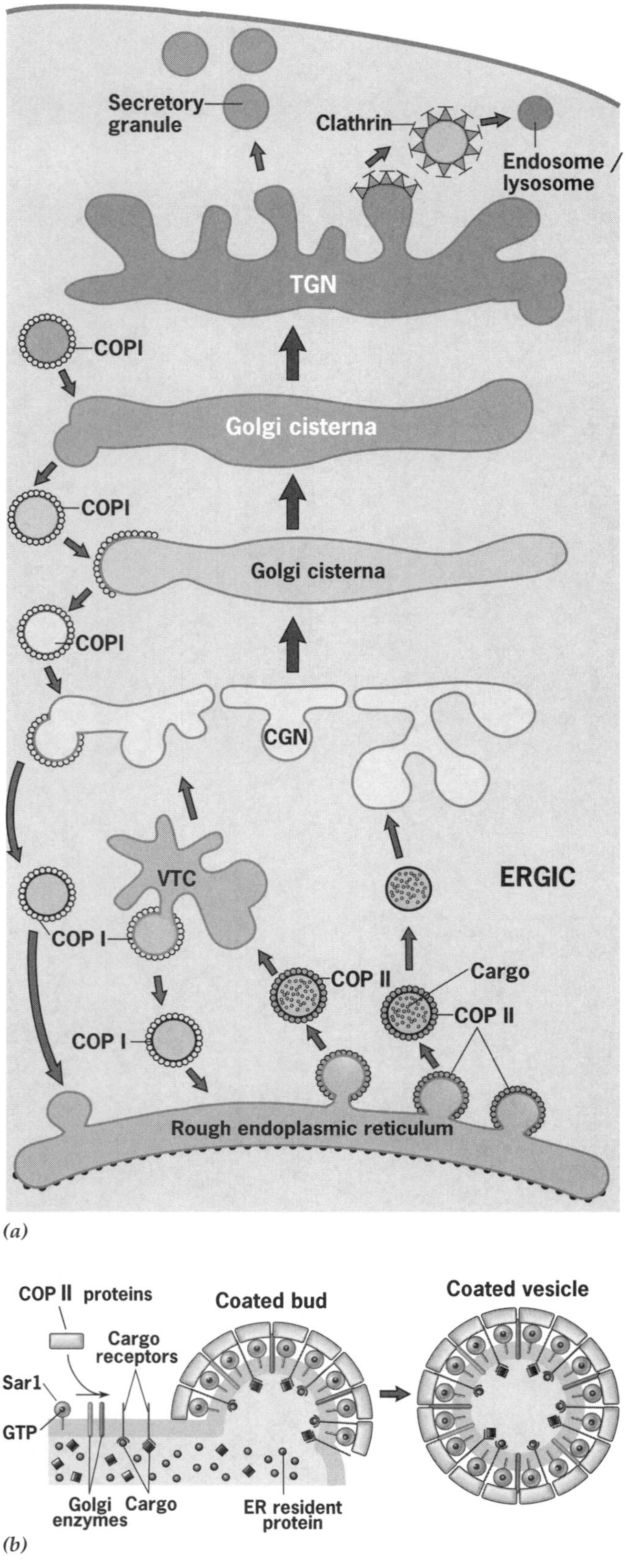

Figure 8.26 Proposed movement of materials by vesicular transport between membranous compartments of the biosynthetic/secretory pathway.

Copyright © 2005 John Wiley & Sons, Inc.

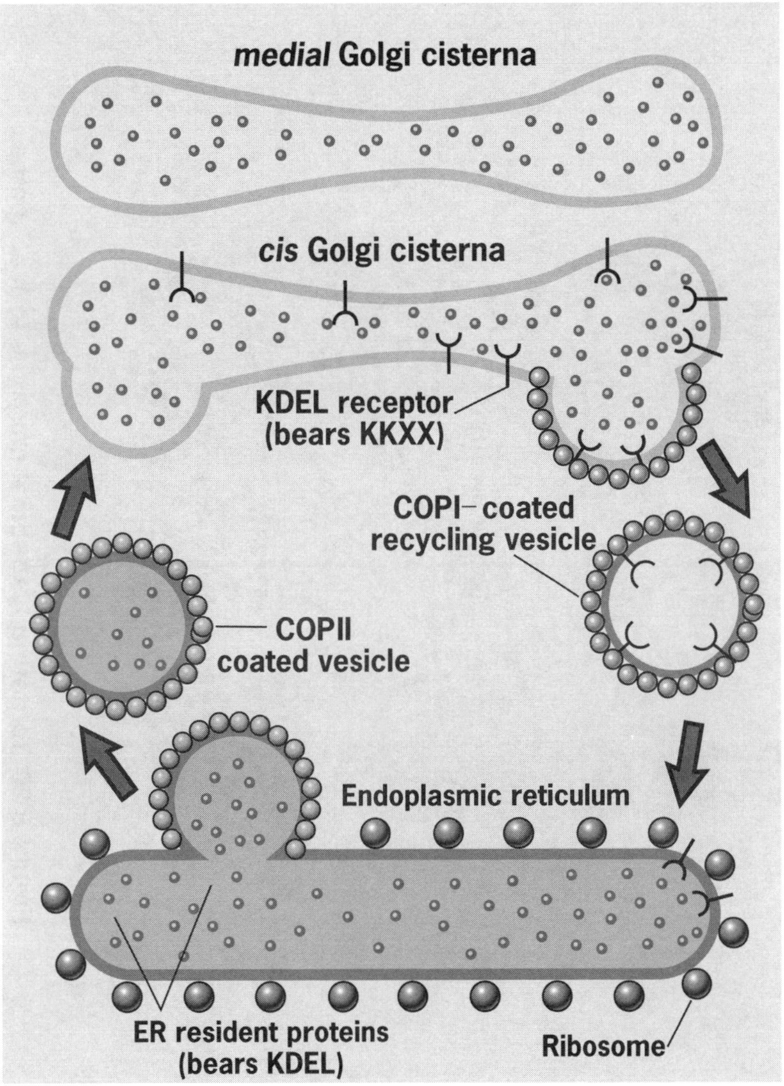

Figure 8.28 Retrieving ER proteins.

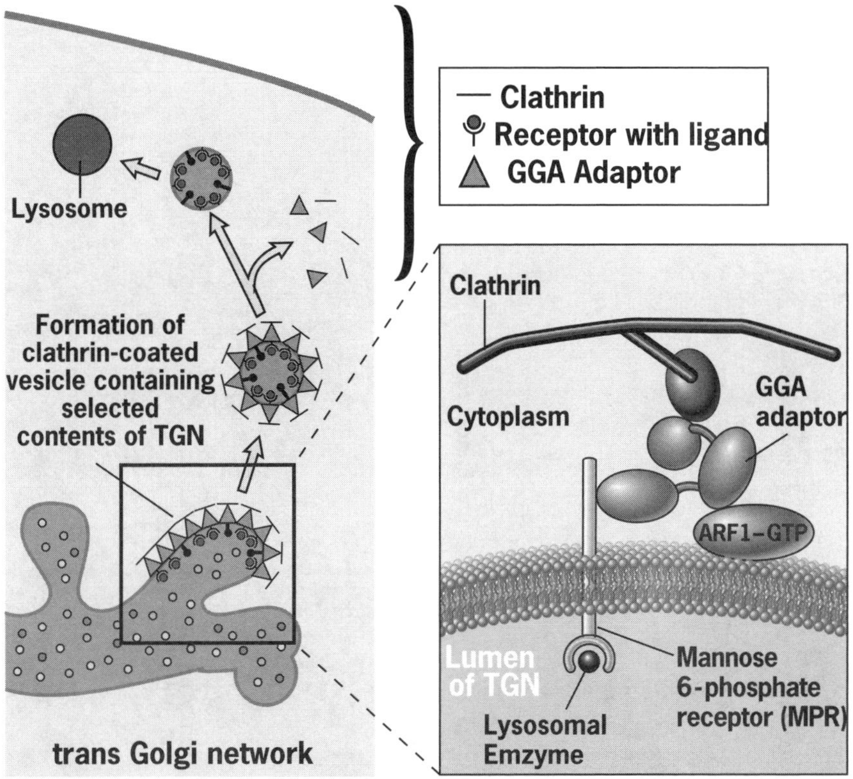

Figure 8.30 The formation of clathrin-coated vesicles at the TGN.

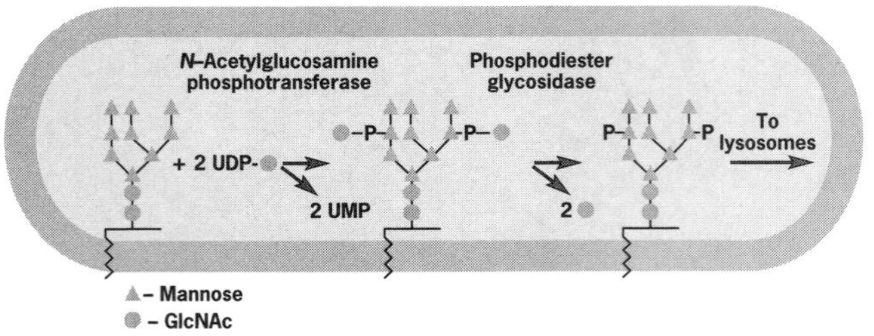

(a)

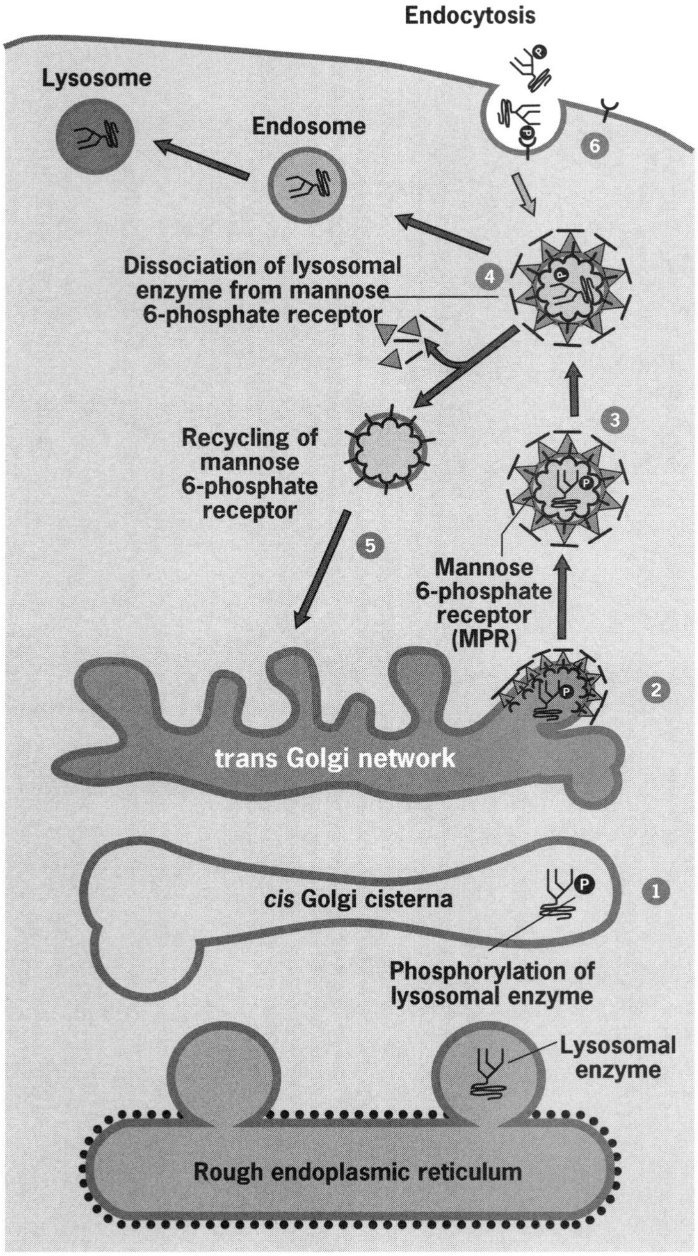

(b)

Figure 8.29 Targeting lysosomal enzymes to lysosomes.

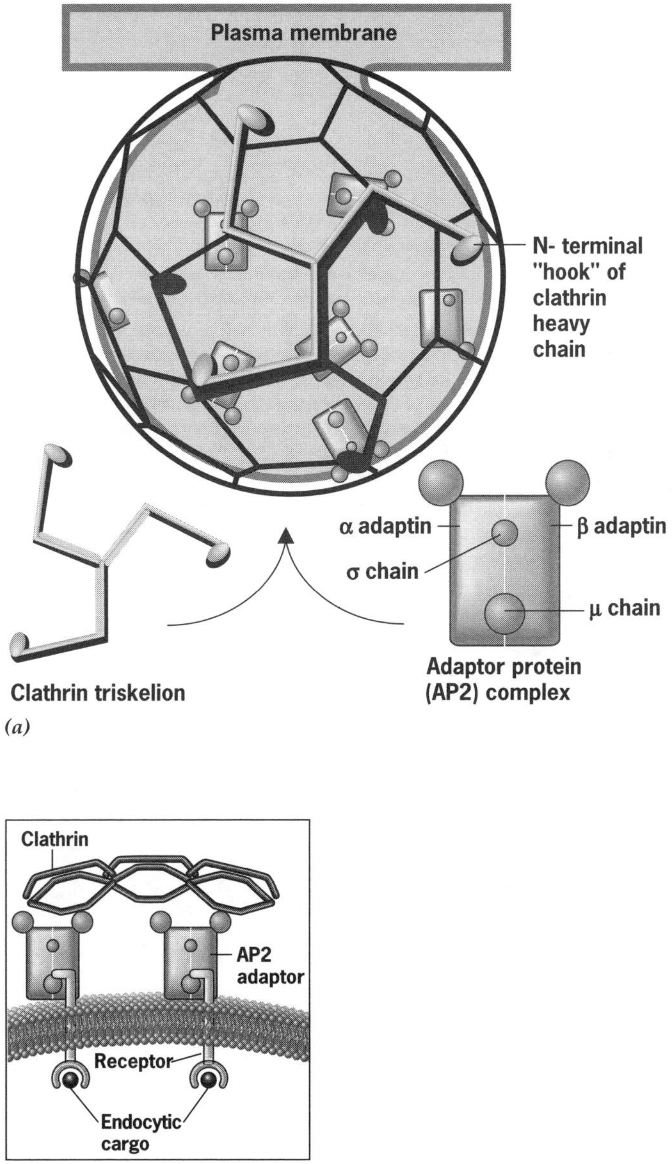

Figure 8.40 Molecular organization of a coated vesicle.

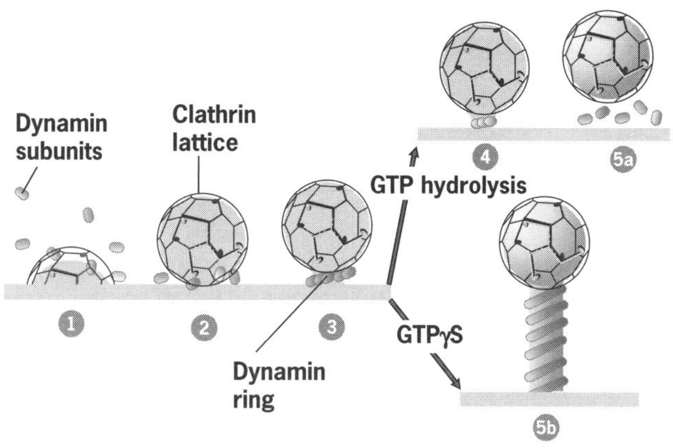

Figure 8.41a The role of dynamin in the formation of clathrin-coated vesicles.

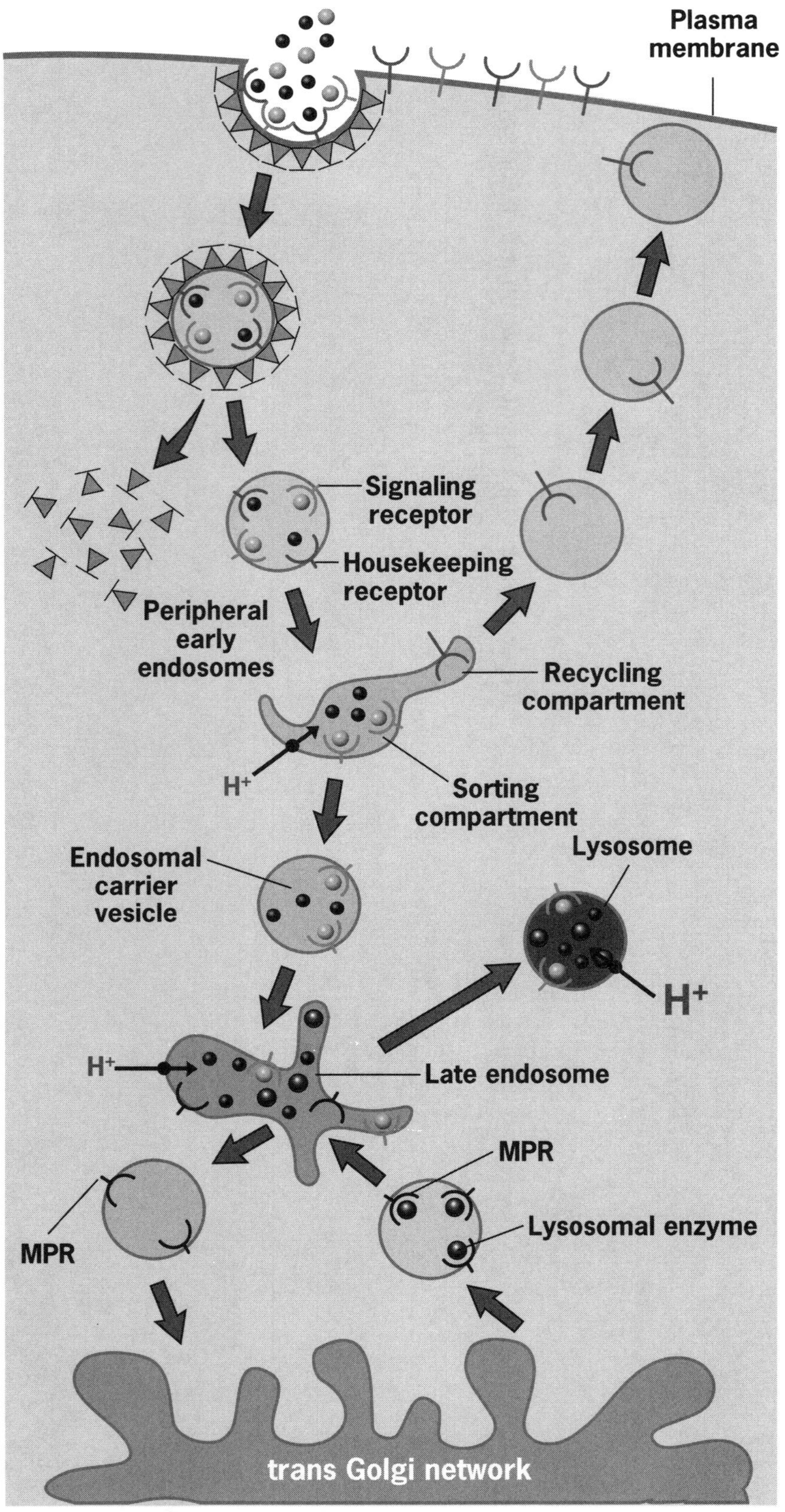

(a)

Figure 8.42a The endocytic pathway.

Copyright © 2005 John Wiley & Sons, Inc.

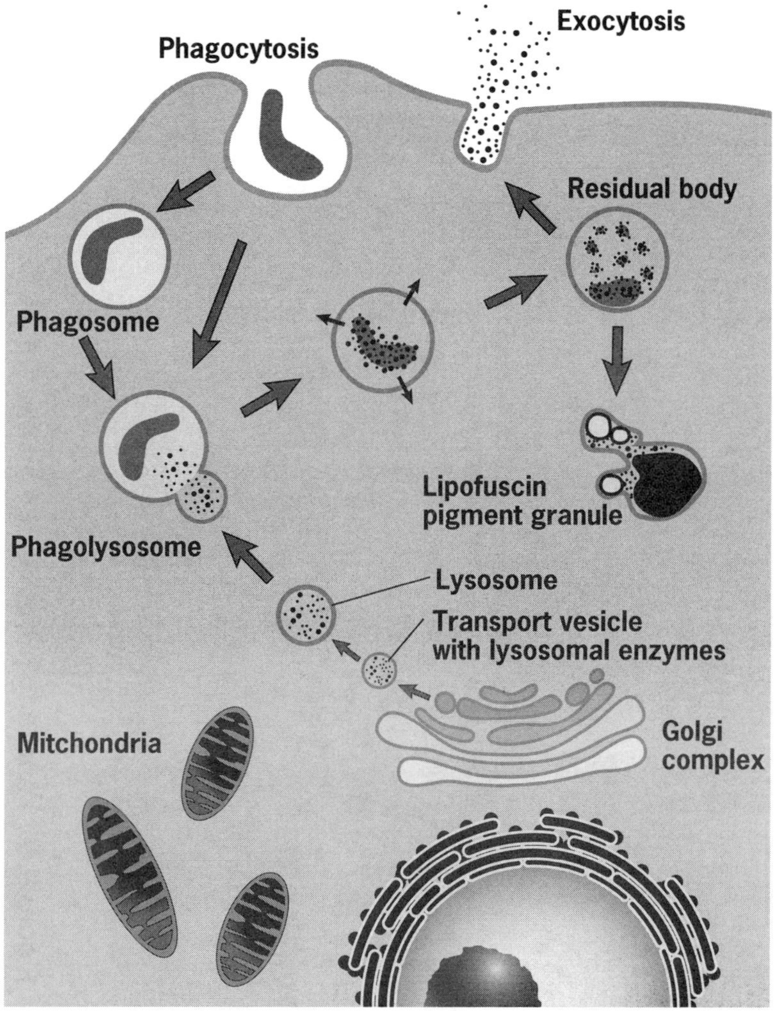

Figure 8.46 A summary of the phagocytic pathway.

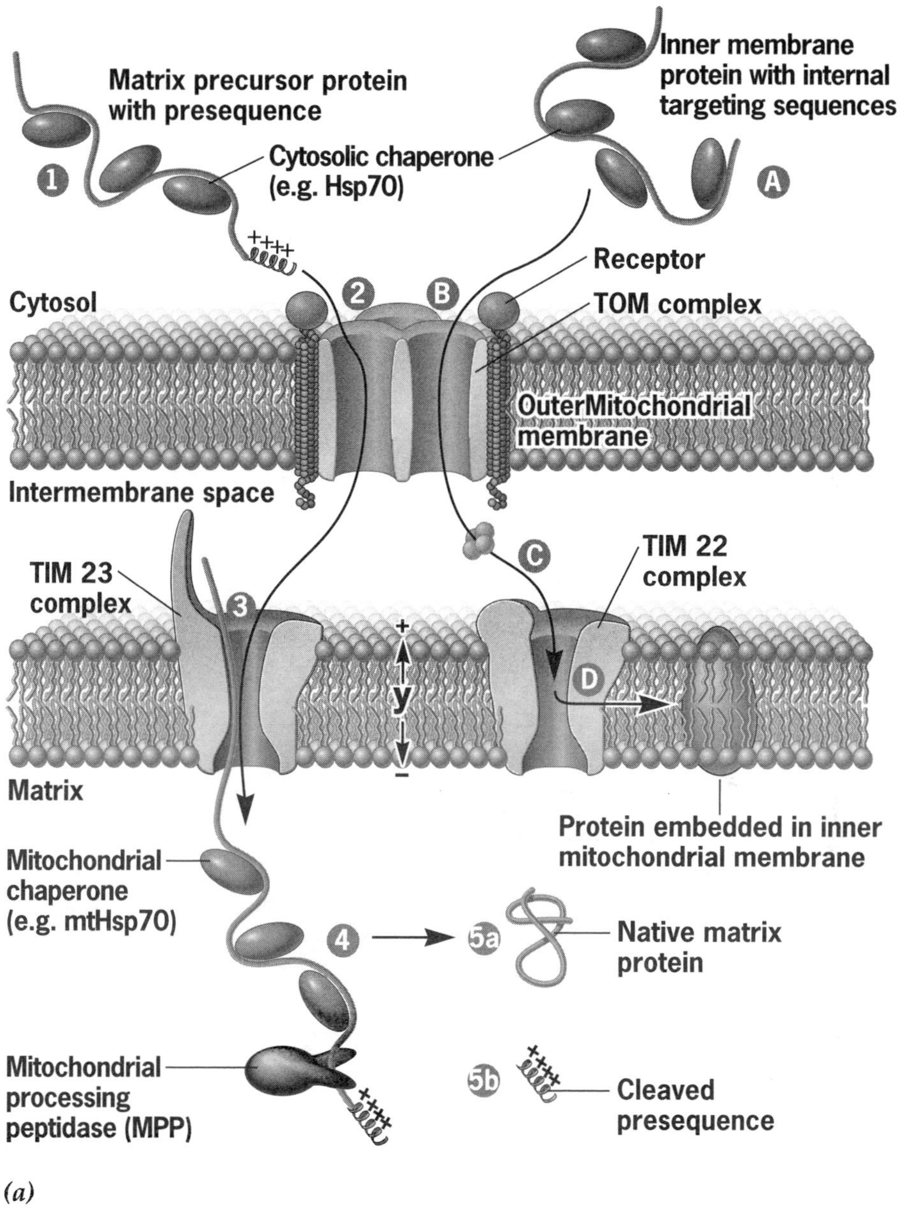

Figure 8.47a Importing proteins into a mitochondrion.

Copyright © 2005 John Wiley & Sons, Inc.

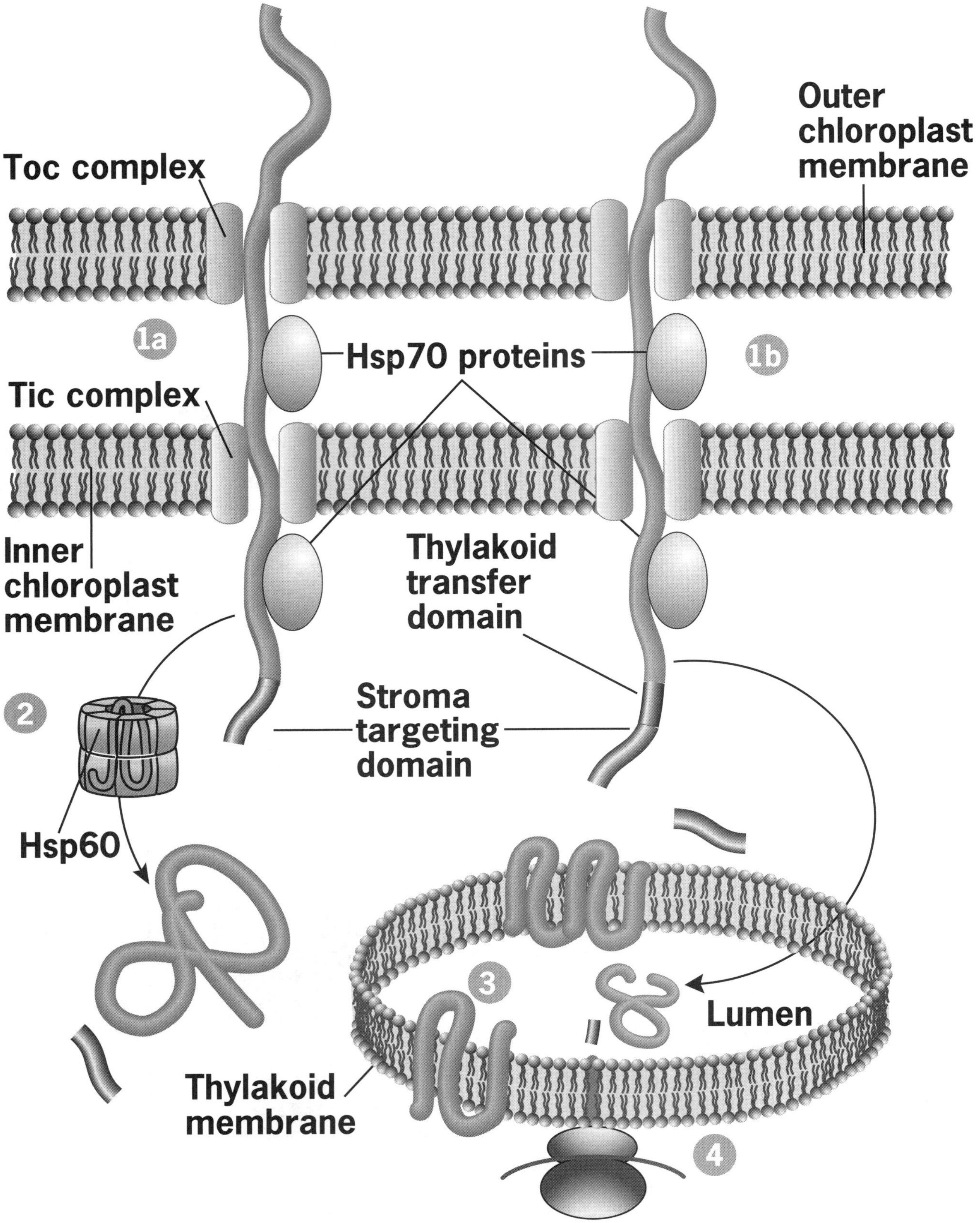

Figure 8.48 Importing proteins into a chloroplast.

Key to Cytoskeletal Functions

(1) Structure and Support (2) Intracellular Transport (3) Contractility and Motility (4) Spatial Organization

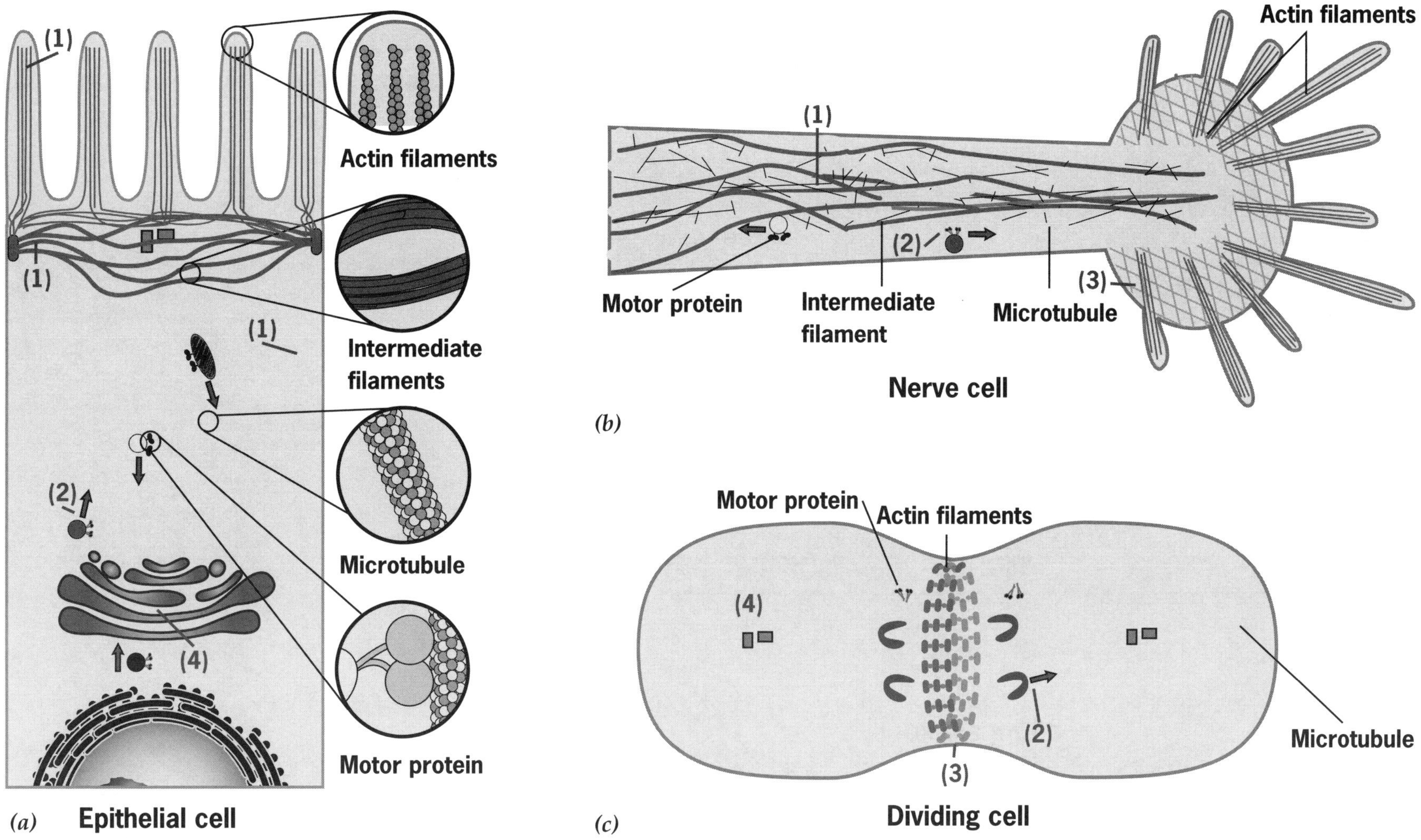

Figure 9.1 An overview of the structure and functions of the cytoskeleton.

Copyright © 2005 John Wiley & Sons, Inc.

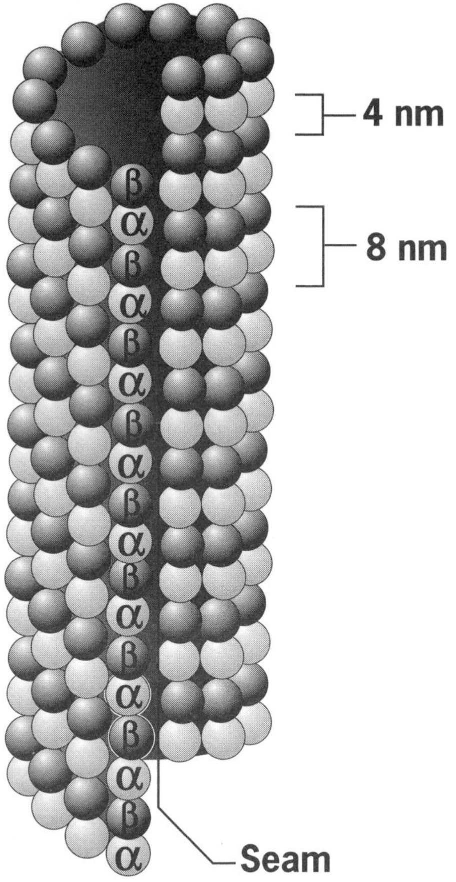

(d)

Figure 9.8d The structure of microtubules.

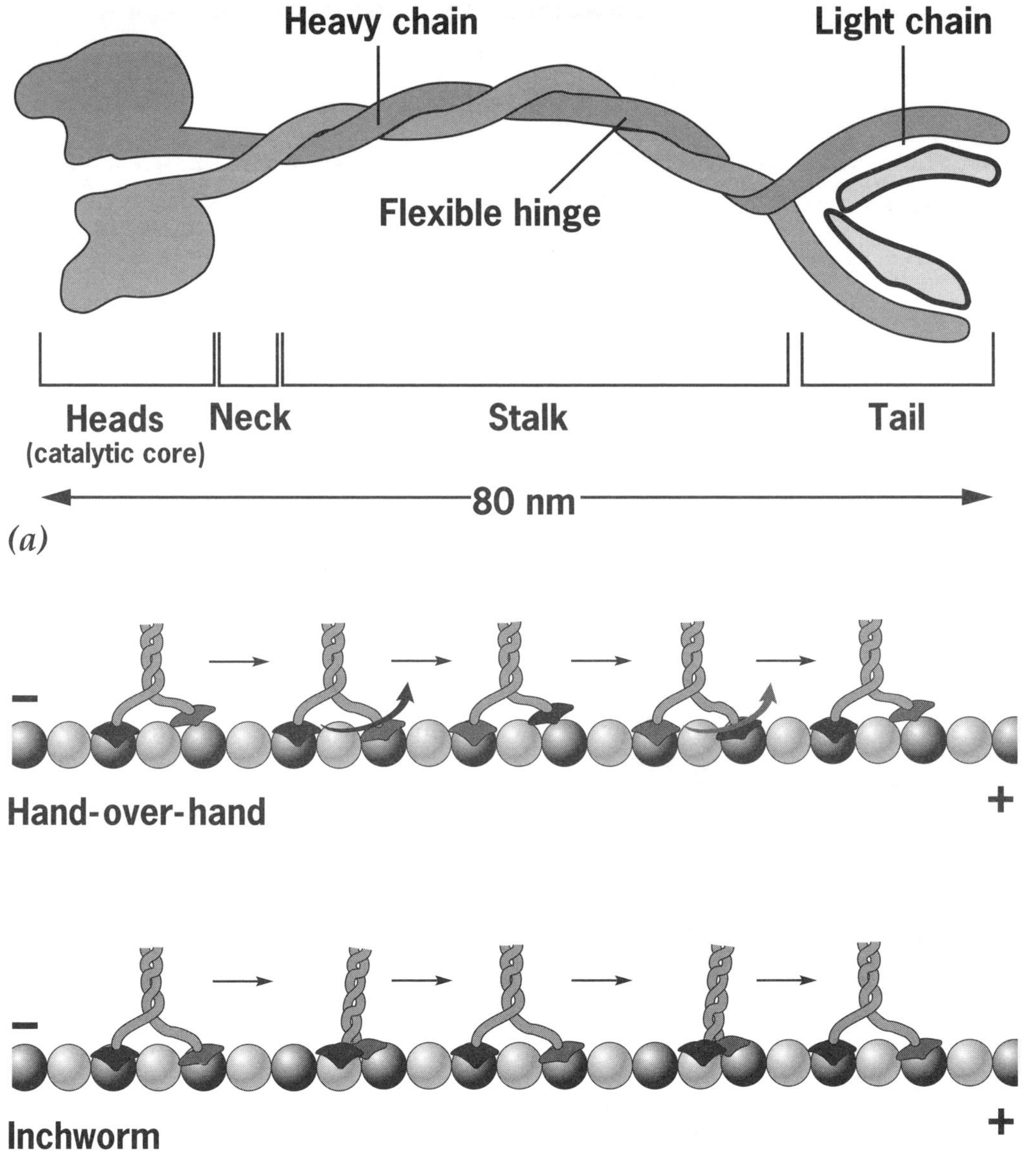

(b)

Figure 9.15a,b Kinesin.

Copyright © 2005 John Wiley & Sons, Inc.

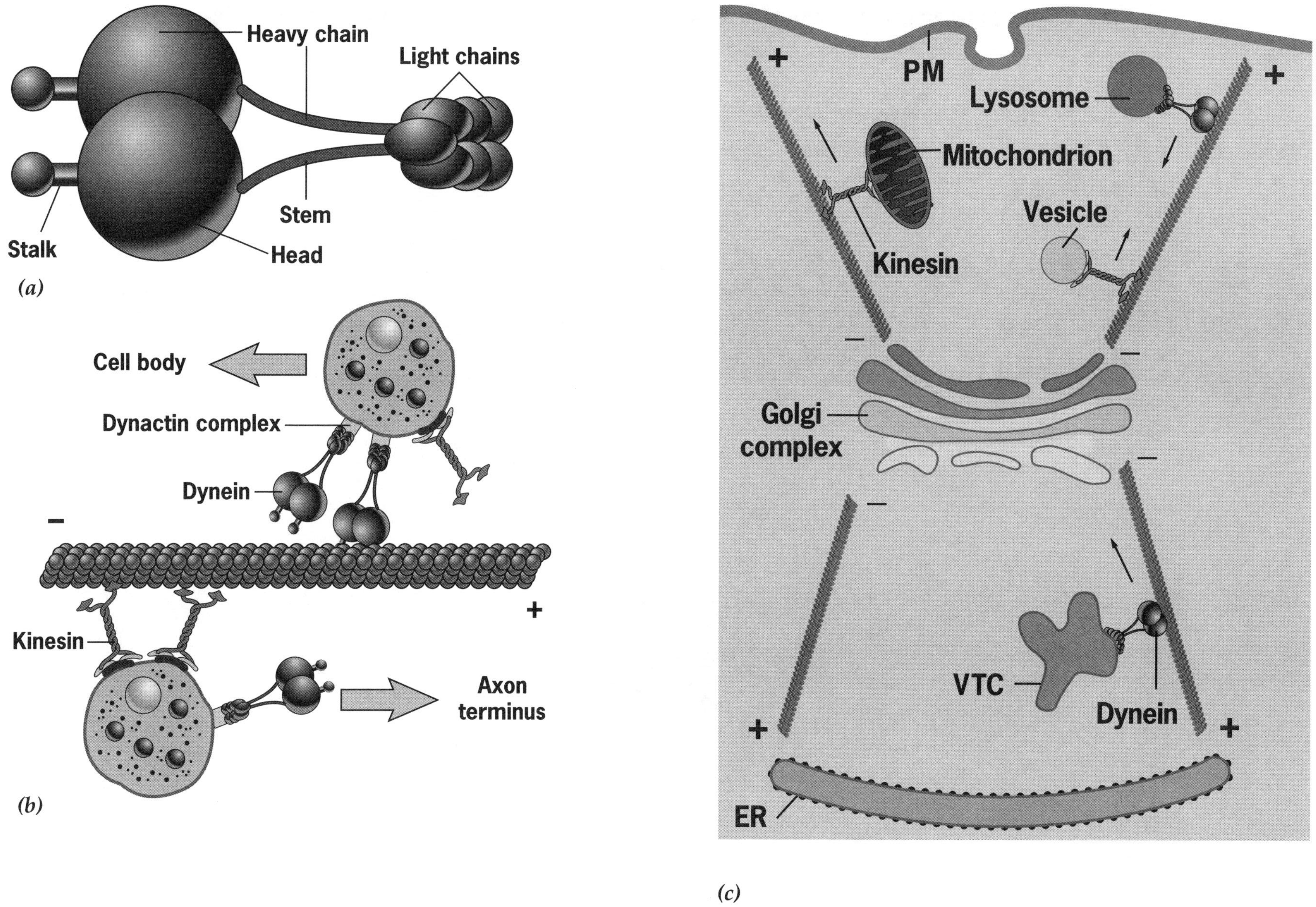

Figure 9.17 Cytoplasmic dynein and organelle transport by microtubule-tracking proteins.

Copyright © 2005 John Wiley & Sons, Inc.

(c)

Figure 9.21c The role of γ-tubulin in centrosome function.

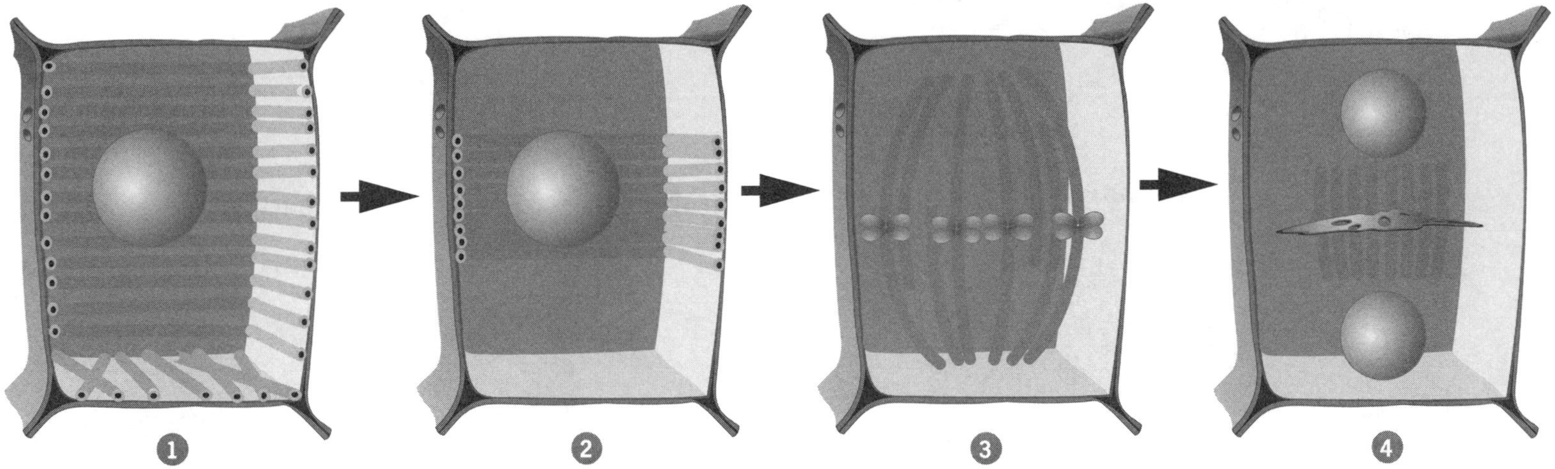

Figure 9.22 Four major arrays of microtubules present during the cell cycle of a plant cell.

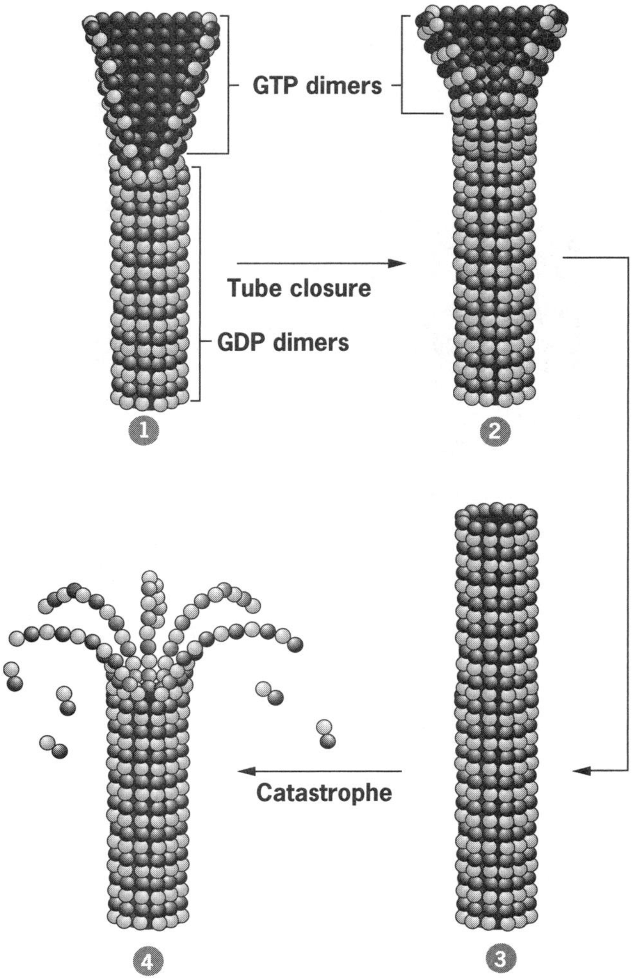

Figure 9.25 The structural cap model of dynamic instability.

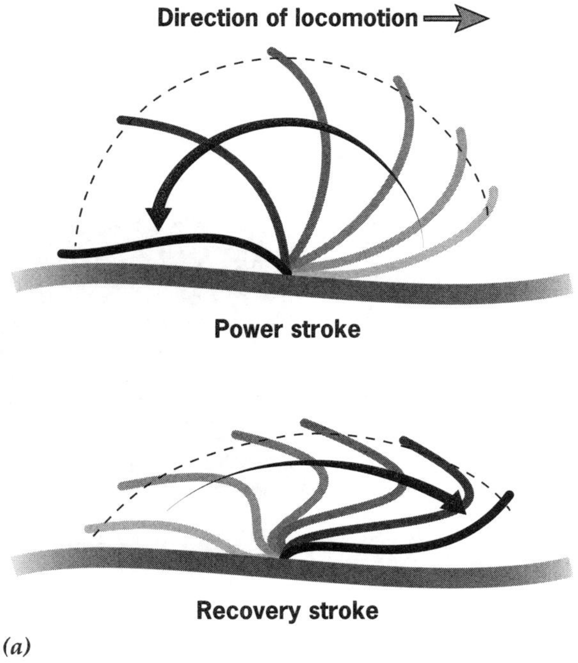

Figure 9.28a Ciliary beat.

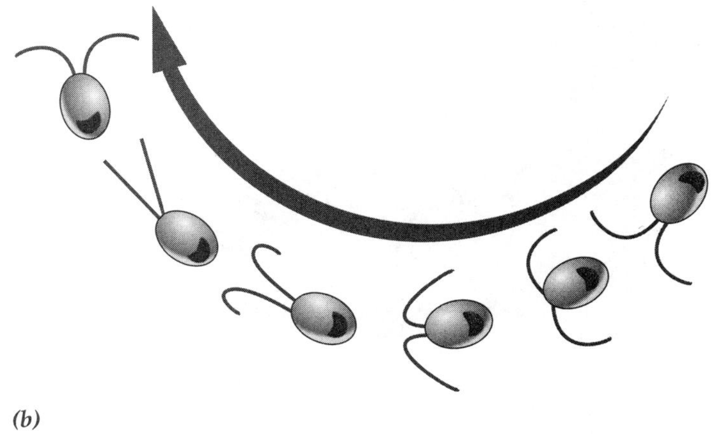

Figure 9.29b Eukaryotic flagella.

Copyright © 2005 John Wiley & Sons, Inc.

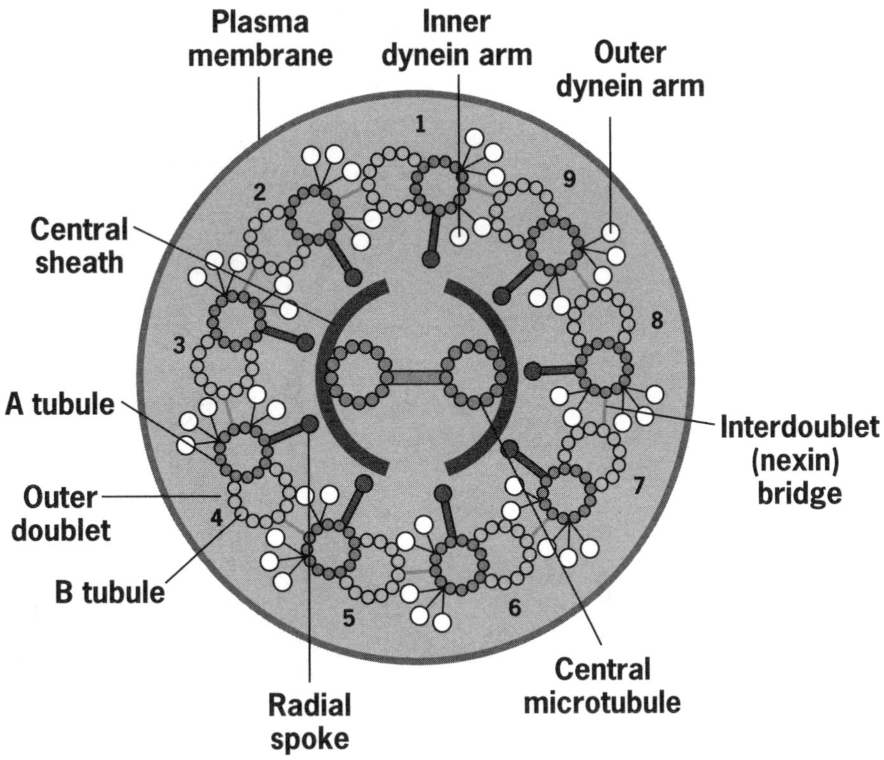

(b)

Figure 9.30b The structure of a ciliary or flagellar axoneme.

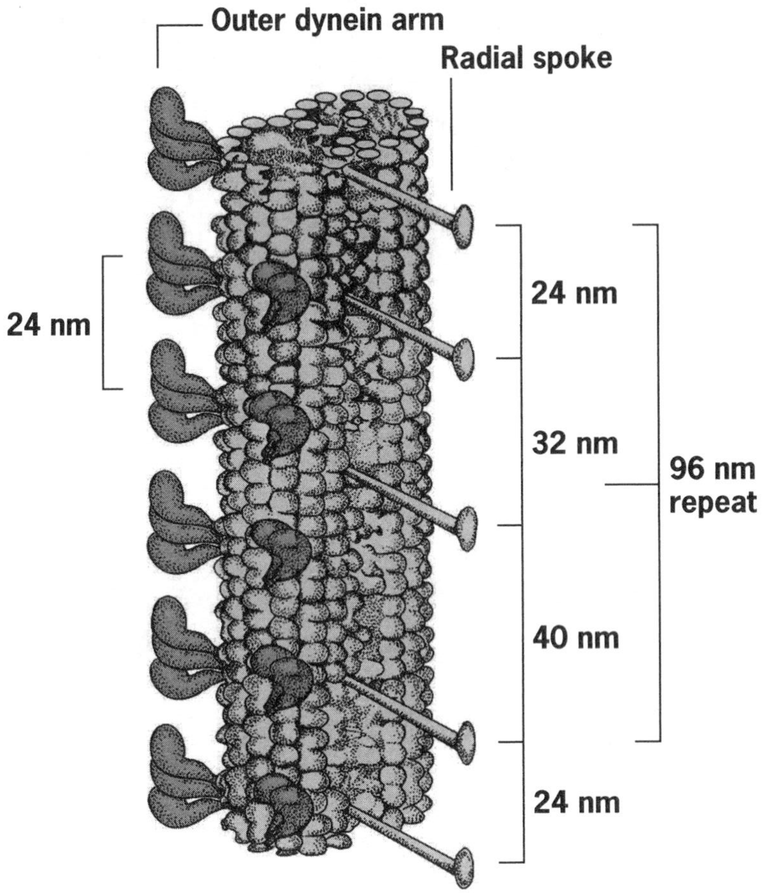

(b)

Figure 9.31b Longitudinal view of an axoneme.

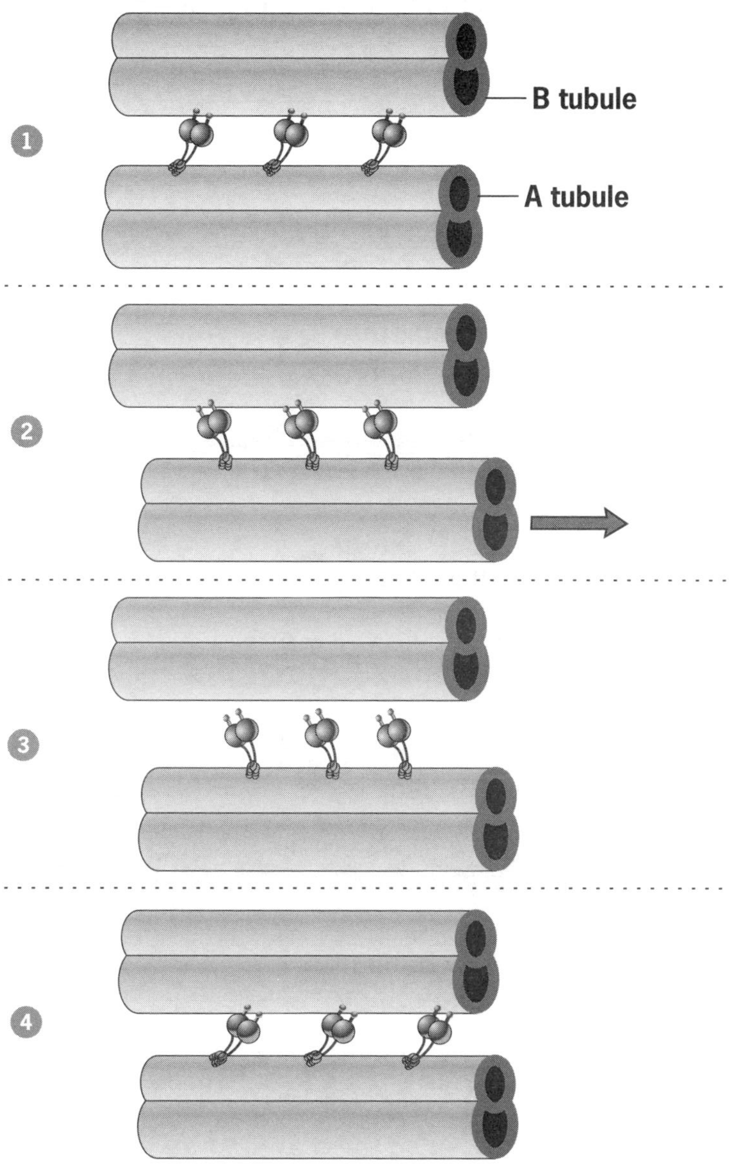

Figure 9.37 Schematic representation of the forces that drive ciliary or flagellar motility.

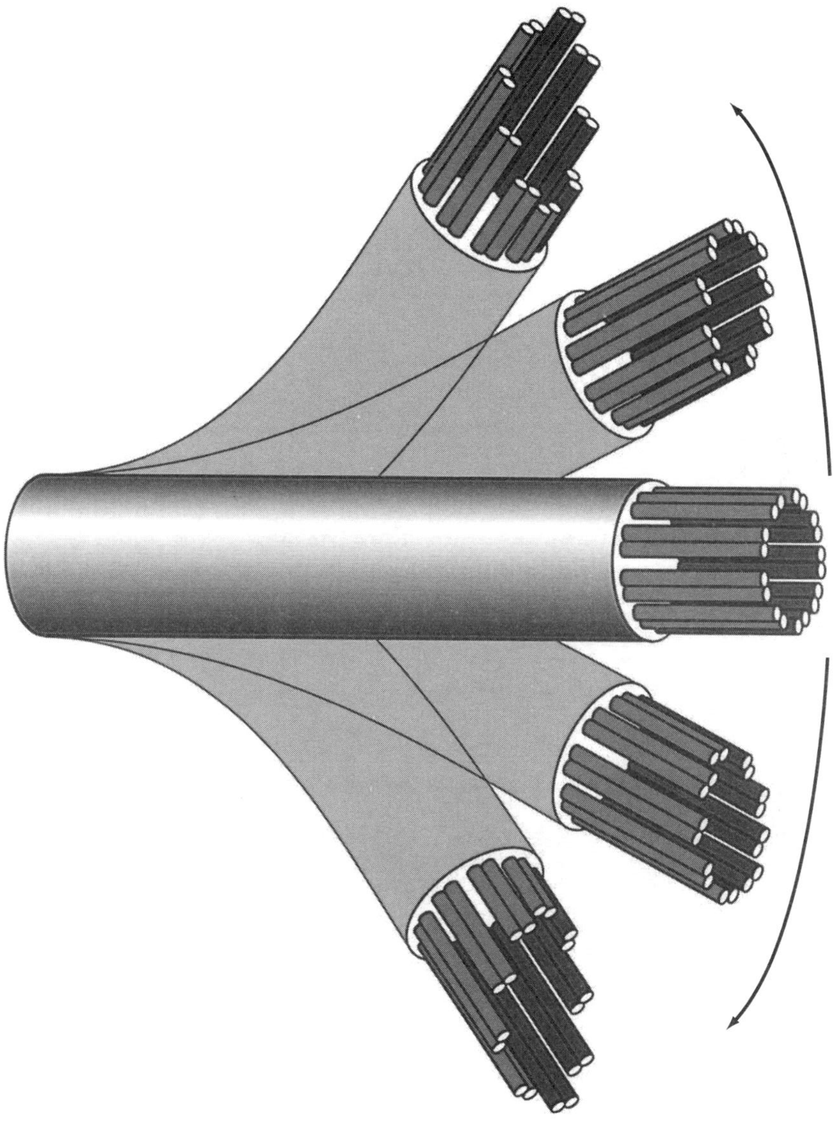

Figure 9.38 The sliding-microtubule mechanism of ciliary or flagellar motility.

Copyright © 2005 John Wiley & Sons, Inc.

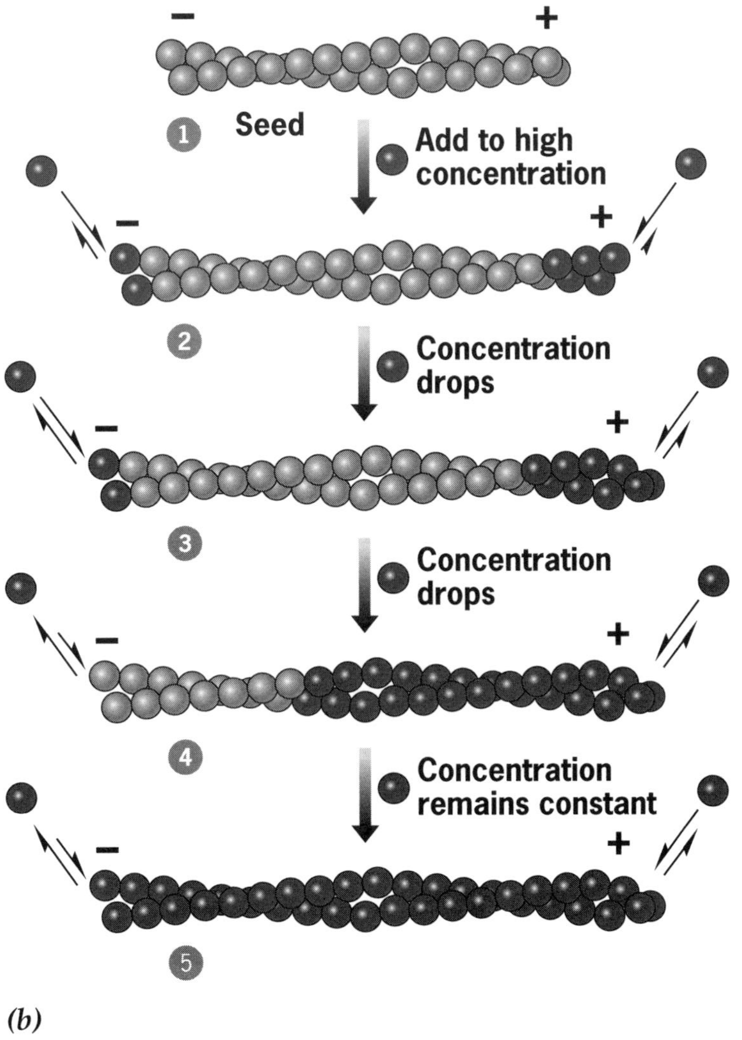

Figure 9.46b Actin assembly in vitro.

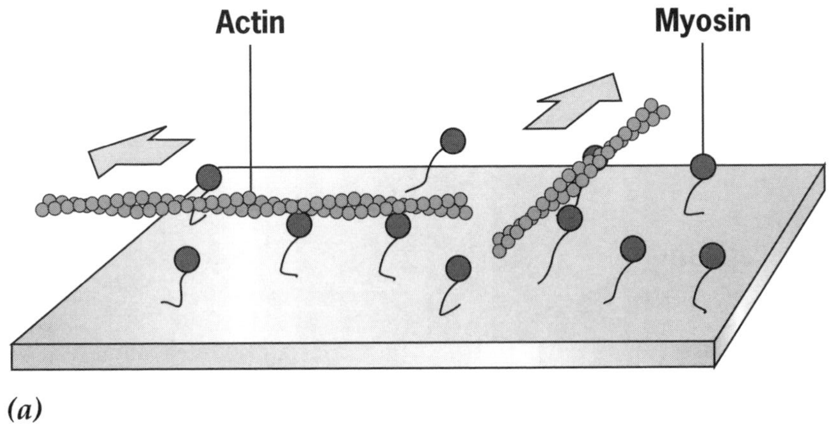

Figure 9.49a In vitro motility assay for myosin.

Copyright © 2005 John Wiley & Sons, Inc.

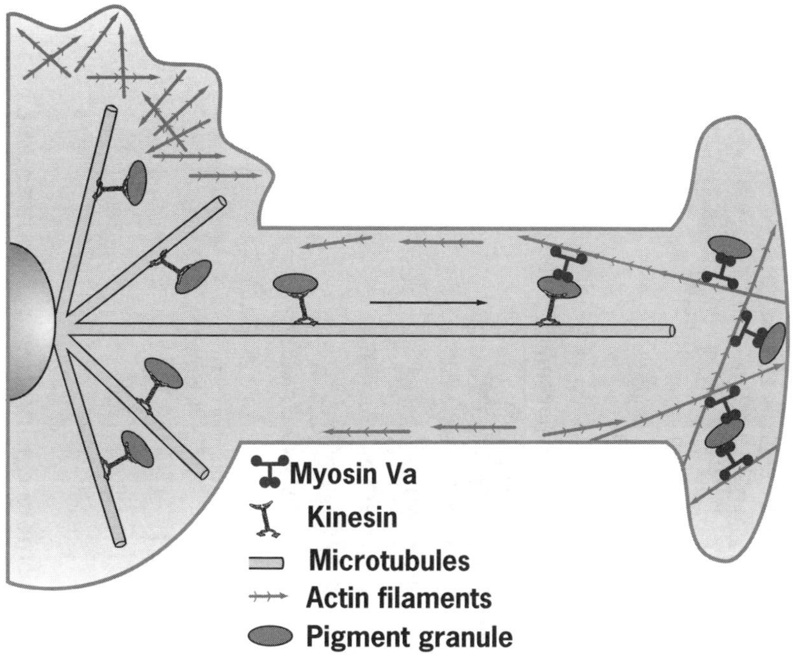

Figure 9.53 The contrasting roles of microtubule- and microfilament-based motors in organelle transport.

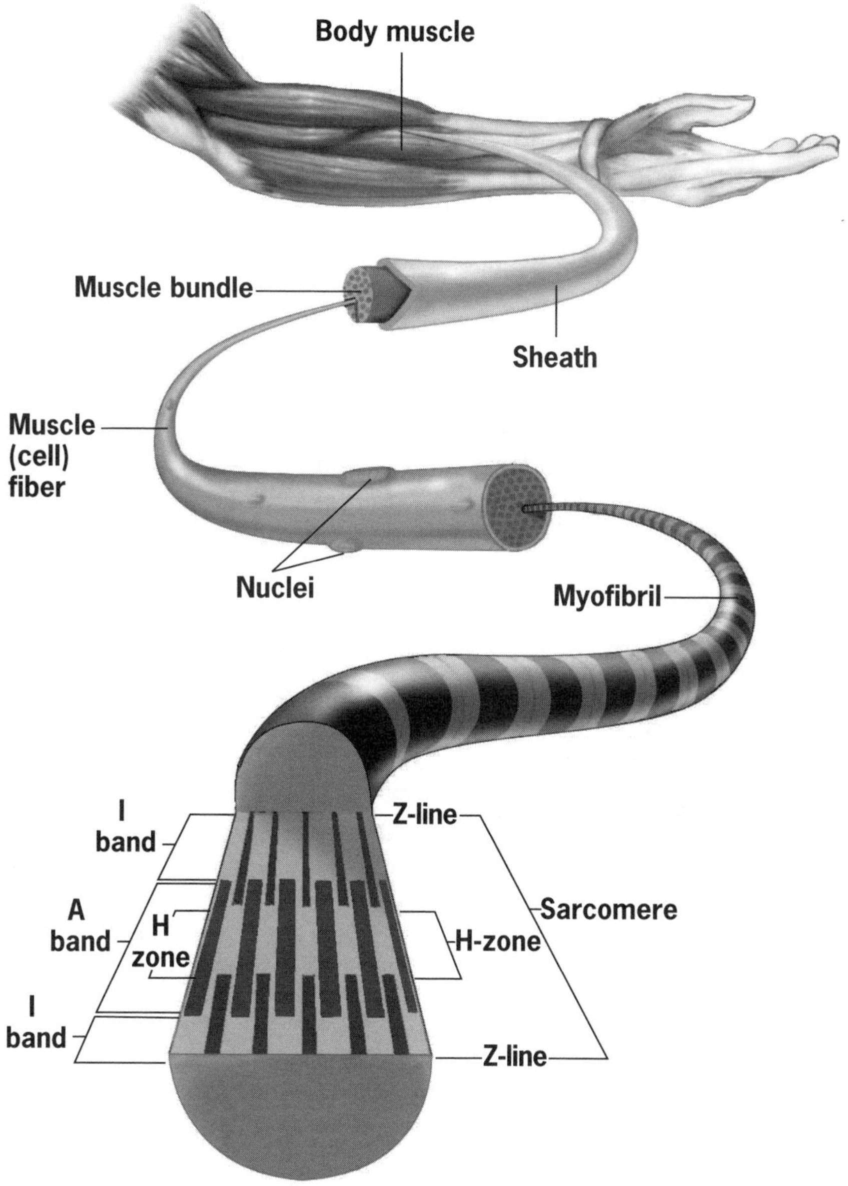

Figure 9.56 The structure of skeletal muscle.

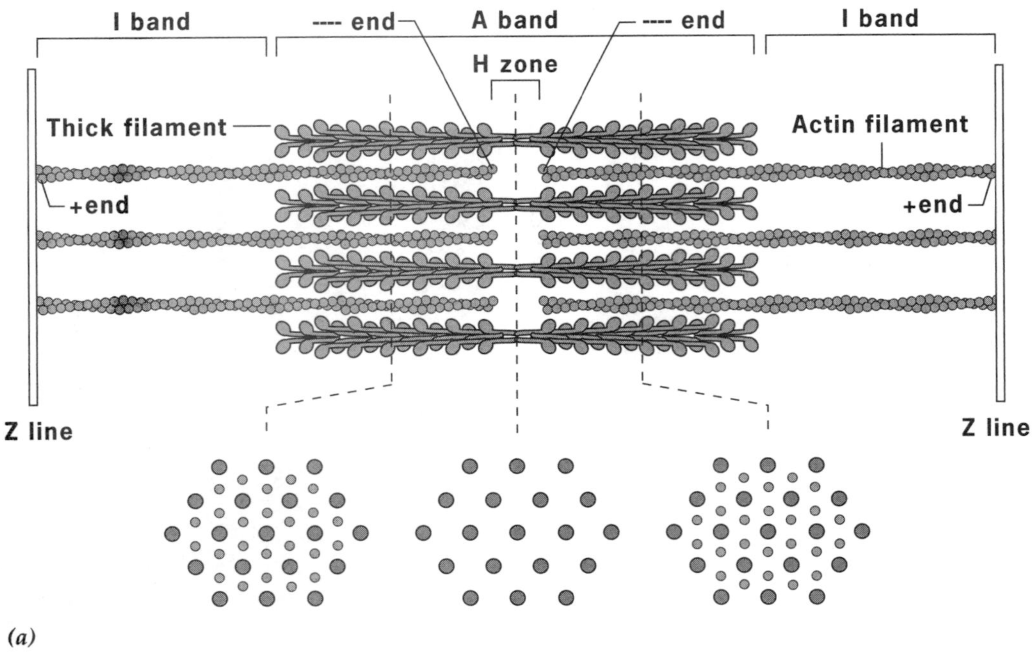

Figure 9.57a The contractile machinery of a sarcomere.

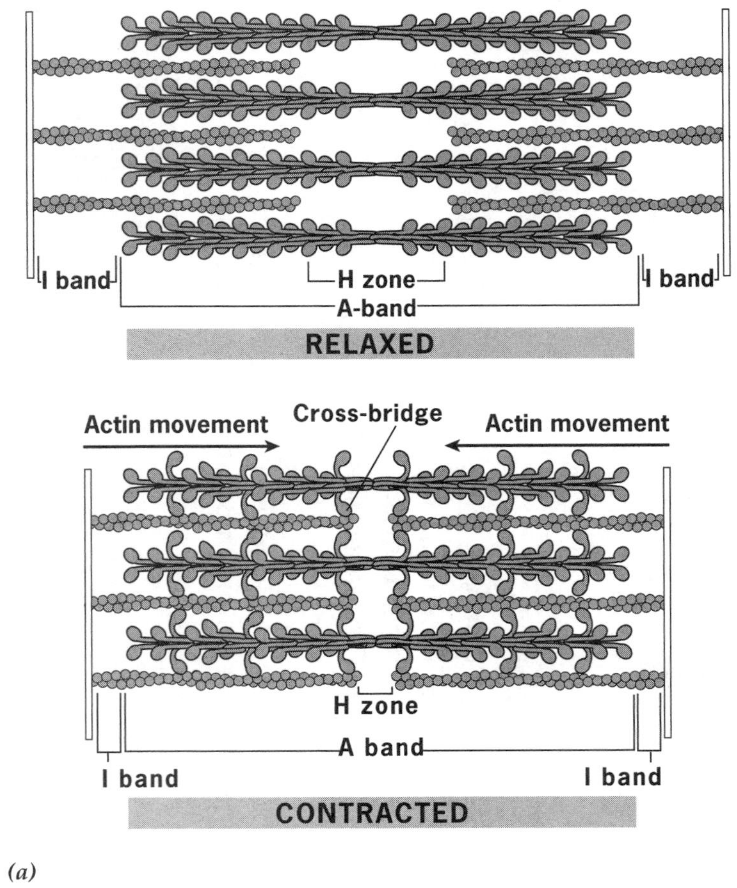

Figure 9.58a The shortening of the sarcomere during muscle contraction.

Copyright © 2005 John Wiley & Sons, Inc.

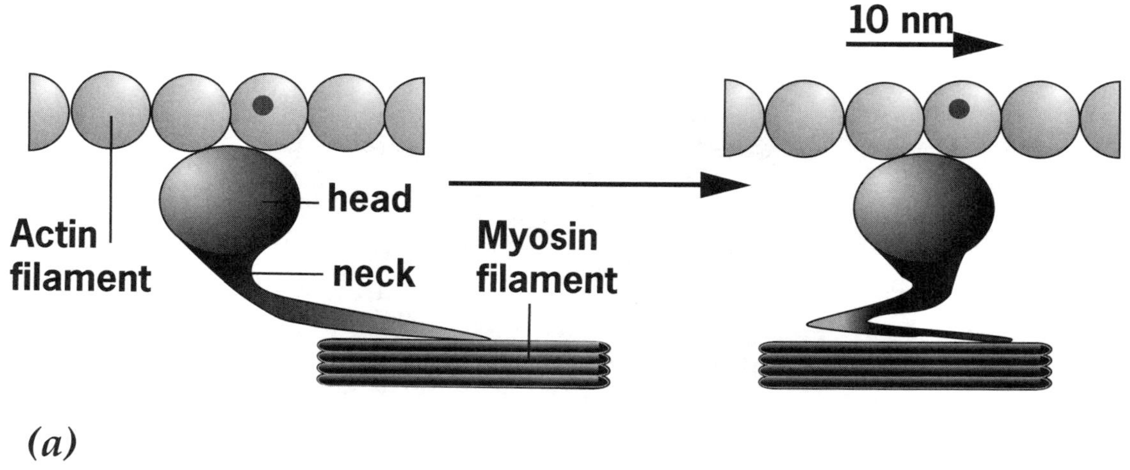

Figure 9.61a Model of the swinging lever arm of a myosin II molecule.

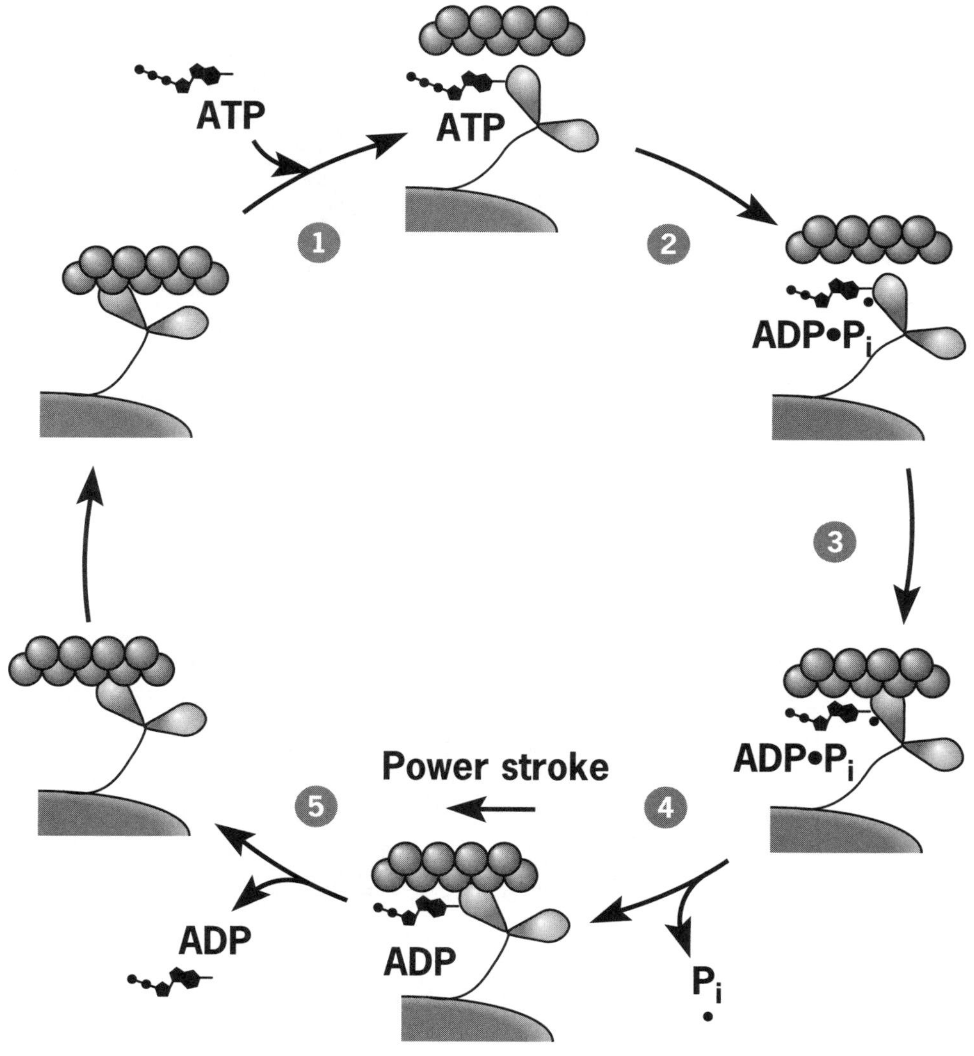

Figure 9.62 A schematic model of the actinomyosin contractile cycle.

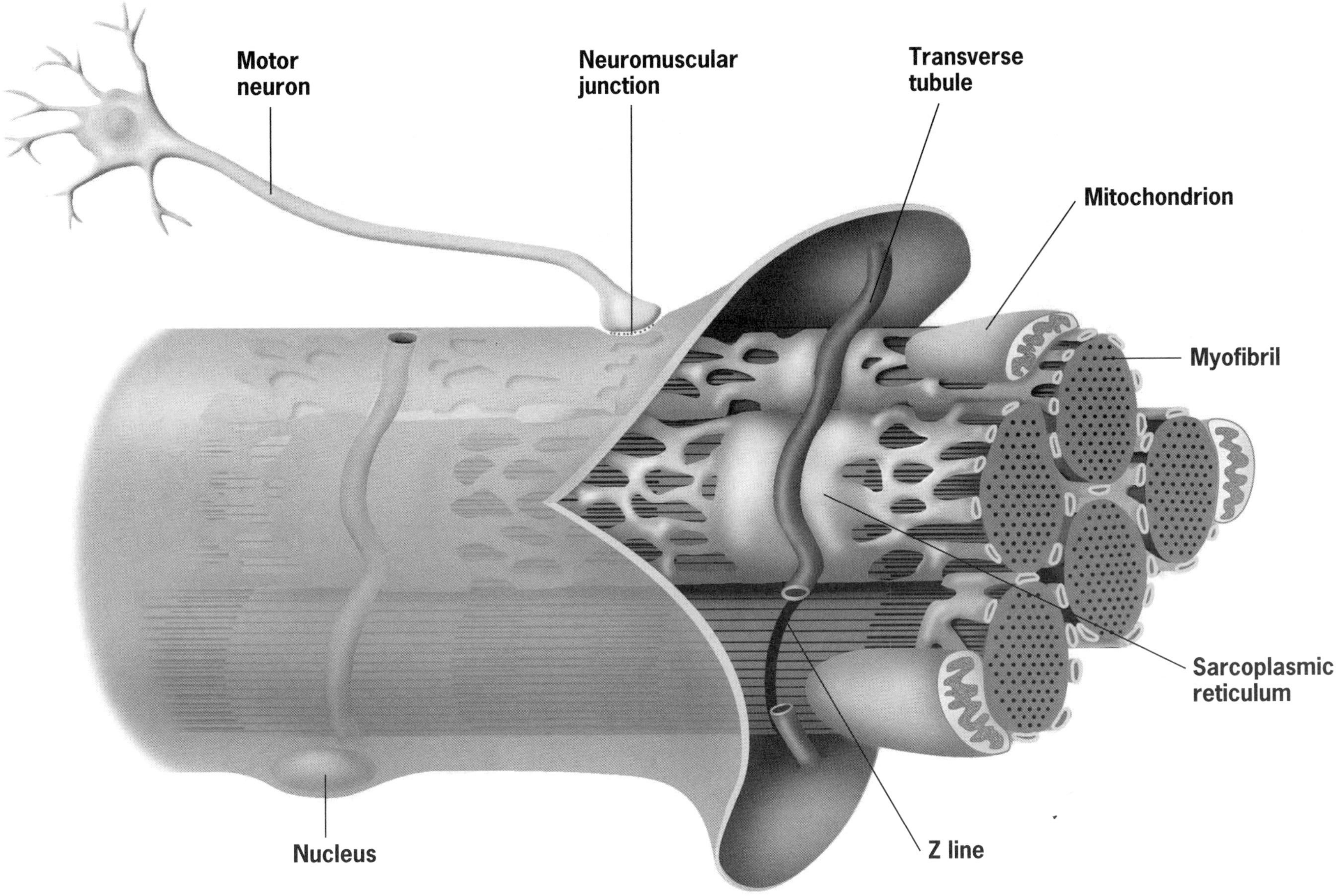

Figure 9.63 The functional anatomy of a muscle fiber.

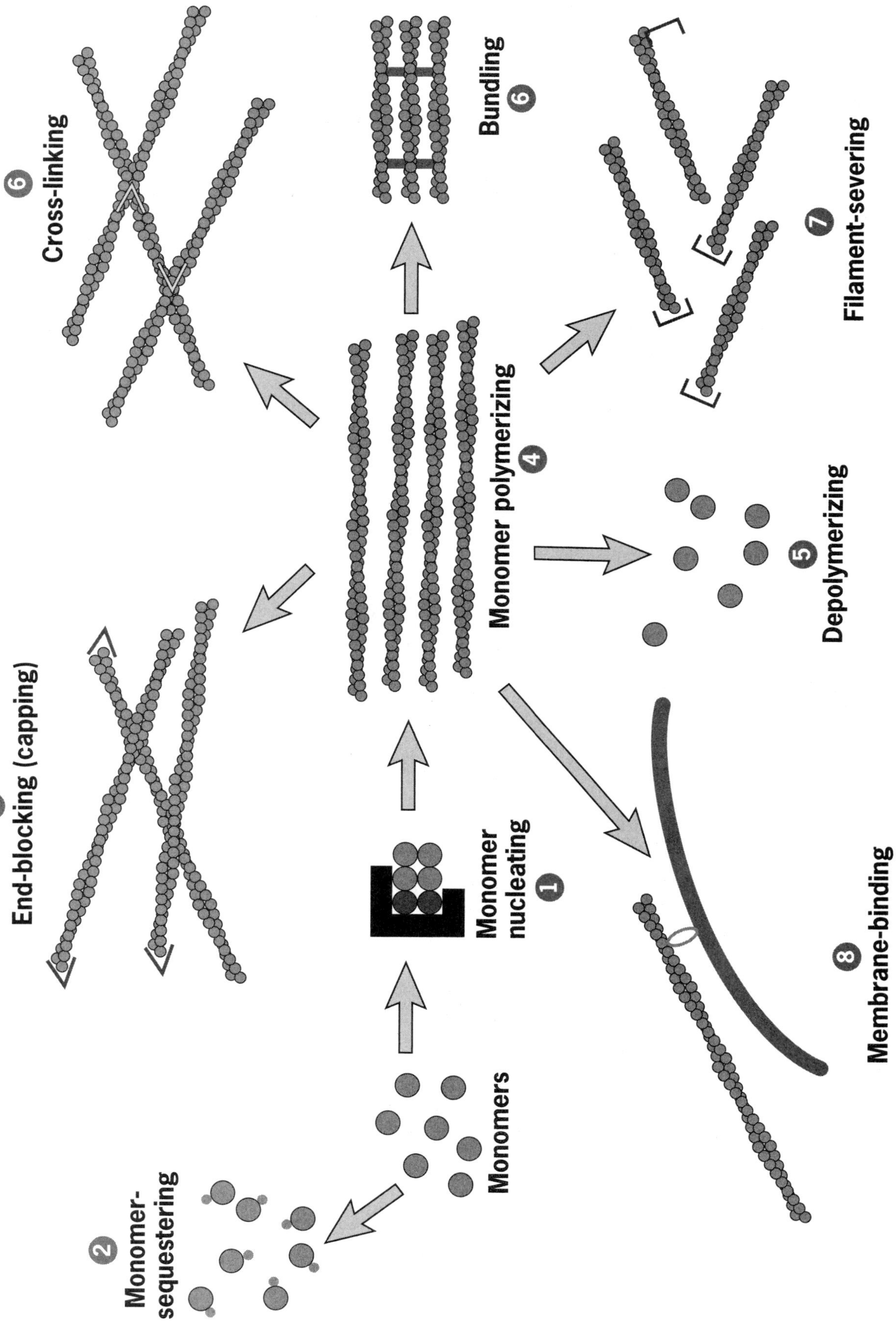

Figure 9.66 The roles of actin-binding proteins.

Copyright © 2005 John Wiley & Sons, Inc.

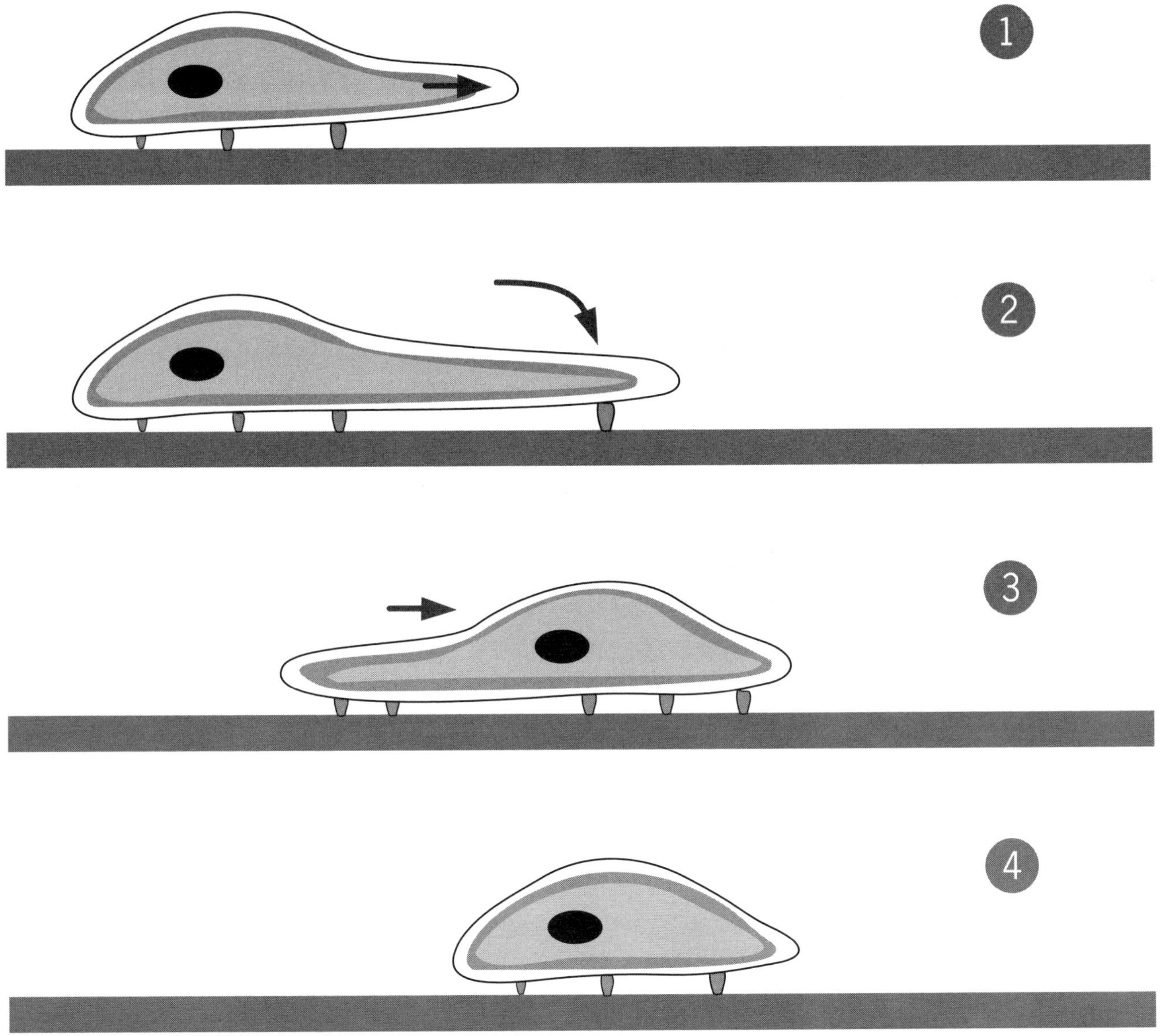

Figure 9.70 The repetitive sequence of activities that occurs as a cell crawls over the substratum.

Copyright © 2005 John Wiley & Sons, Inc.

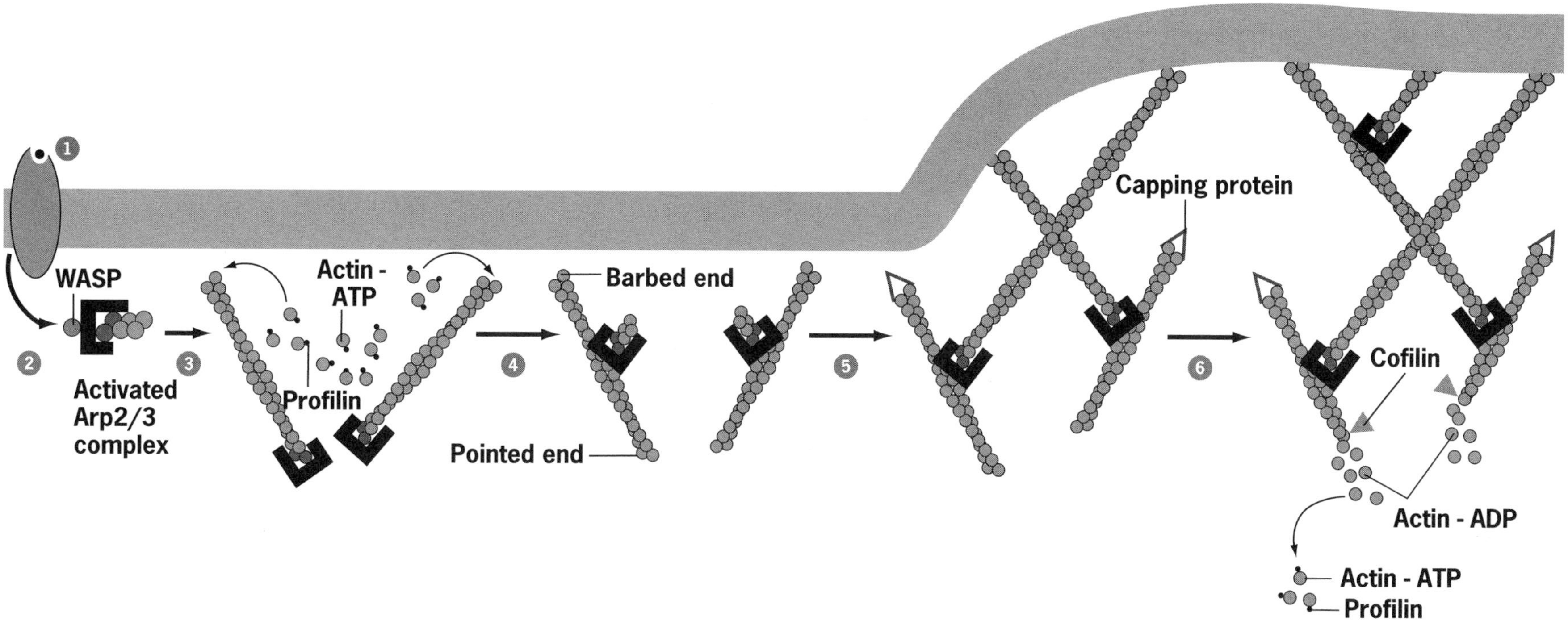

Figure 9.72 A proposed mechanism for the movement of a non-muscle cell in a directed manner.

Copyright © 2005 John Wiley & Sons, Inc.

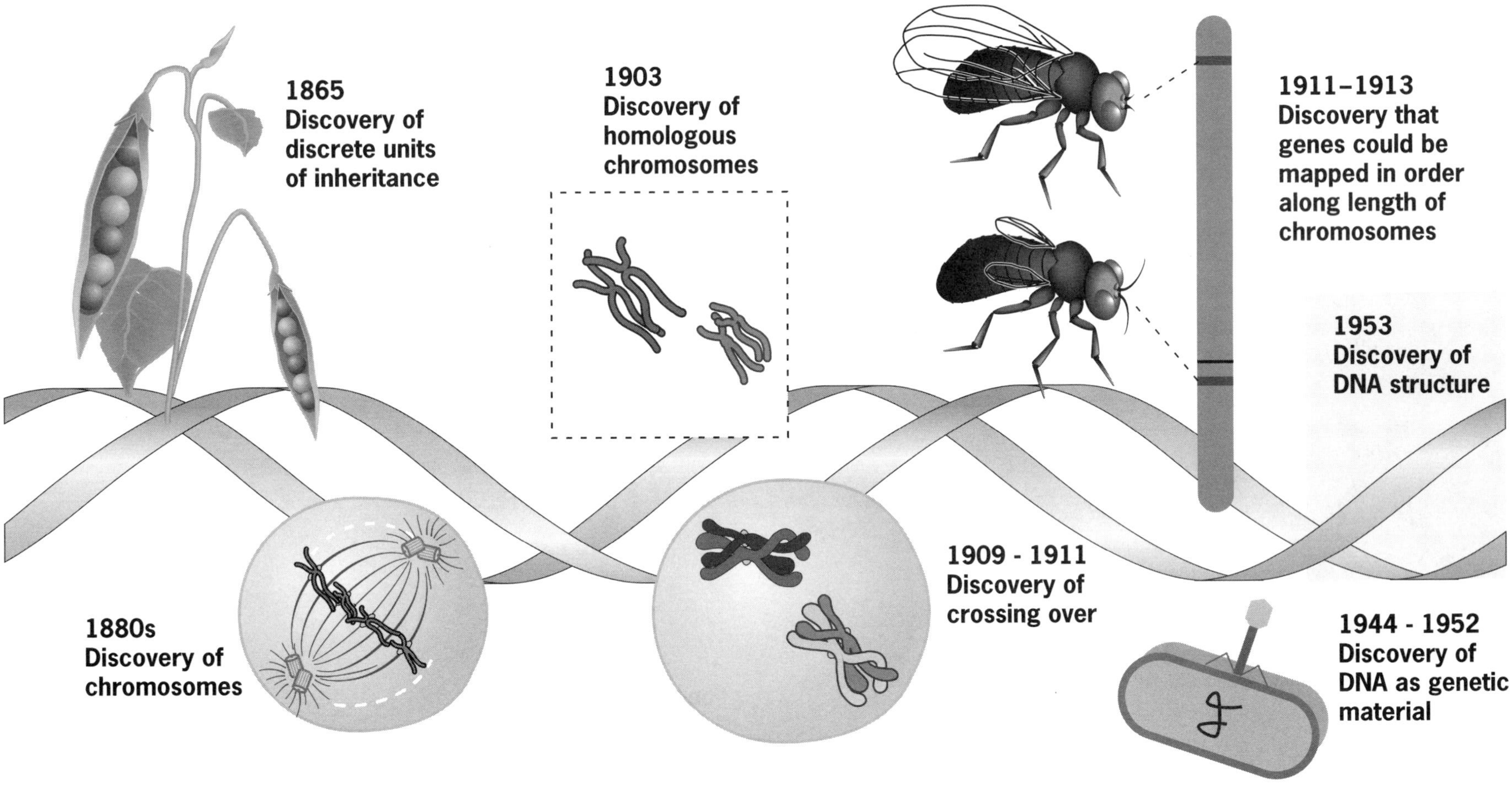

Figure 10.1 An overview depicting several of the most important early discoveries on the nature of the gene.

Copyright © 2005 John Wiley & Sons, Inc.

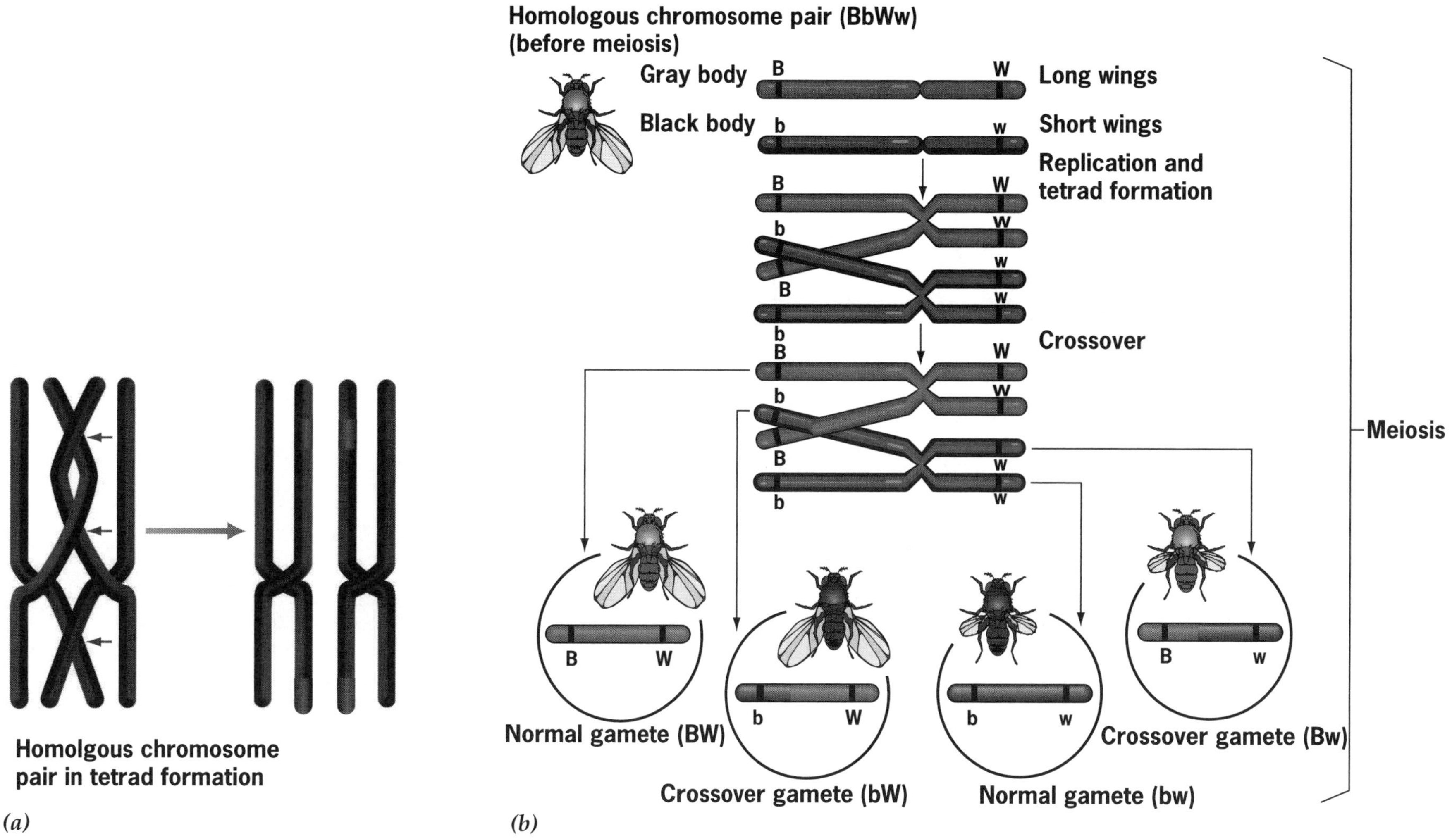

Figure 10.7 Crossing over provides the mechanism for reshuffling alleles between maternal and paternal chromosomes.

Copyright © 2005 John Wiley & Sons, Inc.

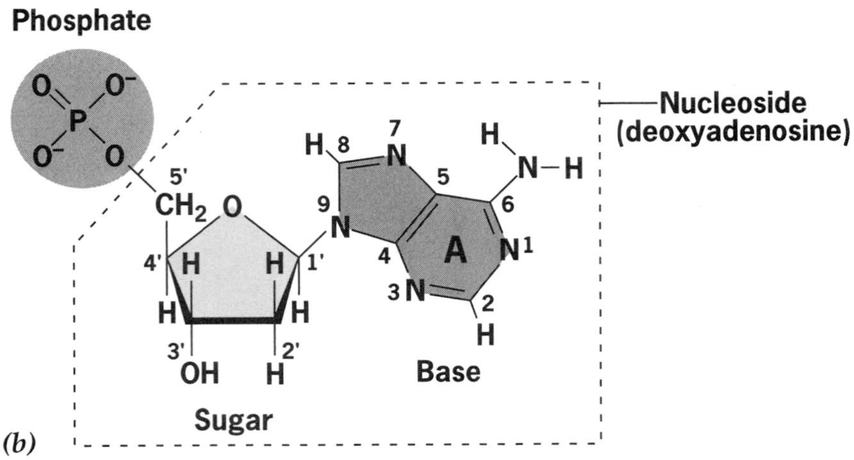

(b)

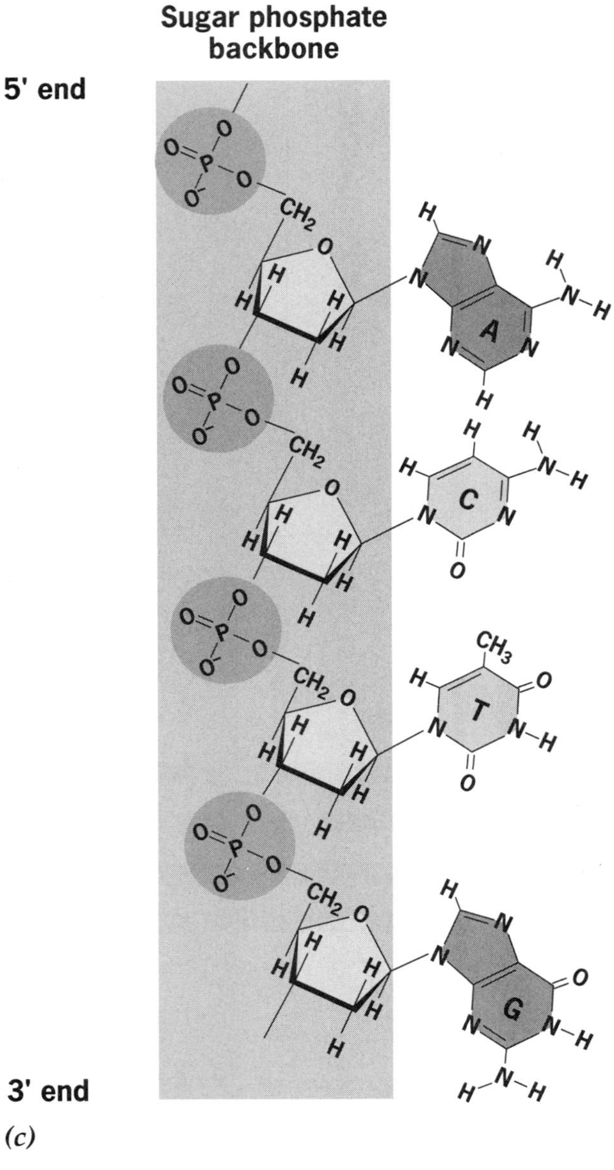

(c)

Figure 10.9b,c The chemical structure of DNA.

Copyright © 2005 John Wiley & Sons, Inc.

20 Å
34 Å
3.4 Å
Minor groove
Major groove
Sugar
Base
Phosphate
(a)

Major groove
A
T
CH_3
51.5°
10.85 Å
Minor groove
Major groove
G
C
51.5°
10.85 Å
Minor groove
(c)

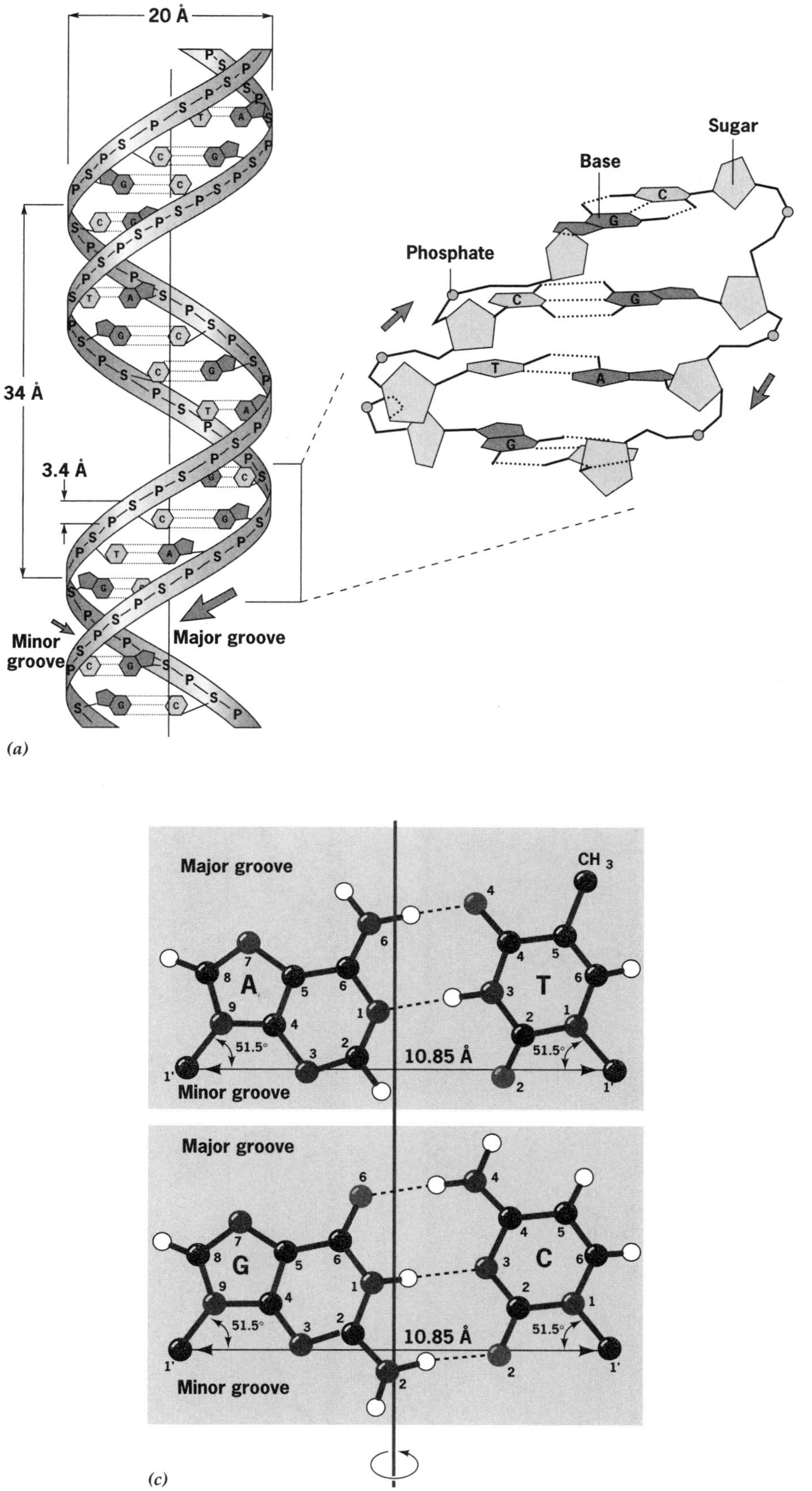

Figure 10.10a,c The double helix.

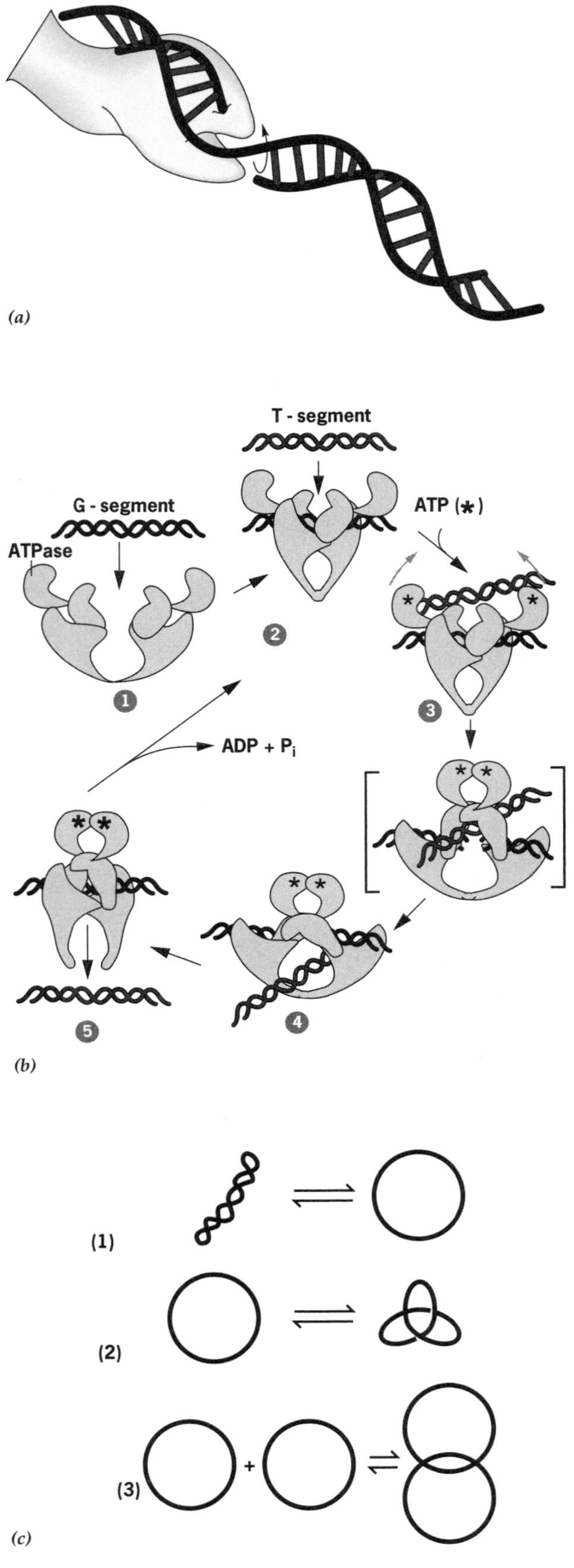

Figure 10.14a-c DNA topoisomerases.

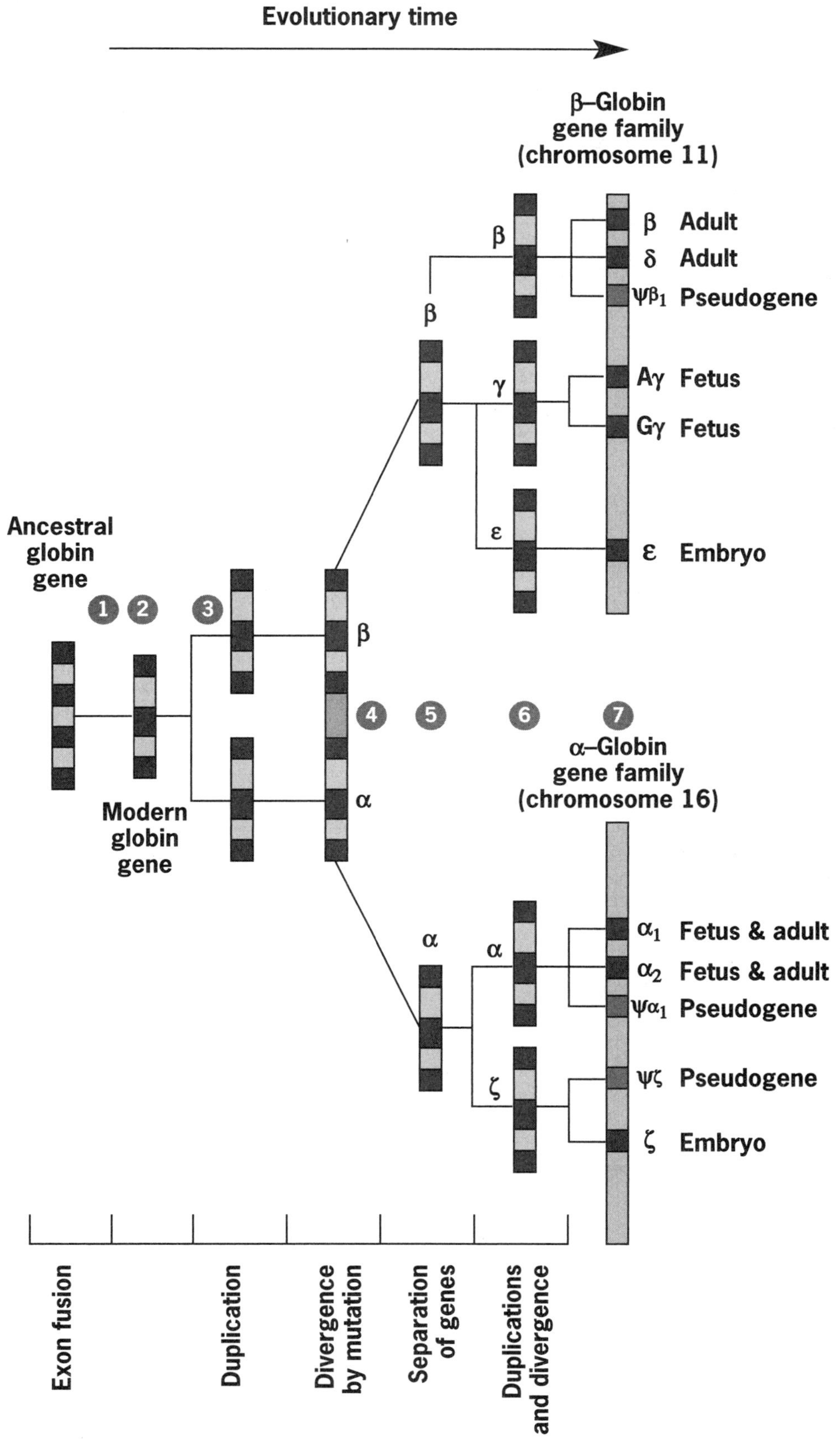

Figure 10.25 A pathway for the evolution of globin genes.

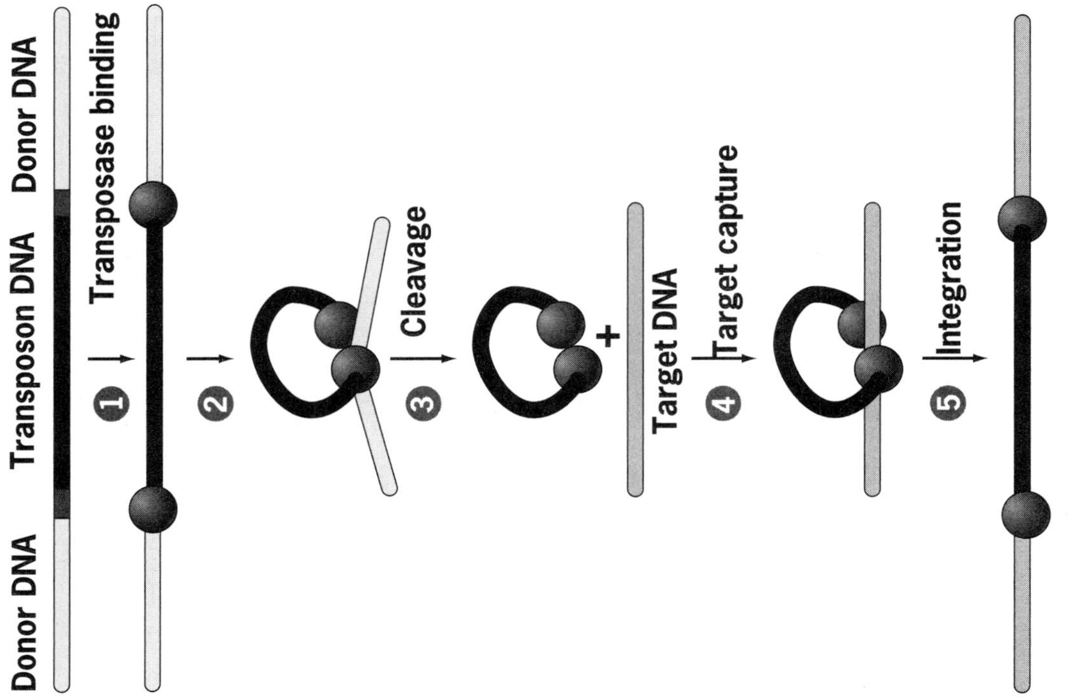

Figure 10.27 Transposition of a bacterial transposon by a "cut-and-paste" mechanism.

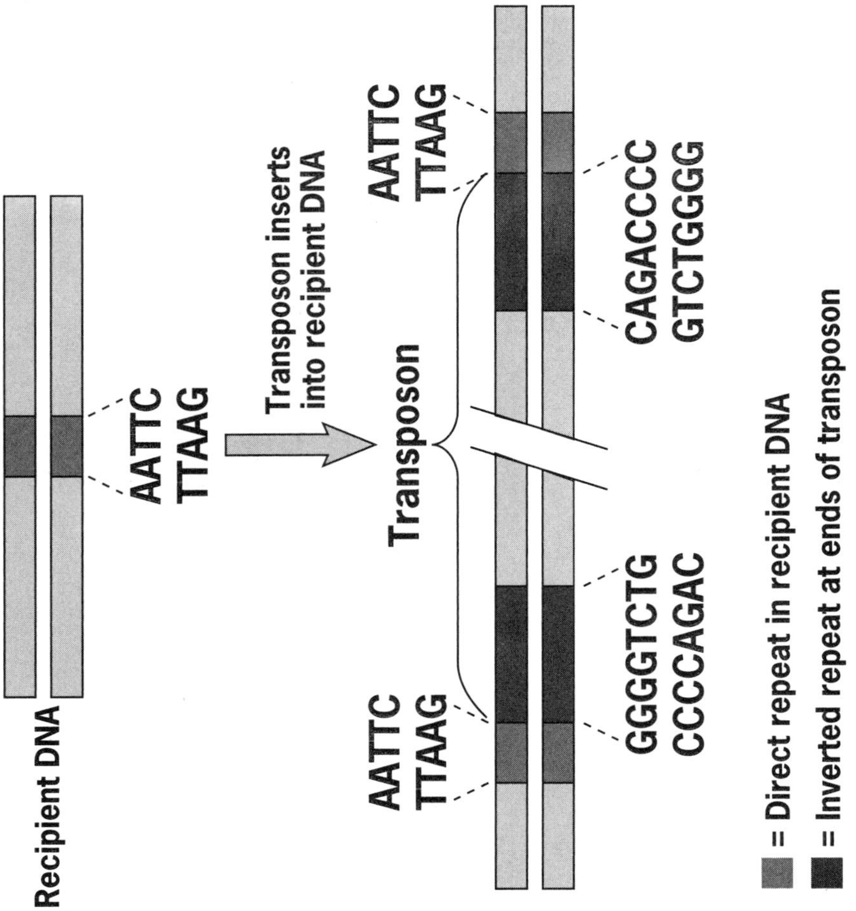

Figure 10.28 Sequence organization of a transposon.

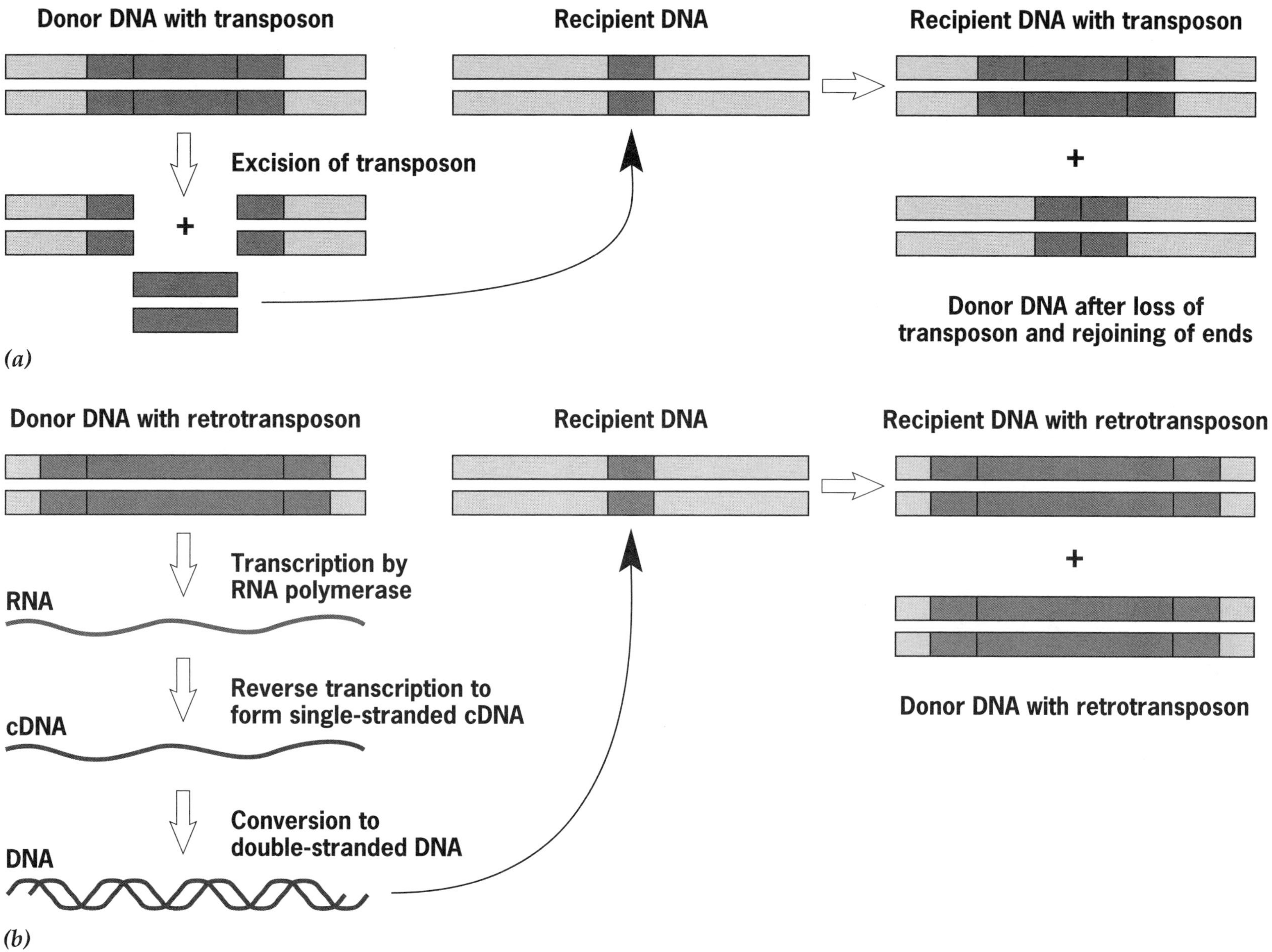

Figure 10.29 Schematic pathways in the movement of transposable elements.

Copyright © 2005 John Wiley & Sons, Inc.

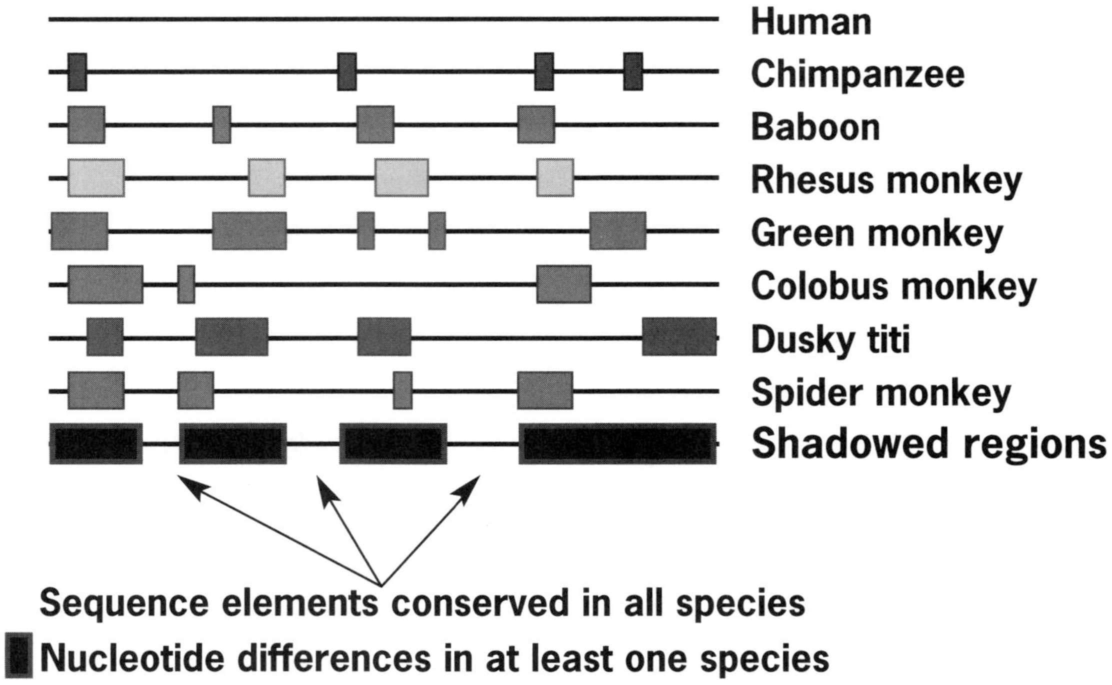

Figure 10.31 Small segments of DNA are highly conserved between humans and related species.

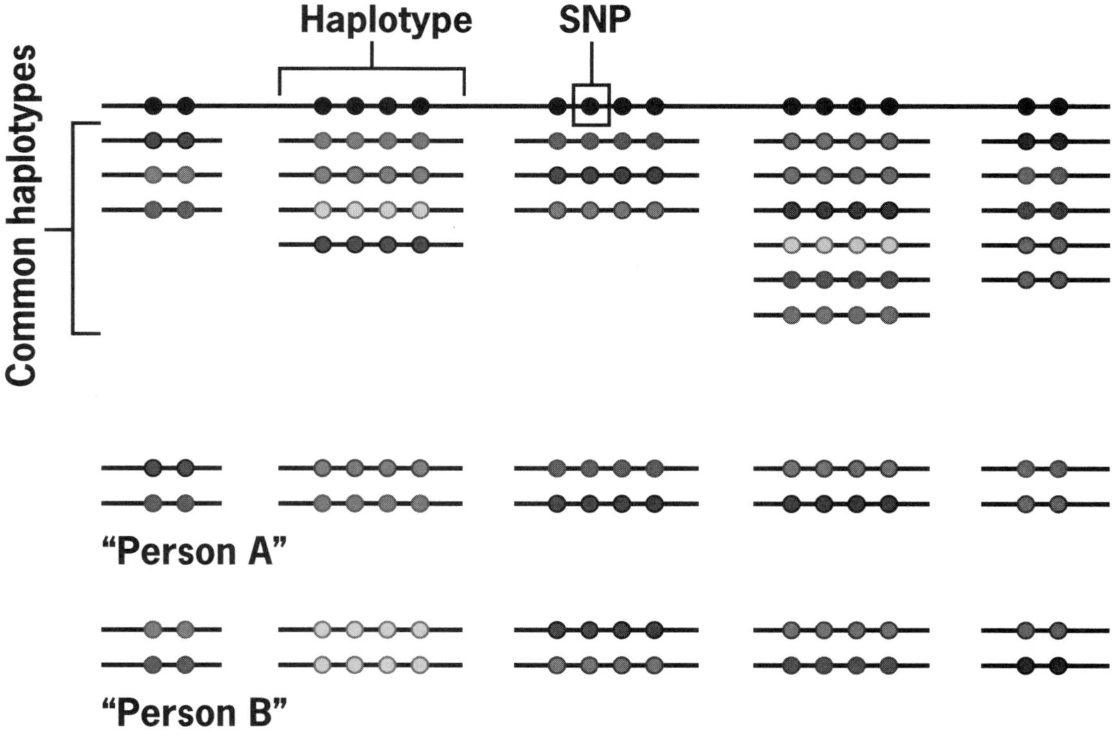

Figure 1 The genome is divided into blocks (haplotypes).

Copyright © 2005 John Wiley & Sons, Inc.

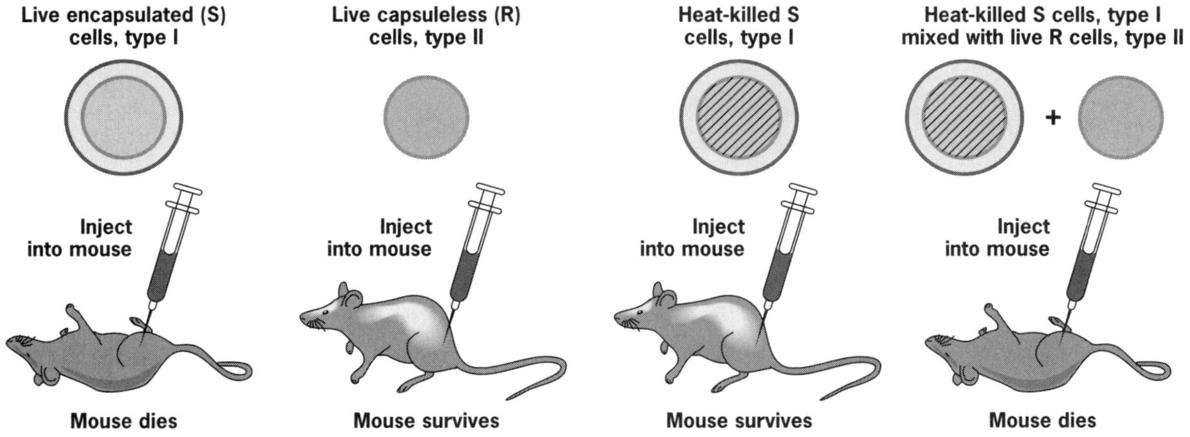

Figure 2 Outline of the experiment by Griffith of the discovery of bacterial transformation.

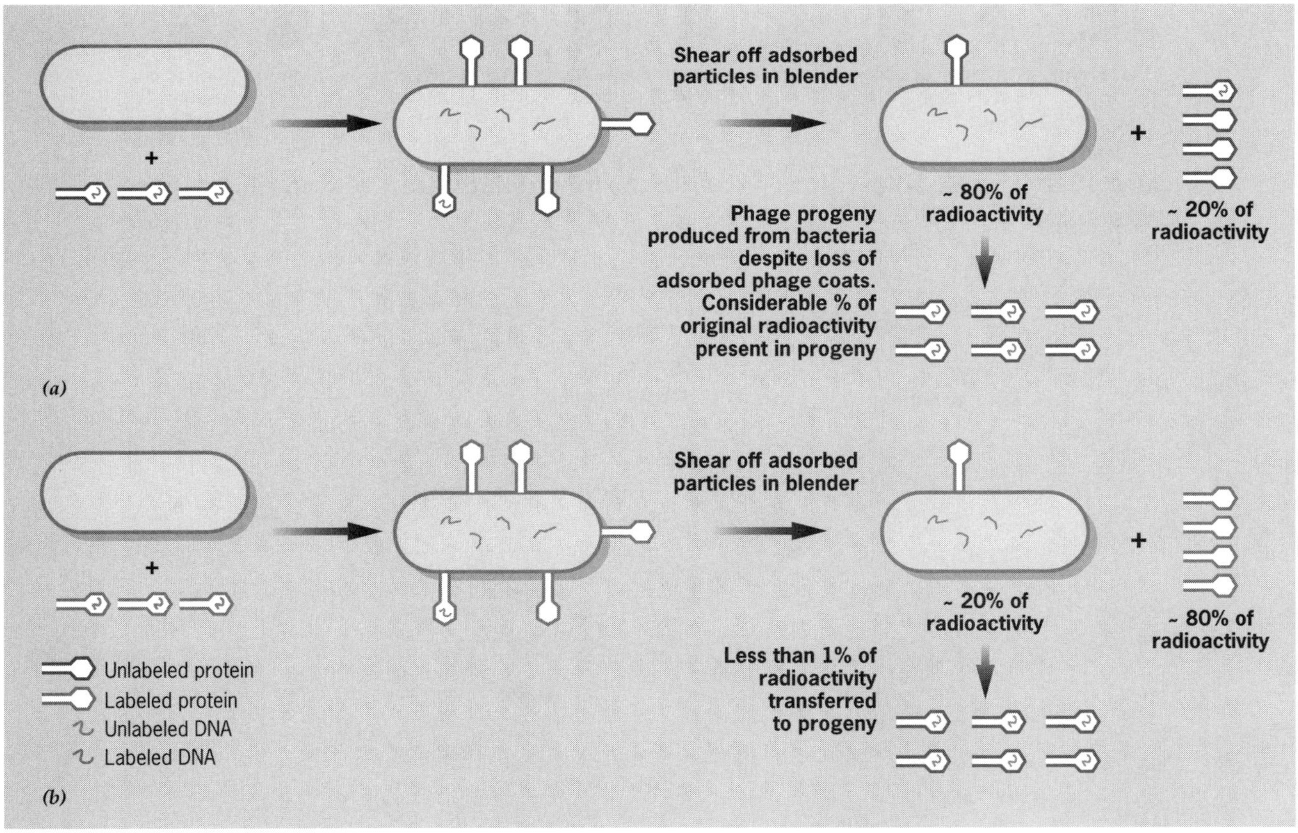

Figure 4 The Hershey-Chase experiment.

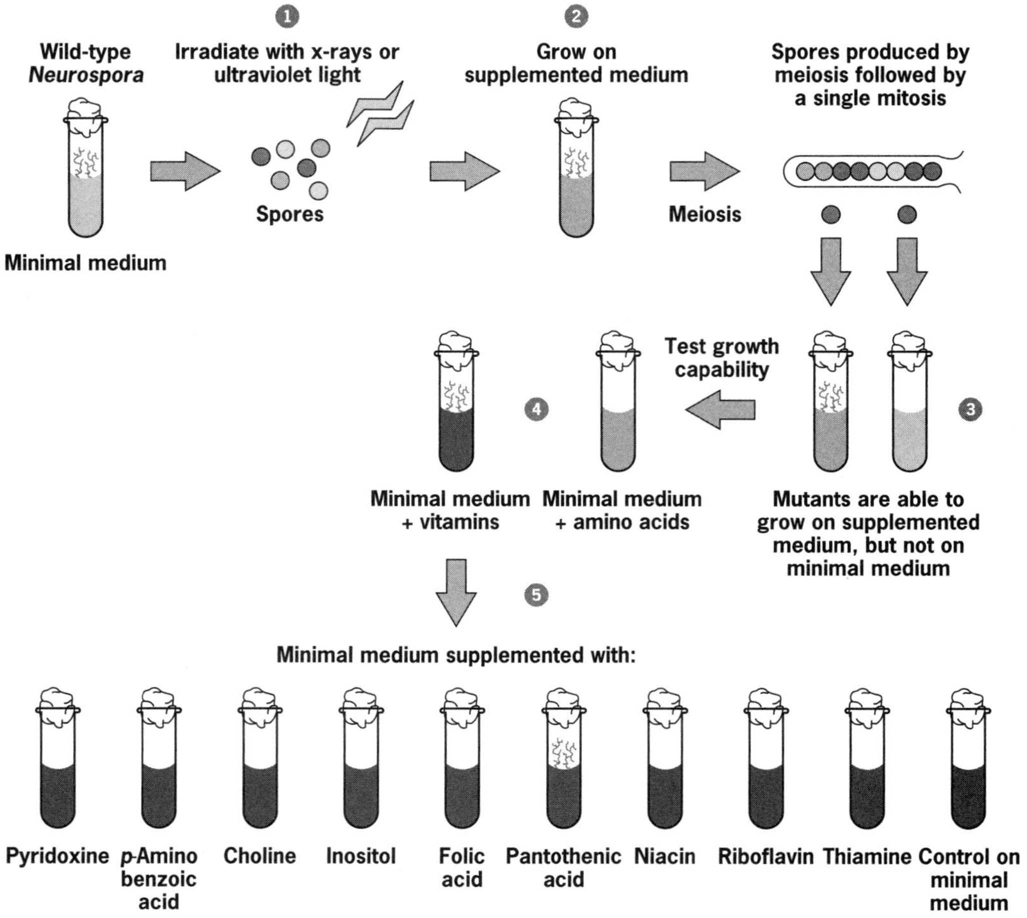

Figure 11.1 The Beadle-Tatum experiment for the isolation of genetic mutants in *Neurospora*.

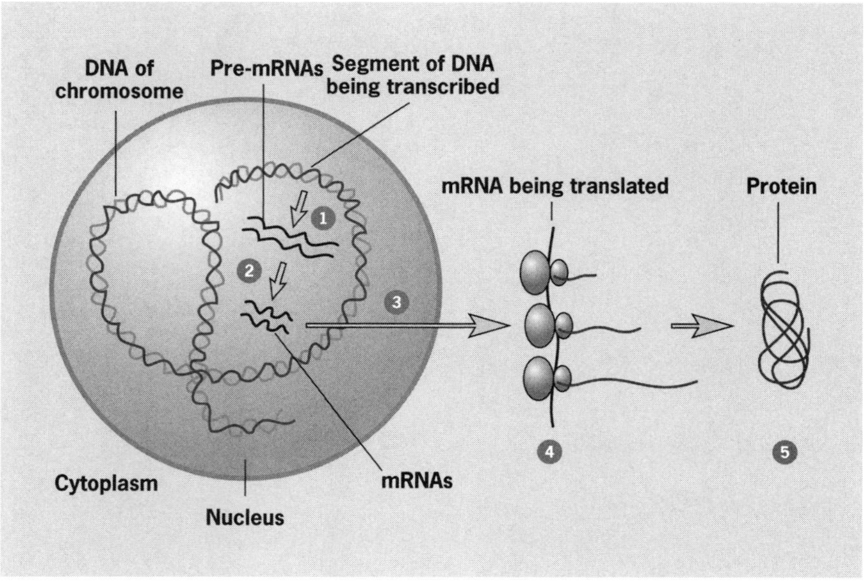

Figure 11.2 An overview of the flow of information in a eukaryotic cell.

Copyright © 2005 John Wiley & Sons, Inc.

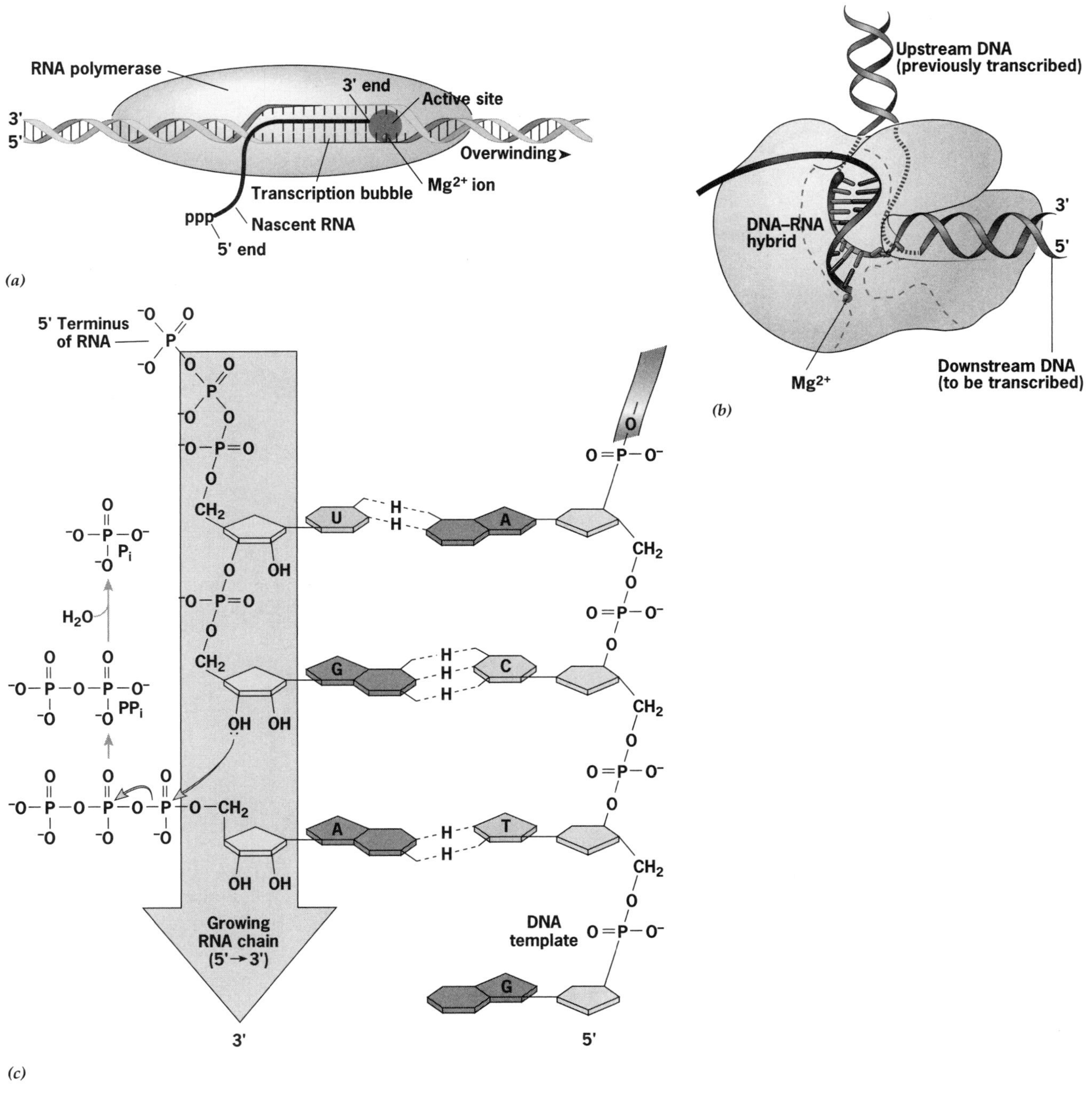

Figure 11.4a-c Chain elongation during transcription.

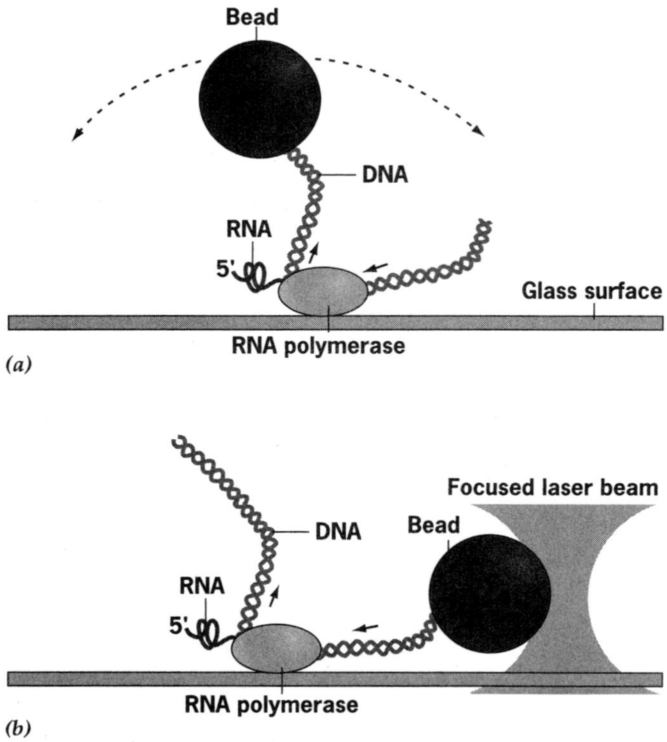

Figure 11.5 Examples of experimental techniques to follow the activities of single RNA polymerase molecules.

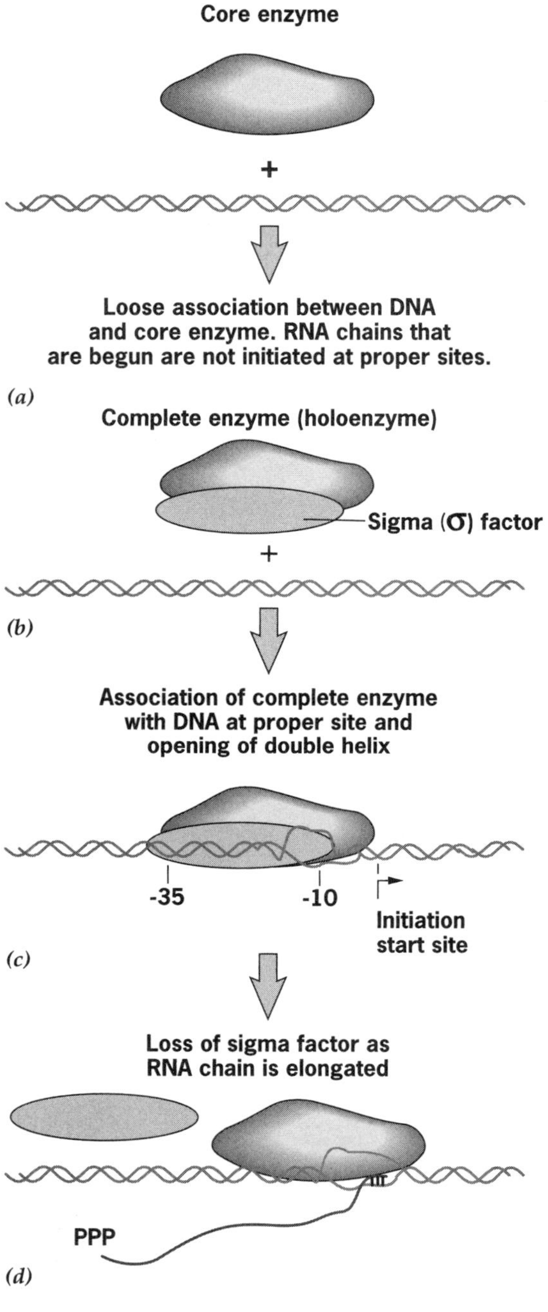

Figure 11.6 Schematic representation of the initiation of transcription in prokaryotes.

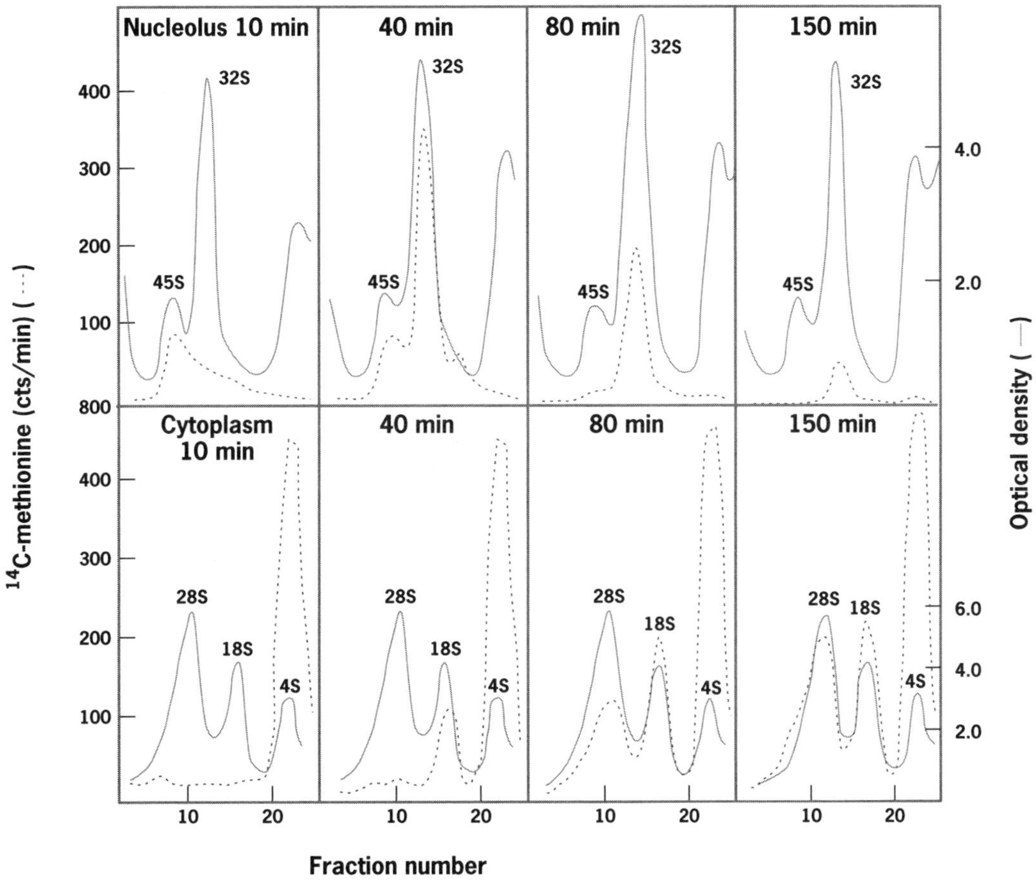

Figure 11.13 Kinetic analysis of rRNA synthesis and processing.

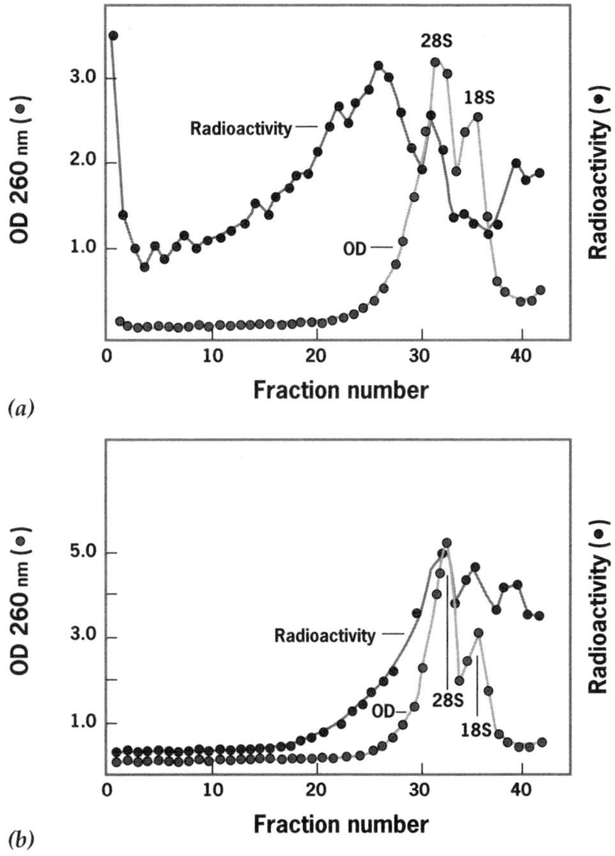

Figure 11.17 The formation of heterogeneous nuclear RNA (hnRNA) and its conversion into smaller mRNAs.

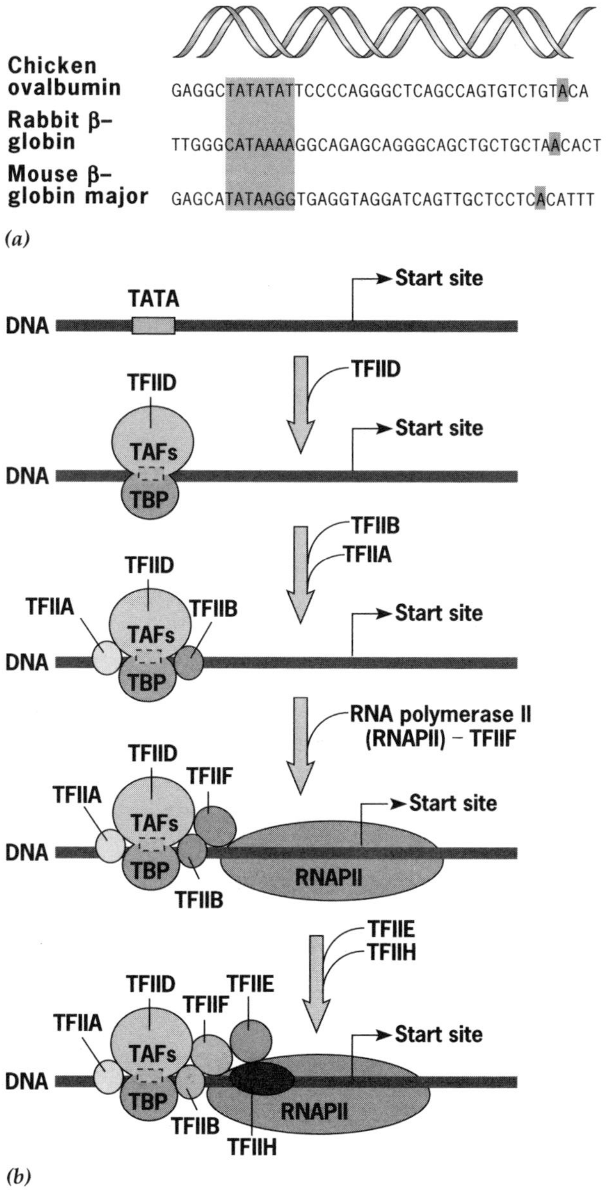

Figure 11.18 Initiation of transcription from a eukaryotic polymerase II promoter.

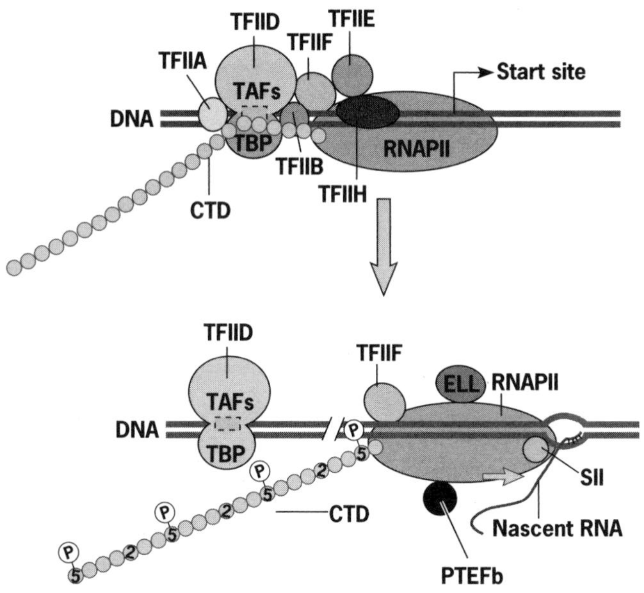

Figure 11.20 Initiation of transcription by RNA polymerase II is associated with phosphorylation of the C-terminal domain (CTD).

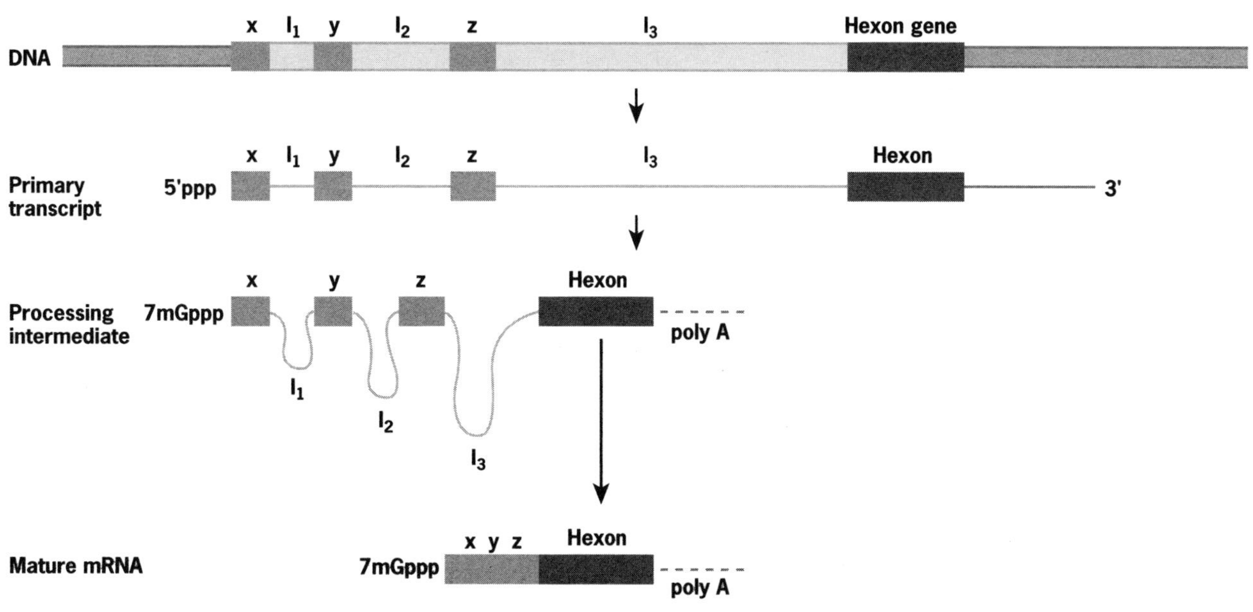

Figure 11.23 The discovery of intervening sequences (introns).

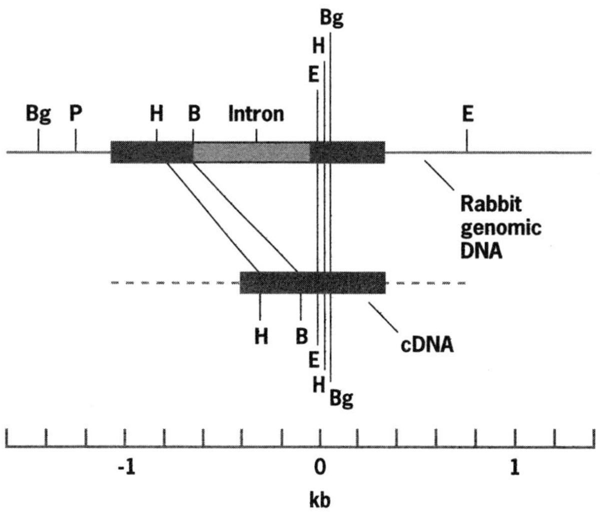

Figure 11.24 The discovery of introns in a eukaryotic gene.

Copyright © 2005 John Wiley & Sons, Inc.

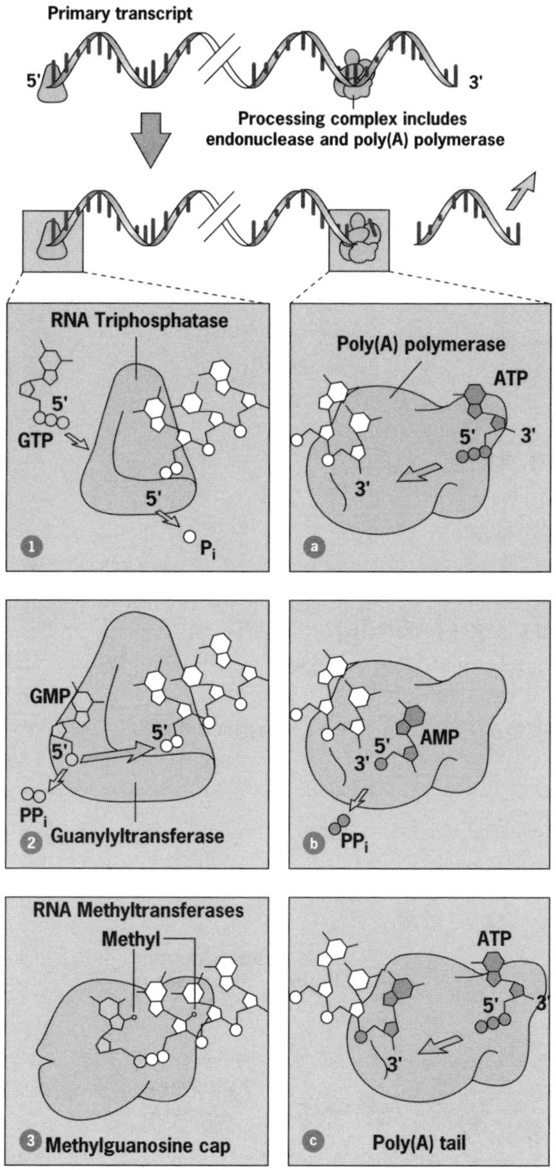

Figure 11.28 Steps in the addition of a 5′ methylguanosine cap and a 3′ poly(A) tail to a pre-mRNA.

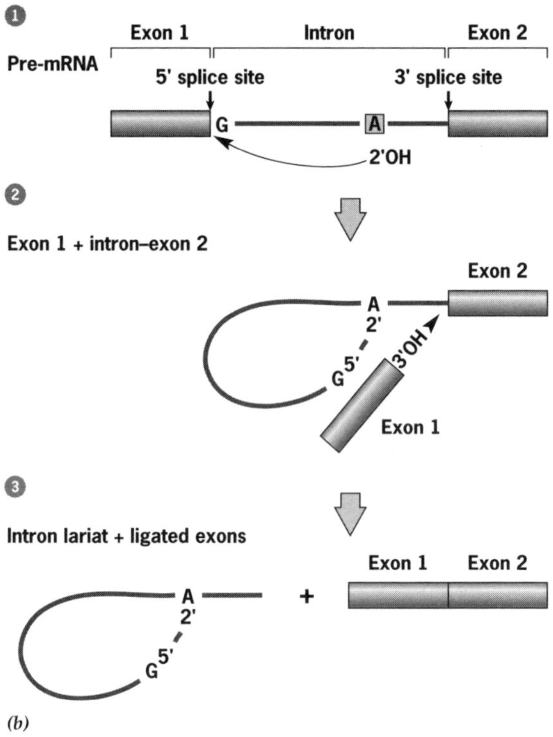

Figure 11.31b The structure and self-splicing pathway of group II introns.

Copyright © 2005 John Wiley & Sons, Inc.

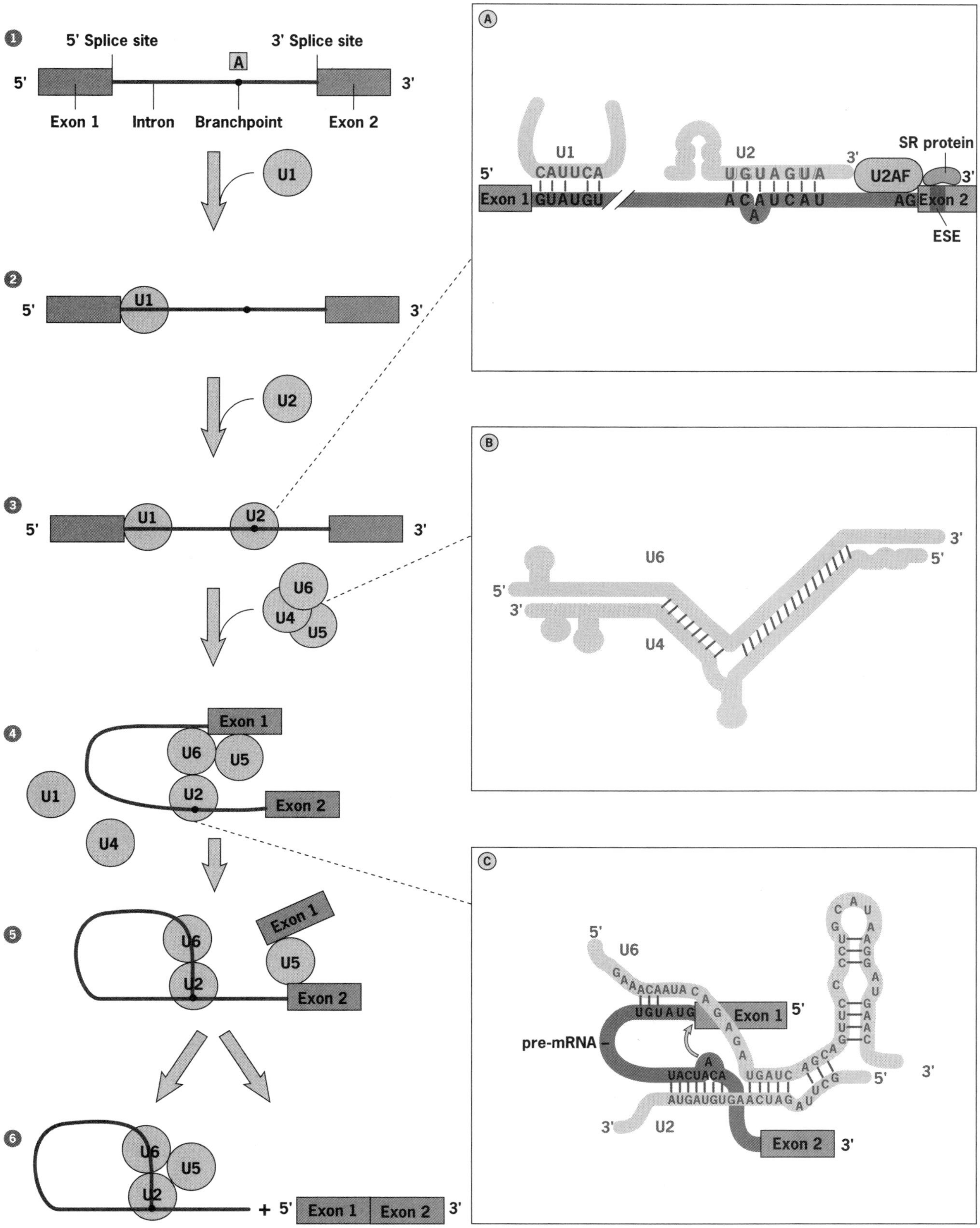

Figure 11.32 Schematic model of the assembly of the splicing machinery and some of the steps that occur during splicing.

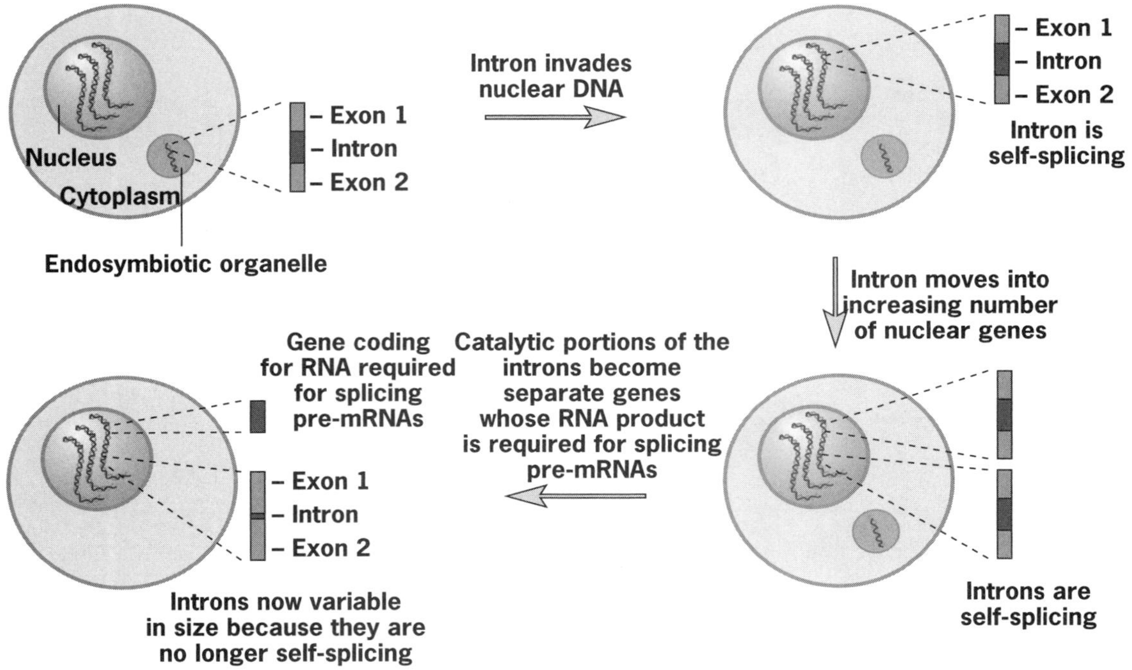

Figure 11.37 A hypothesis that may explain the evolutionary origin of introns in eukaryotic DNA.

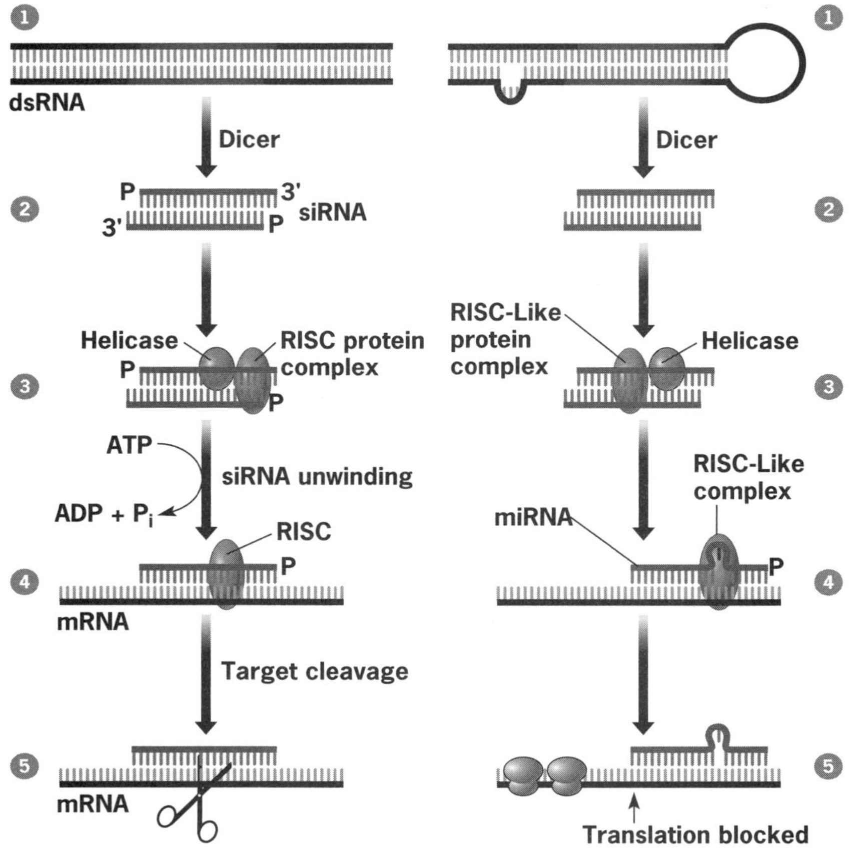

Figure 11.39 The formation and mechanism of action of siRNAs and miRNAs.

Copyright © 2005 John Wiley & Sons, Inc.

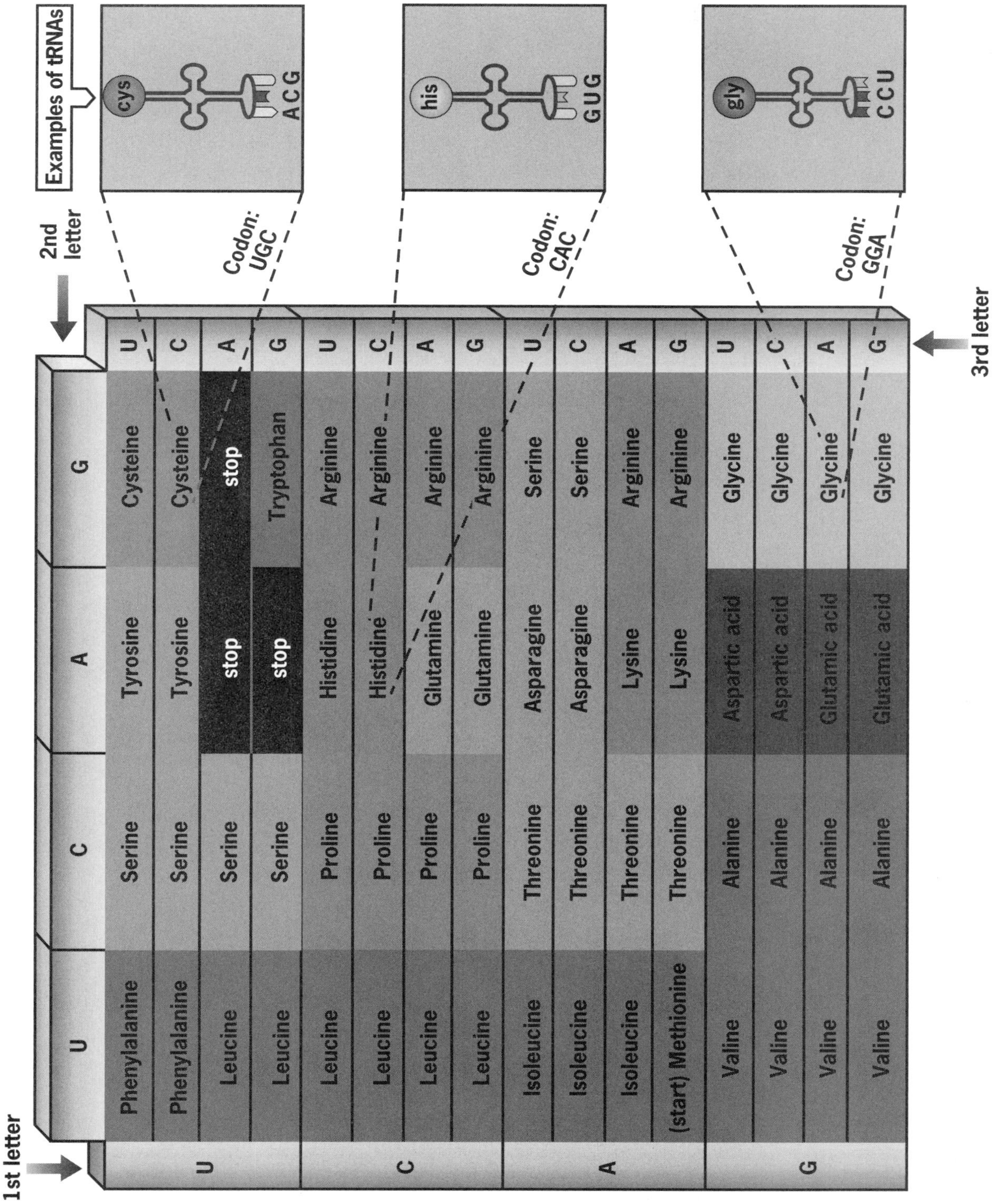

1st letter	2nd: U	2nd: C	2nd: A	2nd: G	3rd letter
U	Phenylalanine	Serine	Tyrosine	Cysteine	U
	Phenylalanine	Serine	Tyrosine	Cysteine	C
	Leucine	Serine	stop	stop	A
	Leucine	Serine	stop	Tryptophan	G
C	Leucine	Proline	Histidine	Arginine	U
	Leucine	Proline	Histidine	Arginine	C
	Leucine	Proline	Glutamine	Arginine	A
	Leucine	Proline	Glutamine	Arginine	G
A	Isoleucine	Threonine	Asparagine	Serine	U
	Isoleucine	Threonine	Asparagine	Serine	C
	Isoleucine	Threonine	Lysine	Arginine	A
	(start) Methionine	Threonine	Lysine	Arginine	G
G	Valine	Alanine	Aspartic acid	Glycine	U
	Valine	Alanine	Aspartic acid	Glycine	C
	Valine	Alanine	Glutamic acid	Glycine	A
	Valine	Alanine	Glutamic acid	Glycine	G

Figure 11.41 The genetic code.

Copyright © 2005 John Wiley & Sons, Inc.

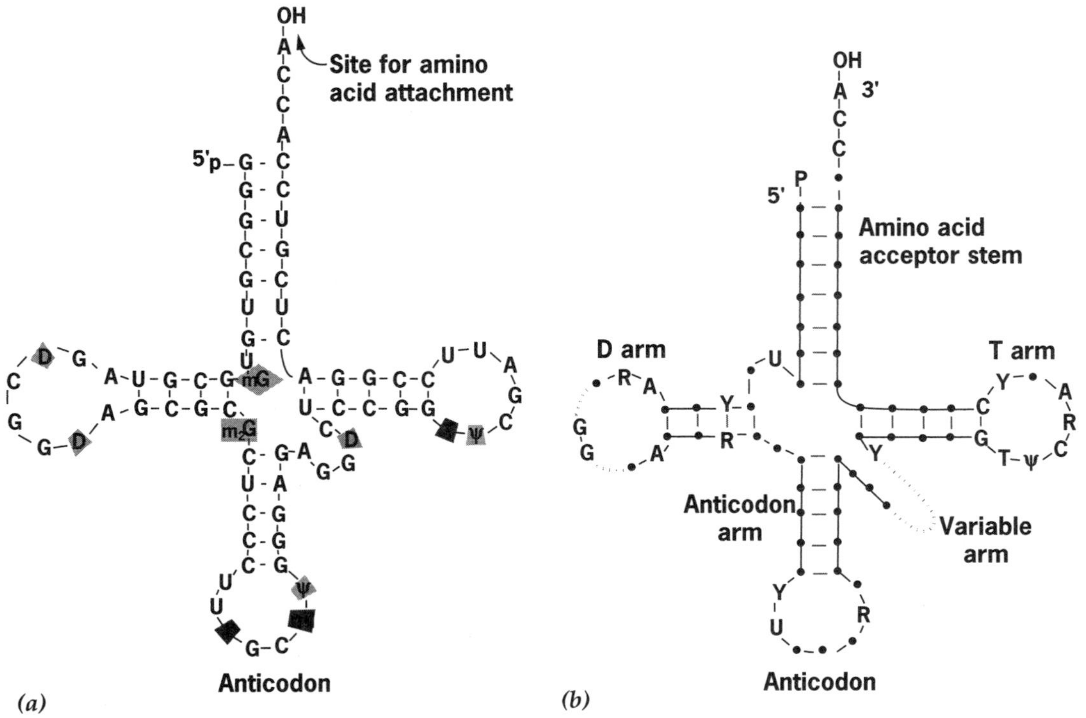

Figure 11.42 Two-dimensional structure of transfer RNAs.

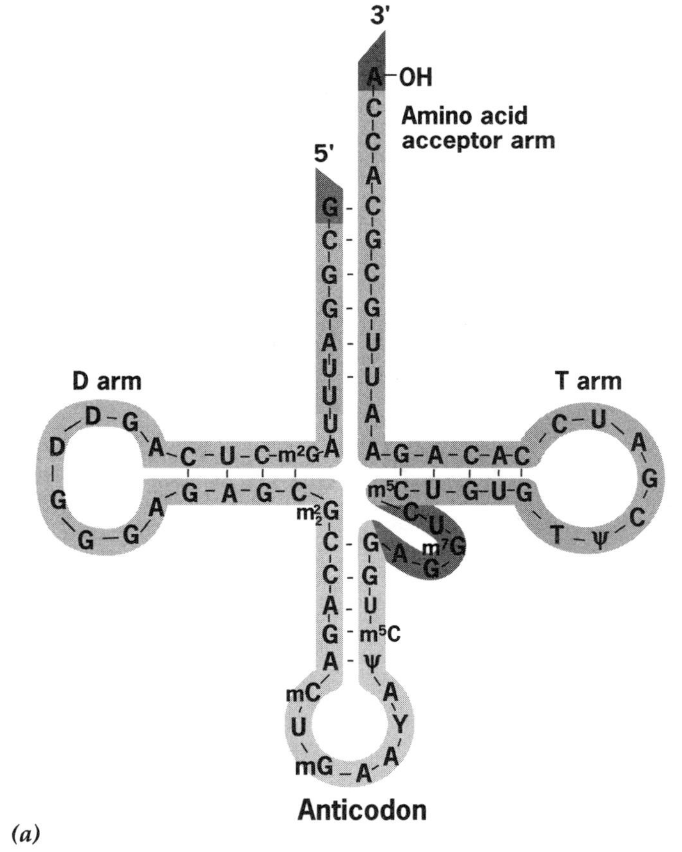

Figure 11.43a The structure of a tRNA.

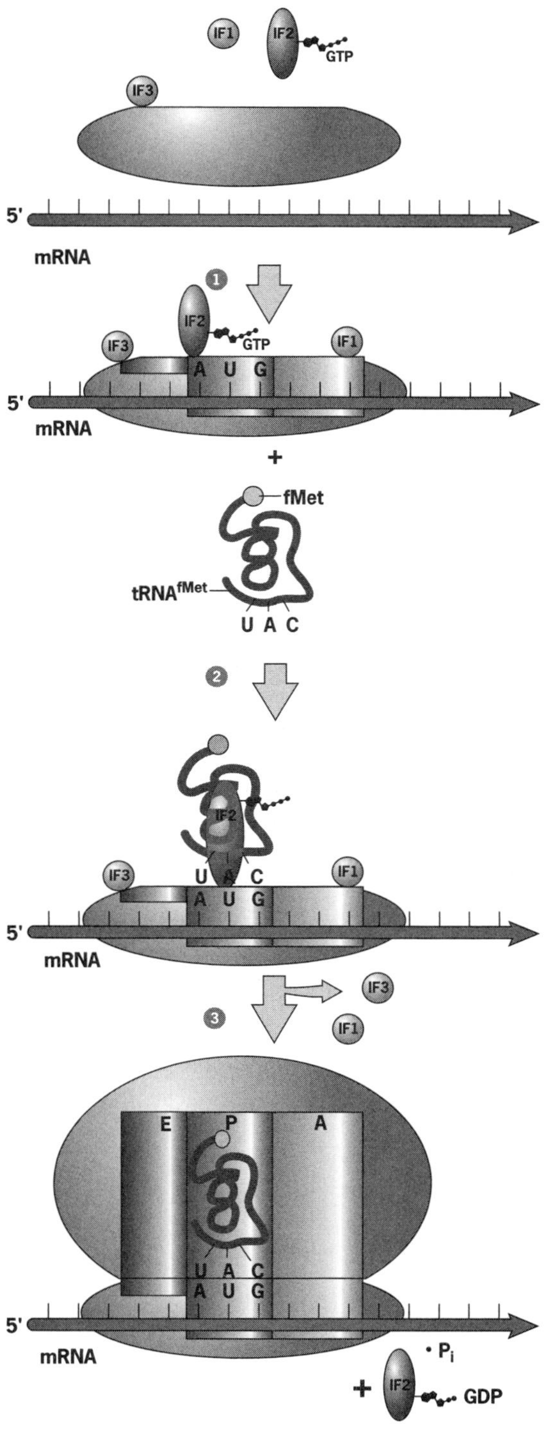

Figure 11.47 Initiation of protein synthesis in prokaryotes.

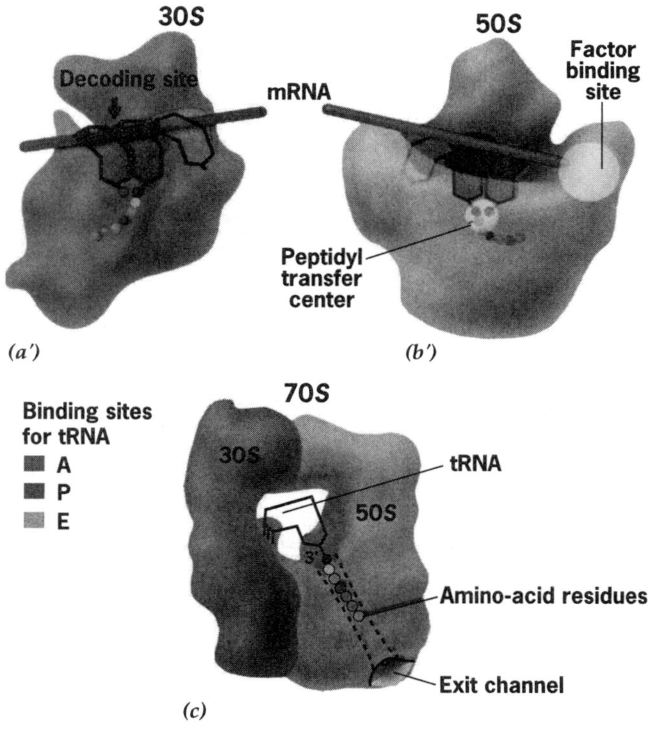

Figure 11.49 Model of the bacterial ribosome based on X-ray crystallographic data, showing tRNAs bound to the A, P, and E sites of the two ribosomal subunits.

Copyright © 2005 John Wiley & Sons, Inc.

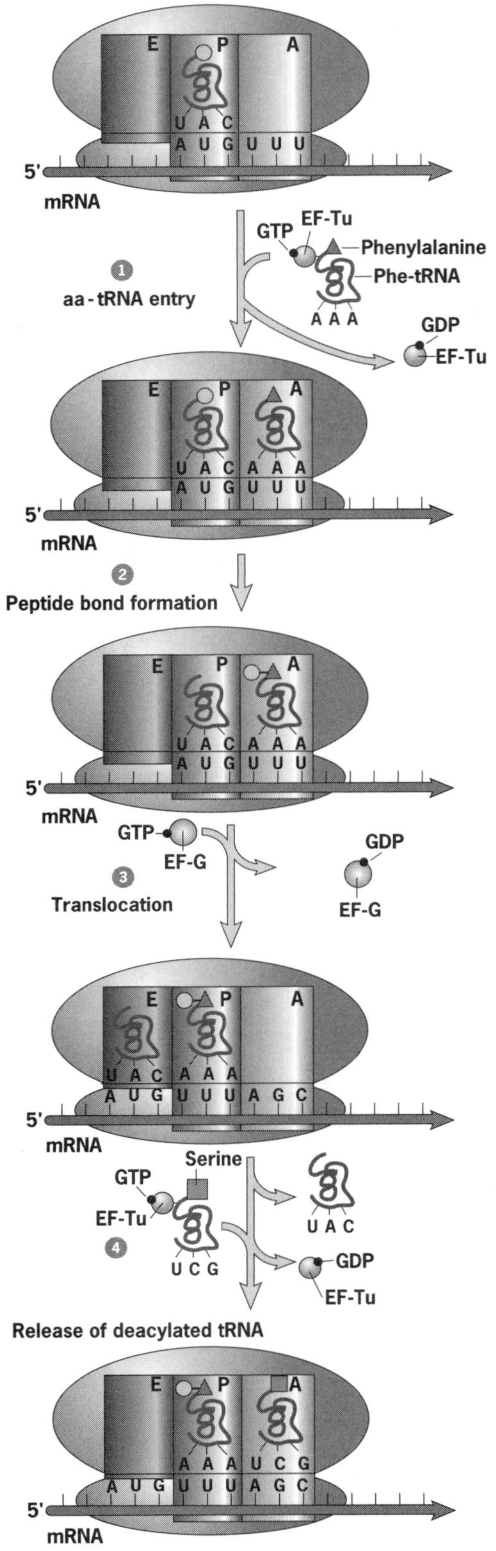

(a)

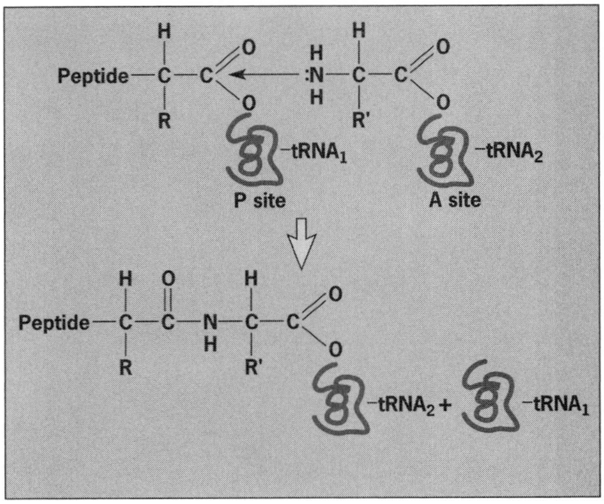

(b)

Figure 11.50 Steps in the elongation of the nascent polypeptide during translation in prokaryotes.

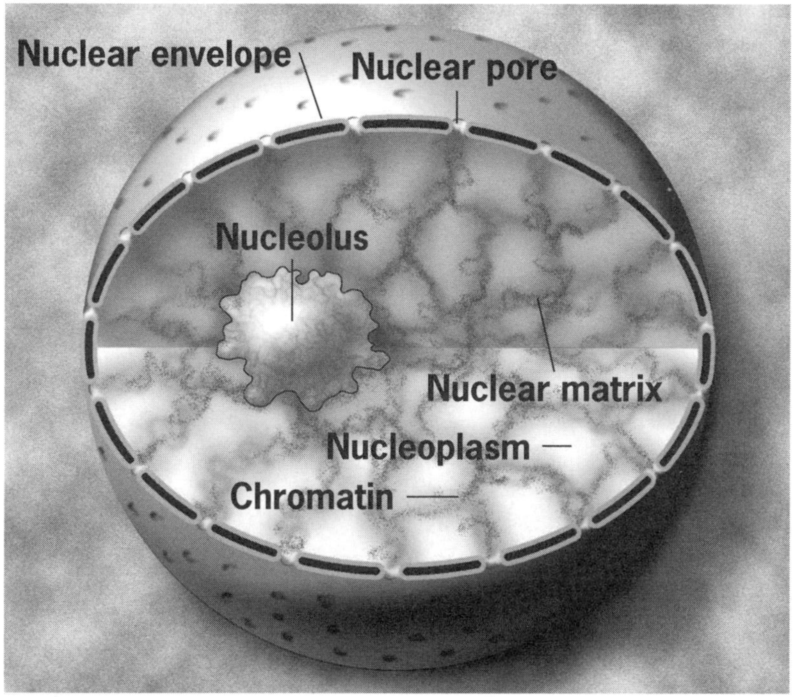

(b)

Figure 12.1b The cell nucleus.

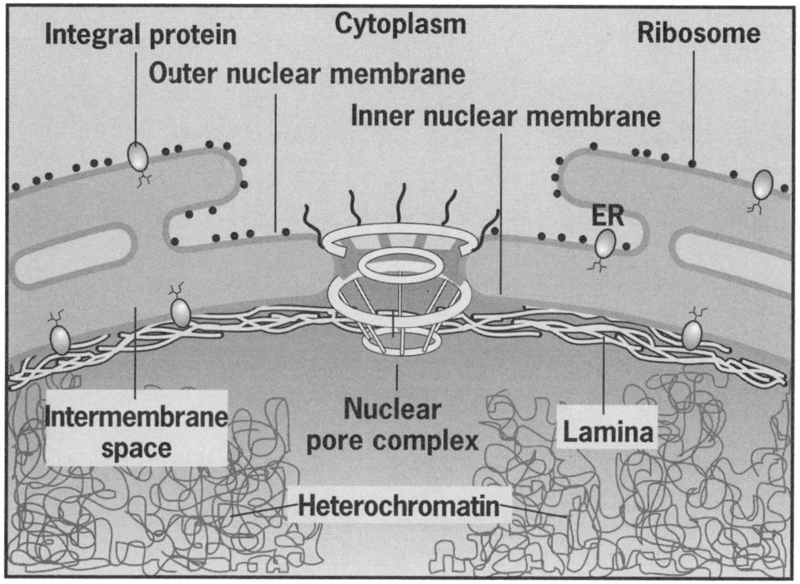

(a)

Figure 12.2a The nuclear envelope.

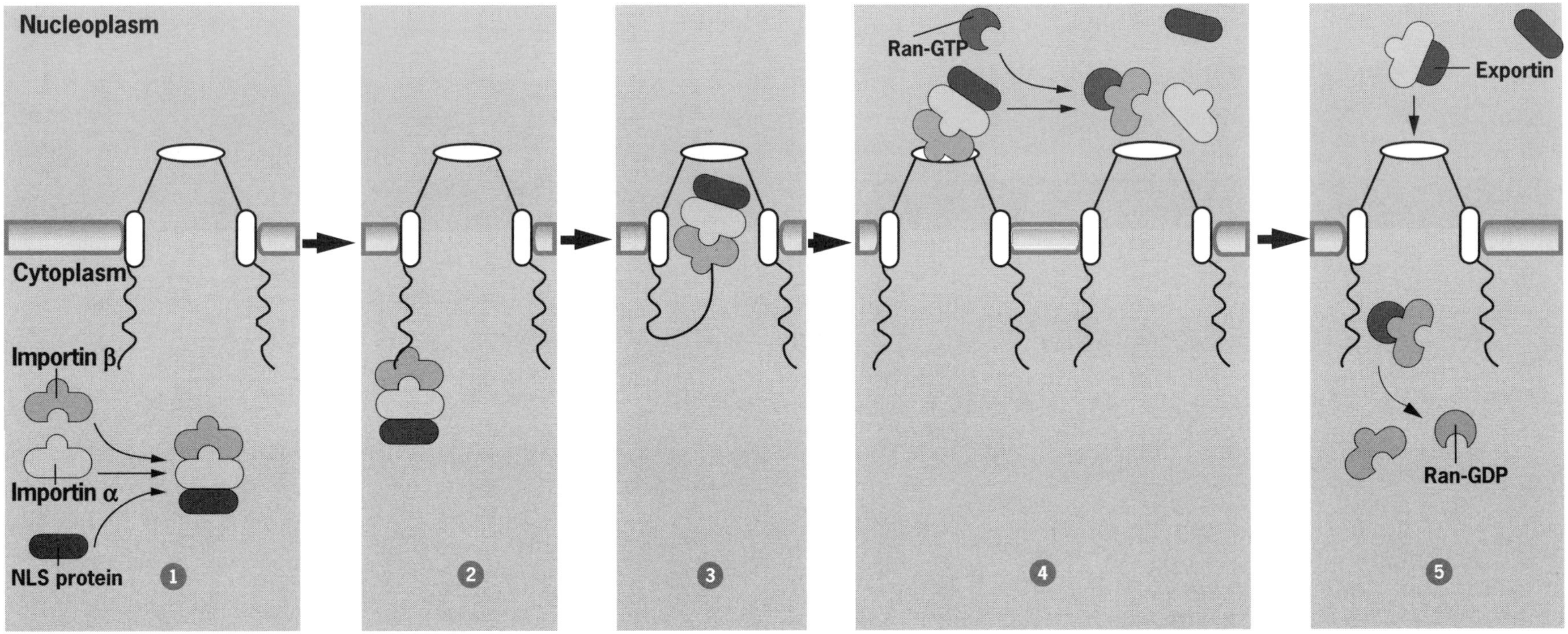

(a)

Figure 12.8a Importing proteins from the cytoplasm into the nucleus.

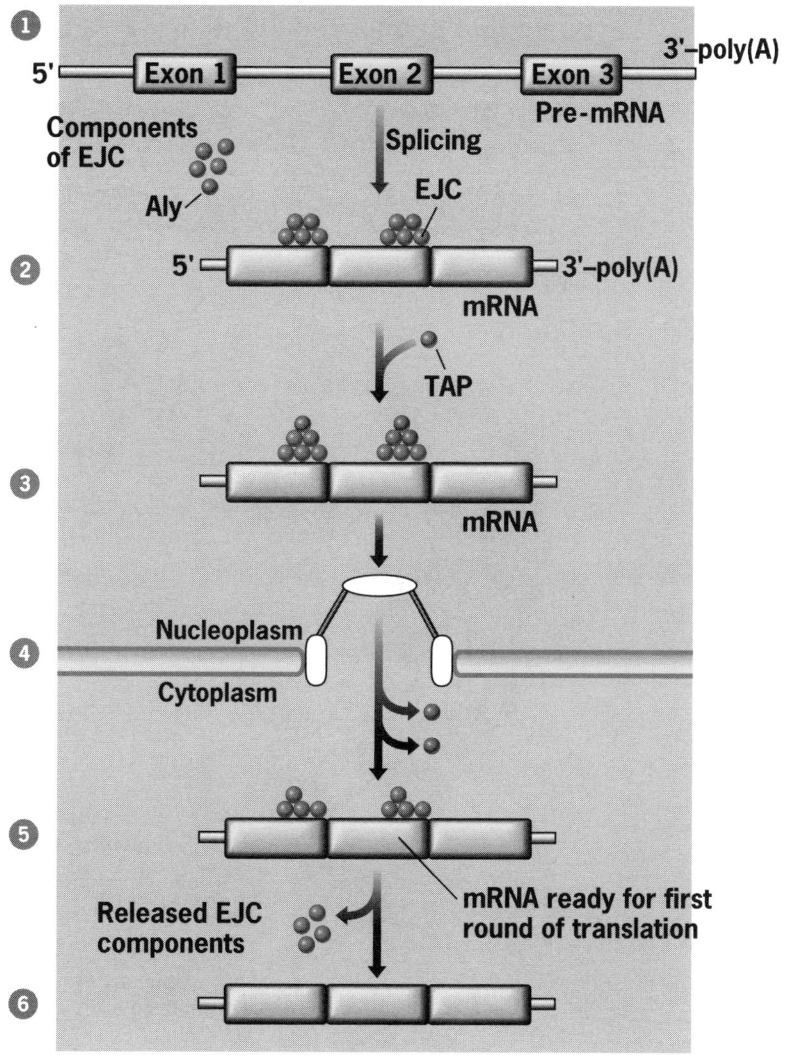

Figure 12.9 Export of mRNPs from the nucleus.

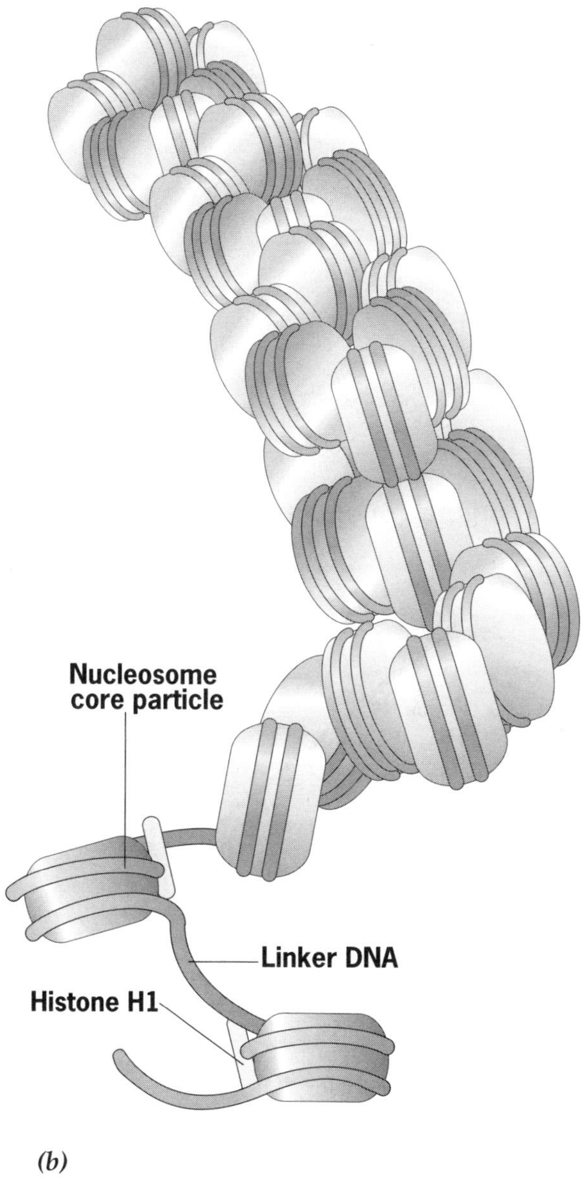

Figure 12.12b The 30-nm fiber: a higher level of chromatin structure.

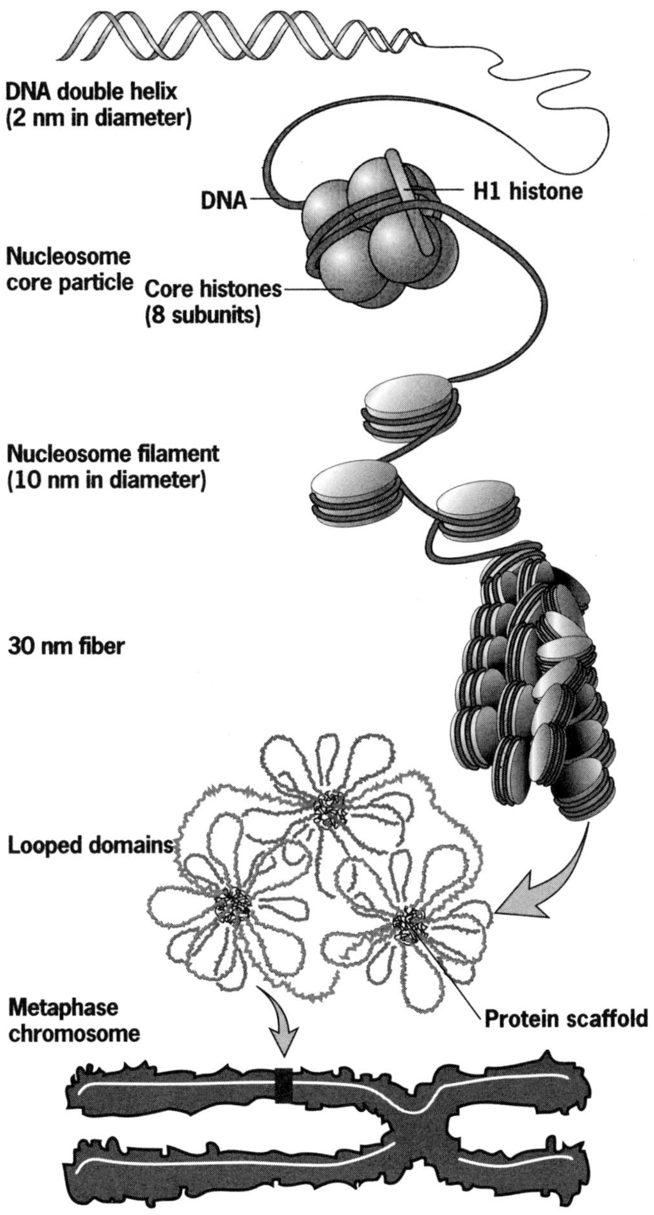

Figure 12.14 Levels of organization of chromatin.

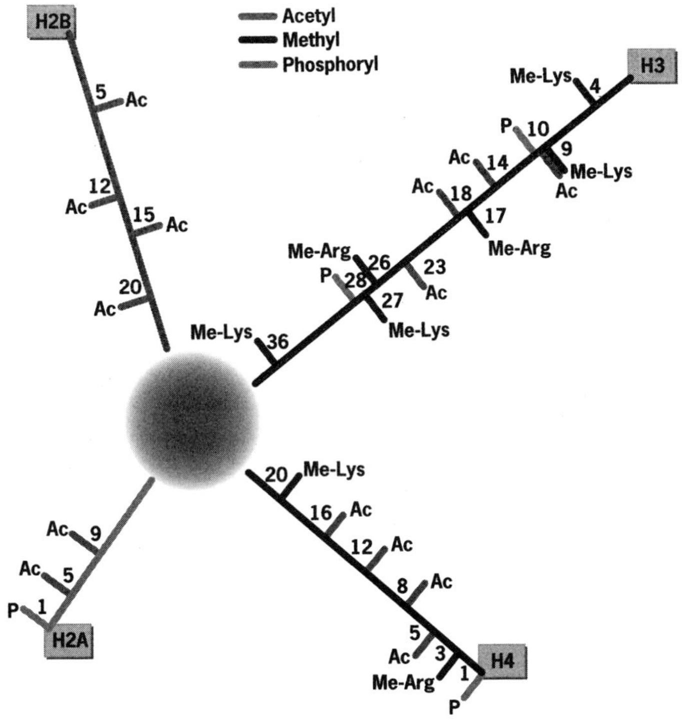

Figure 12.16 Histone modifications and the "histone code."

Copyright © 2005 John Wiley & Sons, Inc.

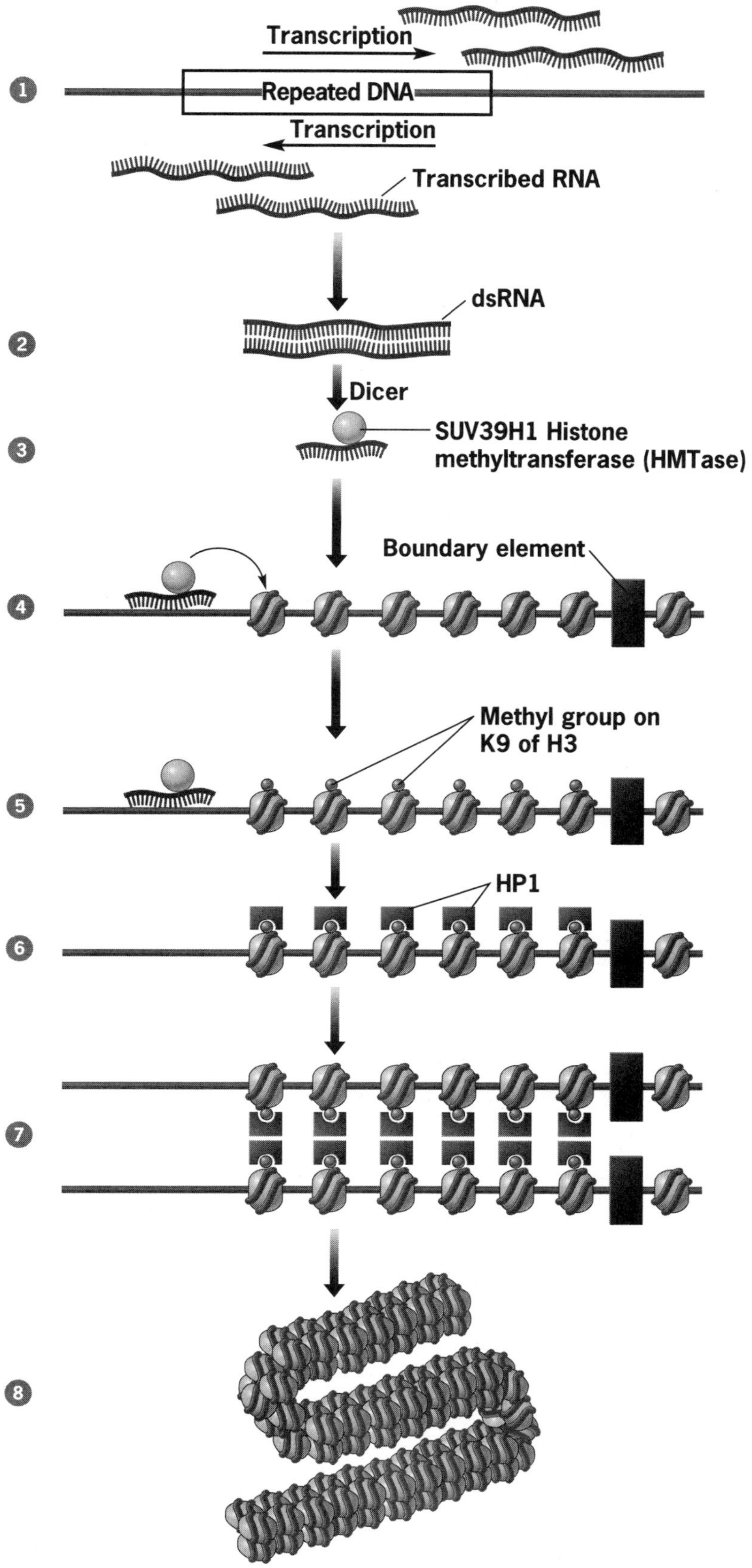

Figure 12.17 A model showing possible events during the formation of heterochromatin.

Copyright © 2005 John Wiley & Sons, Inc.

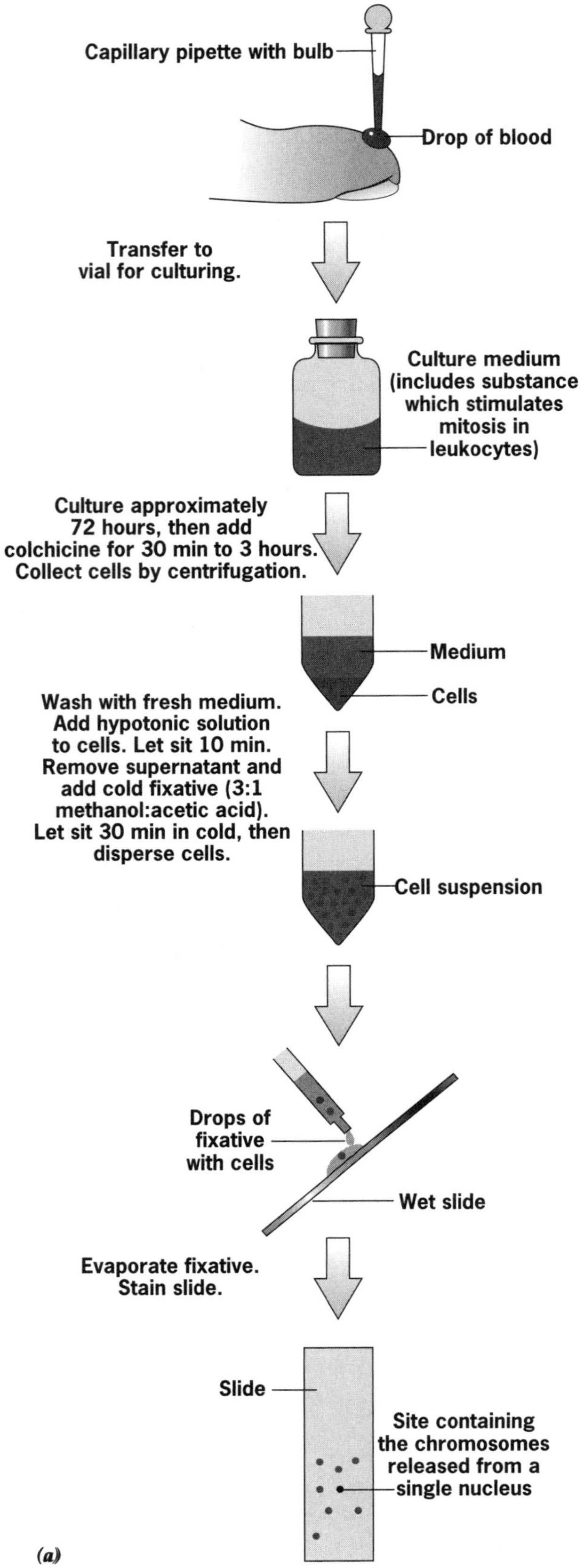
Capillary pipette with bulb
Drop of blood
Transfer to
vial for culturing.
Culture medium
(includes substance
which stimulates
mitosis in
leukocytes)
Culture approximately
72 hours, then add
colchicine for 30 min to 3 hours.
Collect cells by centrifugation.
Medium
Cells
Wash with fresh medium.
Add hypotonic solution
to cells. Let sit 10 min.
Remove supernatant and
add cold fixative (3:1
methanol:acetic acid).
Let sit 30 min in cold, then
disperse cells.
Cell suspension
Drops of
fixative
with cells
Wet slide
Evaporate fixative.
Stain slide.
Slide
Site containing
the chromosomes
released from a
single nucleus
(a)

Figure 12.18a Human mitotic chromosomes and karyotypes.

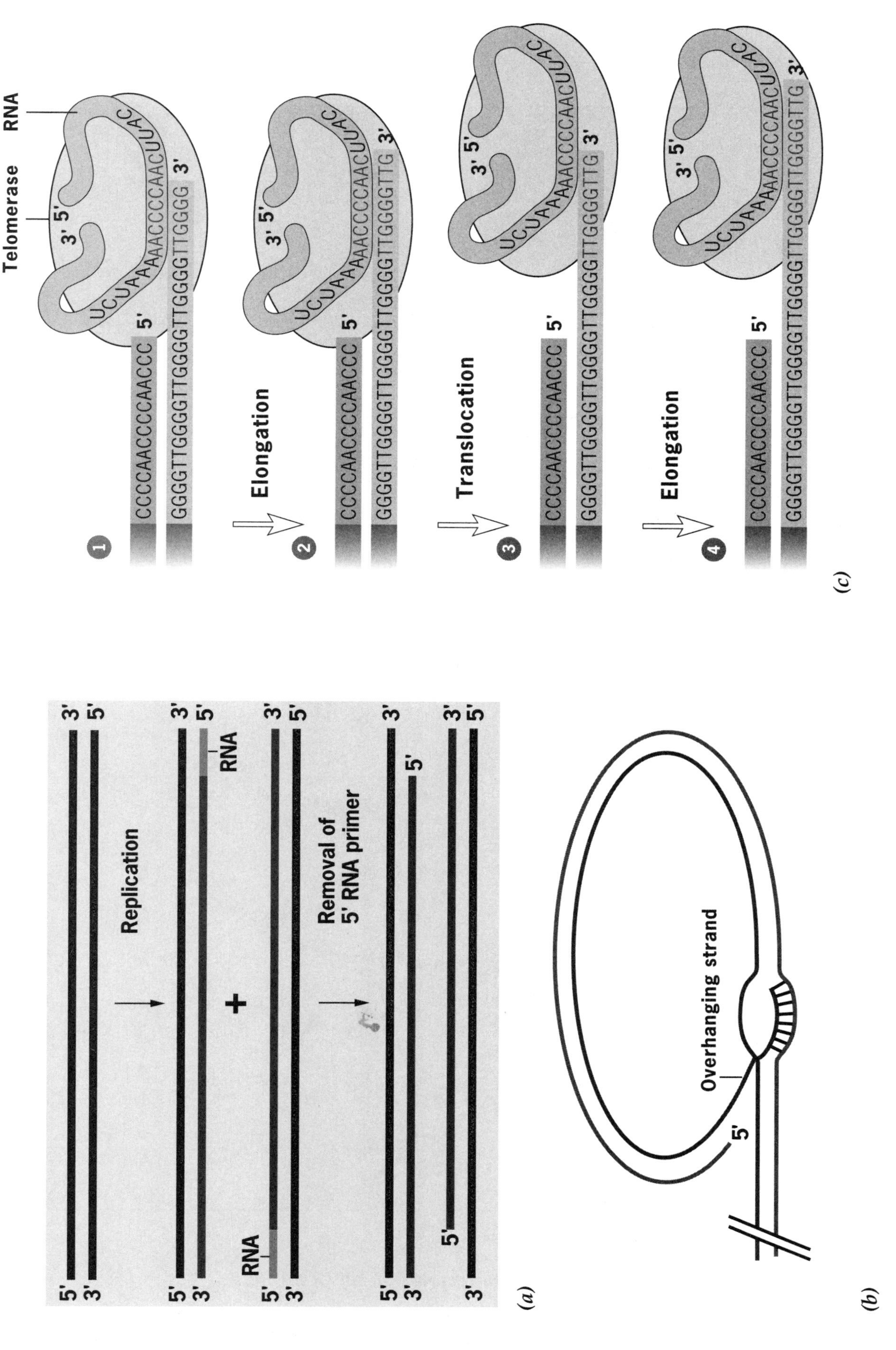

Figure 12.20 The end-replication problem and the role of telomerase.

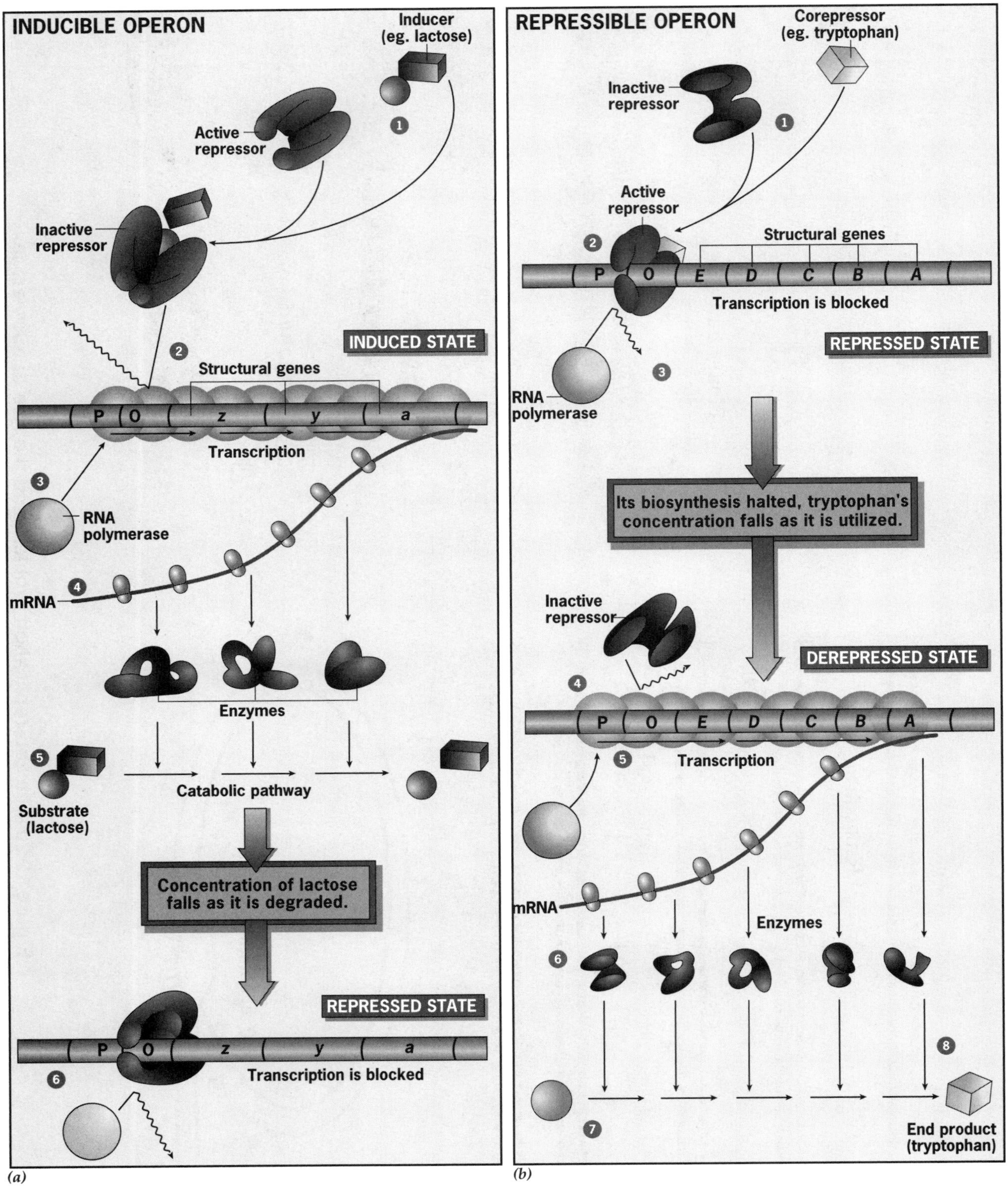

Figure 12.29 Gene regulation by operons.

Copyright © 2005 John Wiley & Sons, Inc.

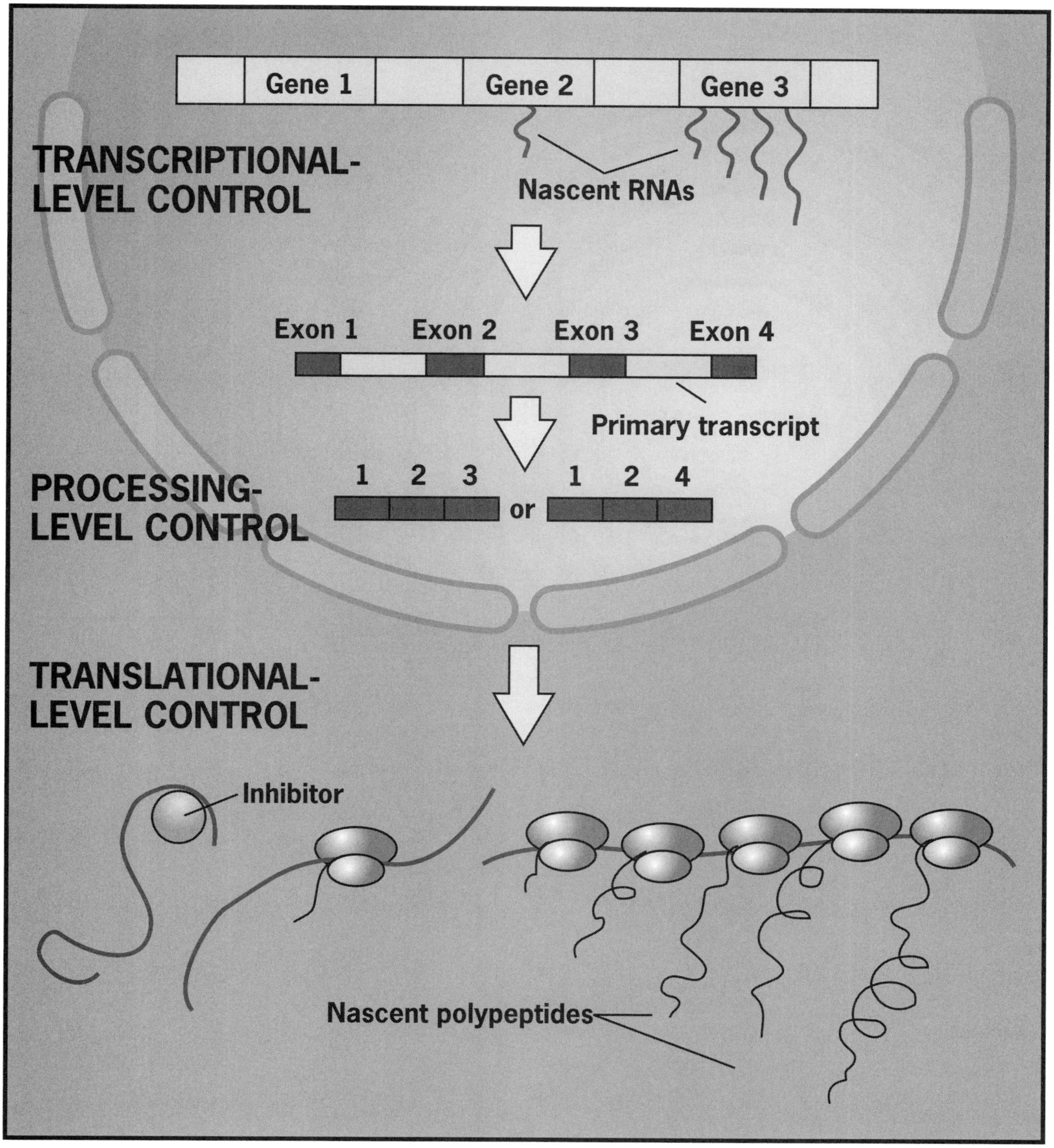

Figure 12.32 Overview of the levels of control of gene expression.

Copyright © 2005 John Wiley & Sons, Inc.

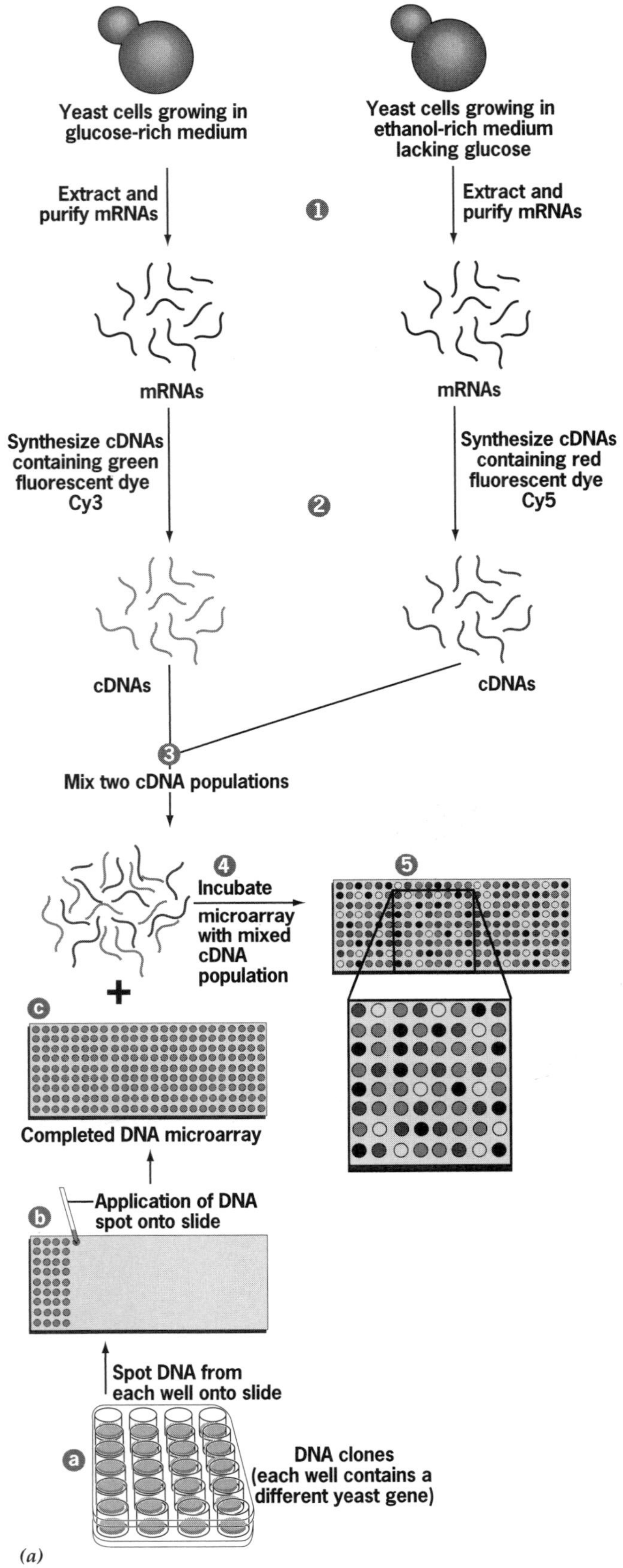

Figure 12.34a The production of DNA microarrays and their use in monitoring gene transcription.

Copyright © 2005 John Wiley & Sons, Inc.

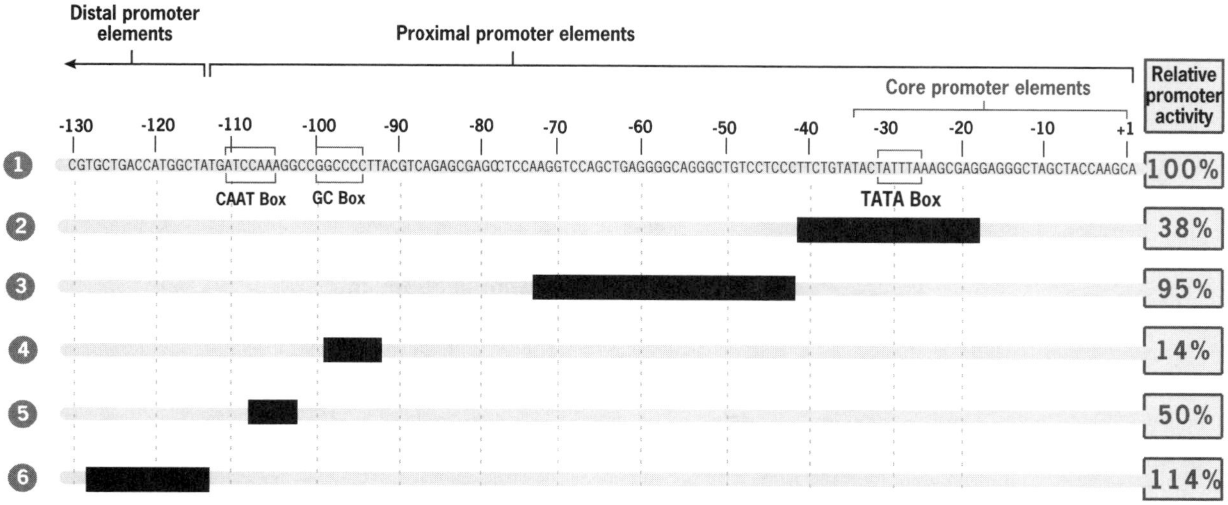

Figure 12.41 Identifying promoter sequences required for transcription.

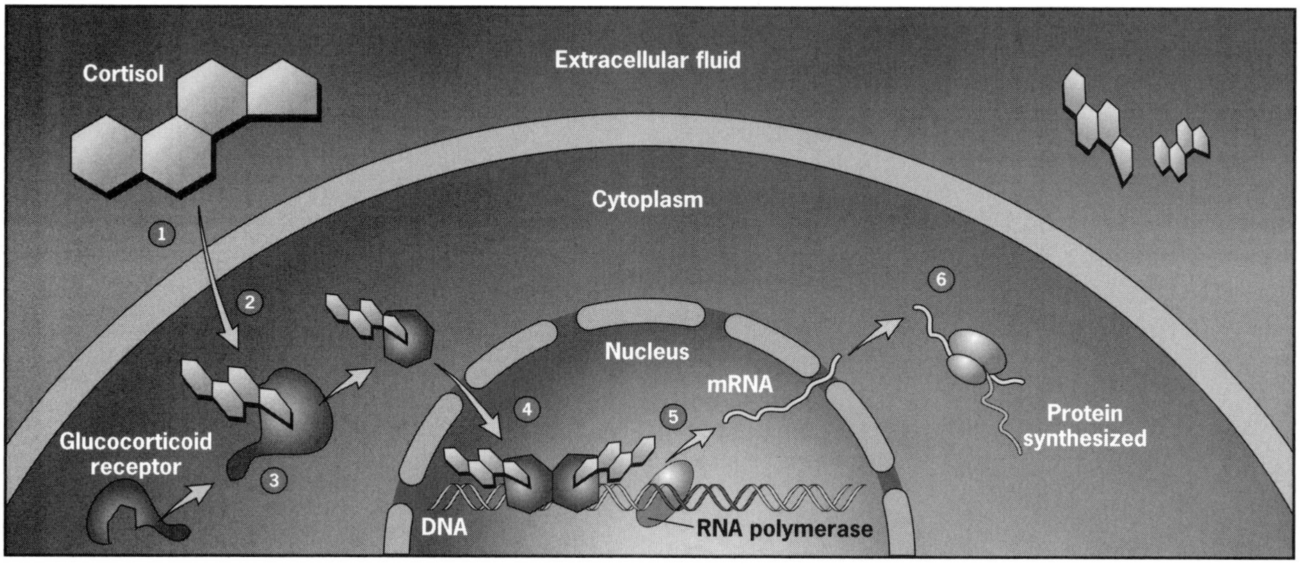

Figure 12.44 Activation of a gene by a steroid hormone, such as the glucocorticoid cortisol.

Copyright © 2005 John Wiley & Sons, Inc.

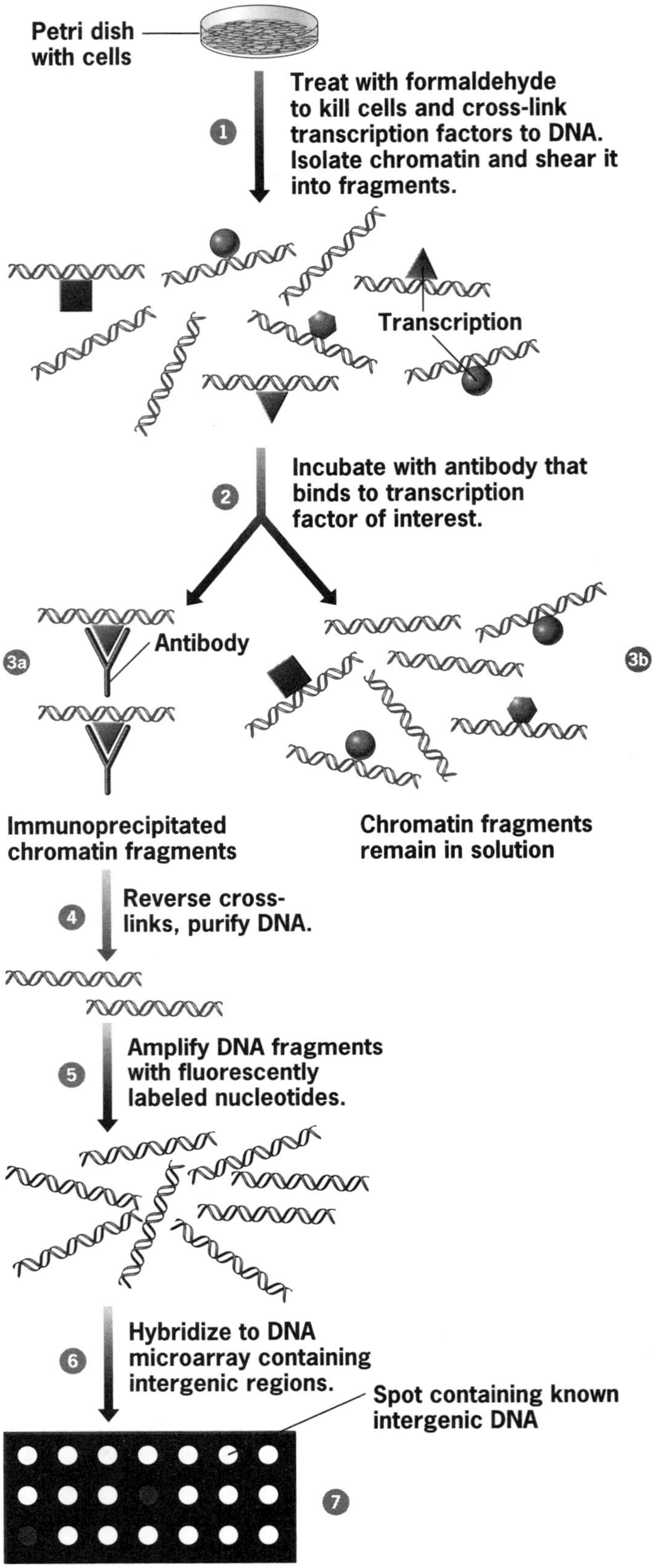
Petri dish
with cells
1
Treat with formaldehyde
to kill cells and cross-link
transcription factors to DNA.
Isolate chromatin and shear it
into fragments.
Transcription
2
Incubate with antibody that
binds to transcription
factor of interest.
3a
Antibody
3b
Immunoprecipitated
chromatin fragments
Chromatin fragments
remain in solution
4
Reverse cross-
links, purify DNA.
5
Amplify DNA fragments
with fluorescently
labeled nucleotides.
6
Hybridize to DNA
microarray containing
intergenic regions.
Spot containing known
intergenic DNA
7

Figure 12.42 Use of chromatin immunoprecipitation (ChIP) and microarray analysis to identify transcription-factor binding sites on a global scale.

Copyright © 2005 John Wiley & Sons, Inc.

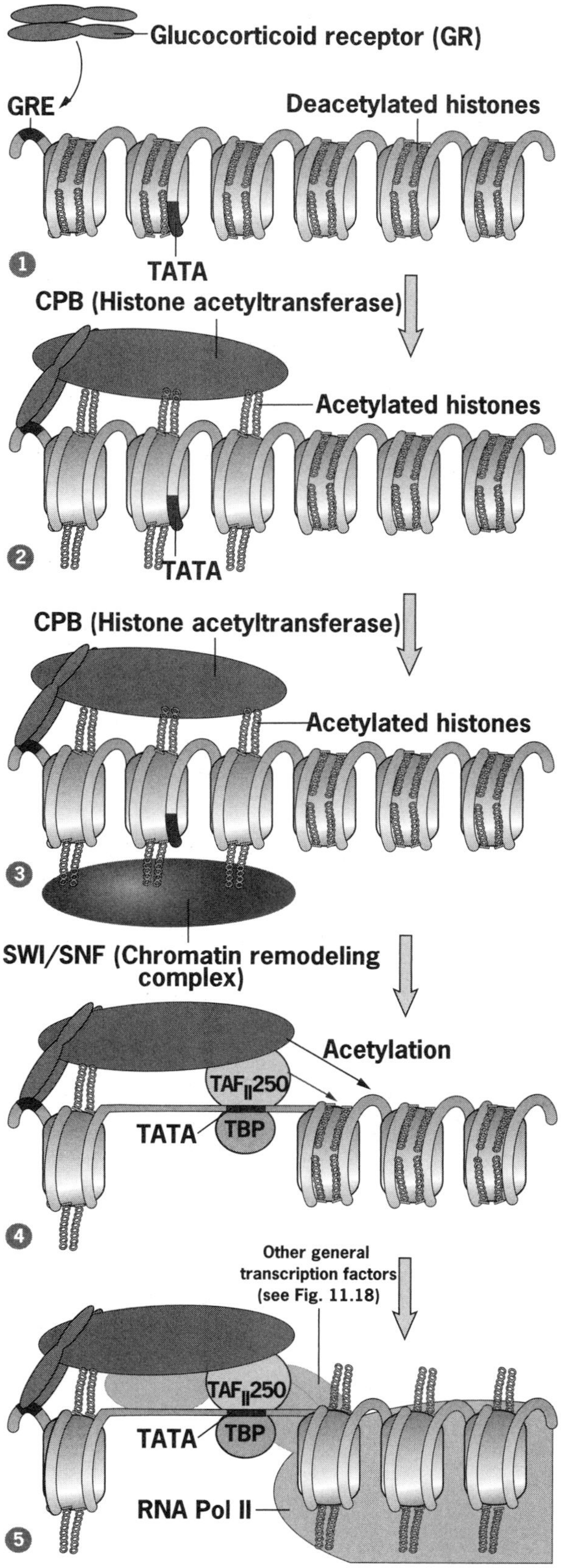

Figure 12.46 A model for the events that occur at a promoter following the binding of a transcriptional activator.

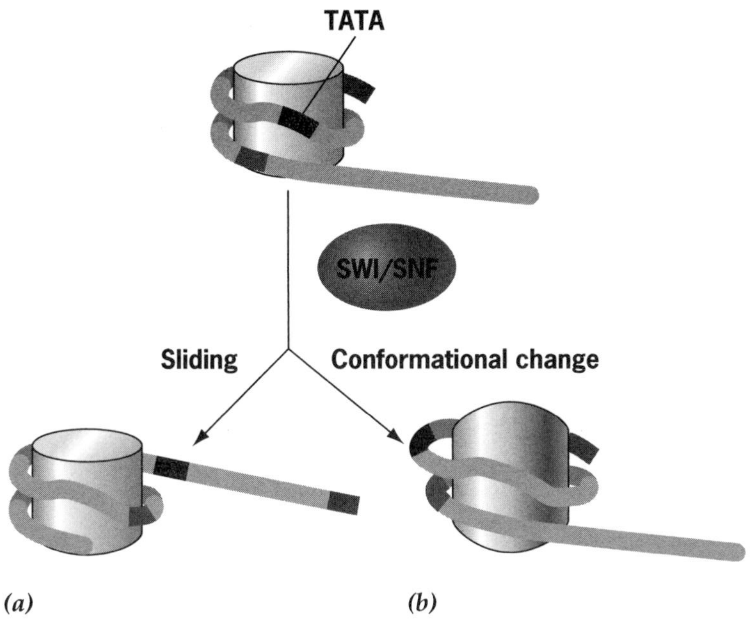

Figure 12.47 Models for the mechanism of action of chromatin remodeling complexes, such as SWI/SNF.

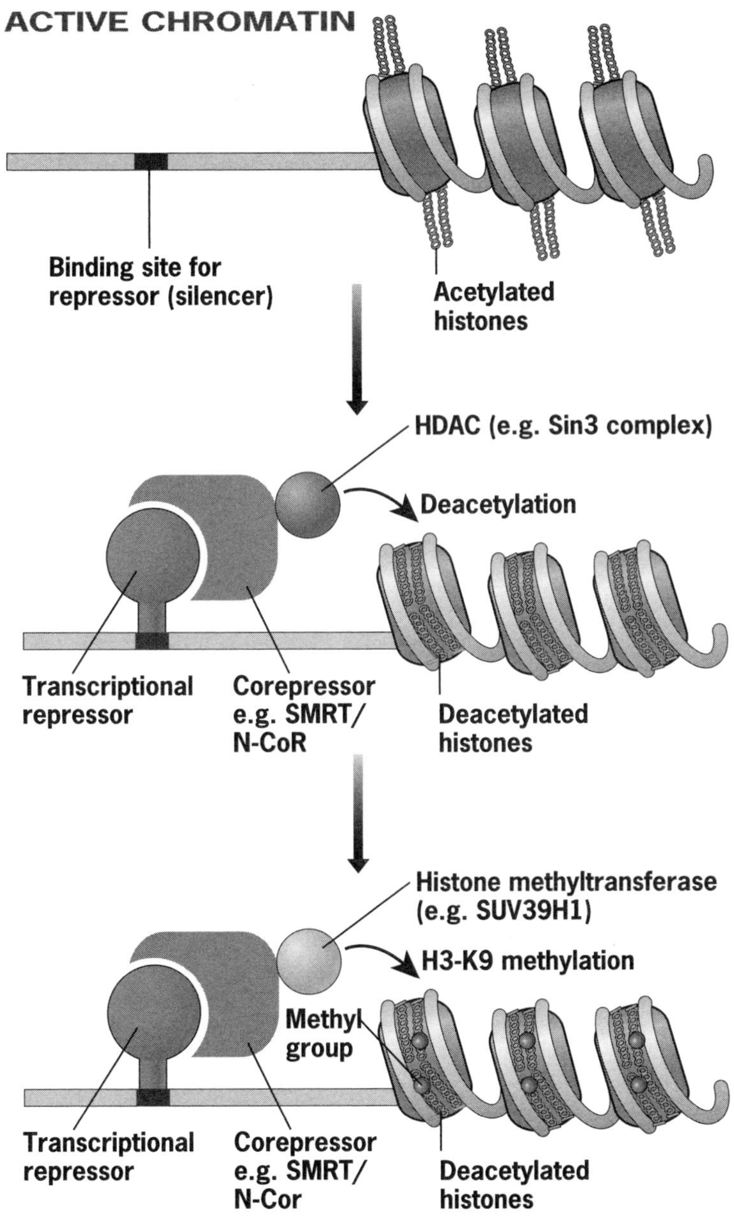

Figure 12.48 A model for transcriptional repression.

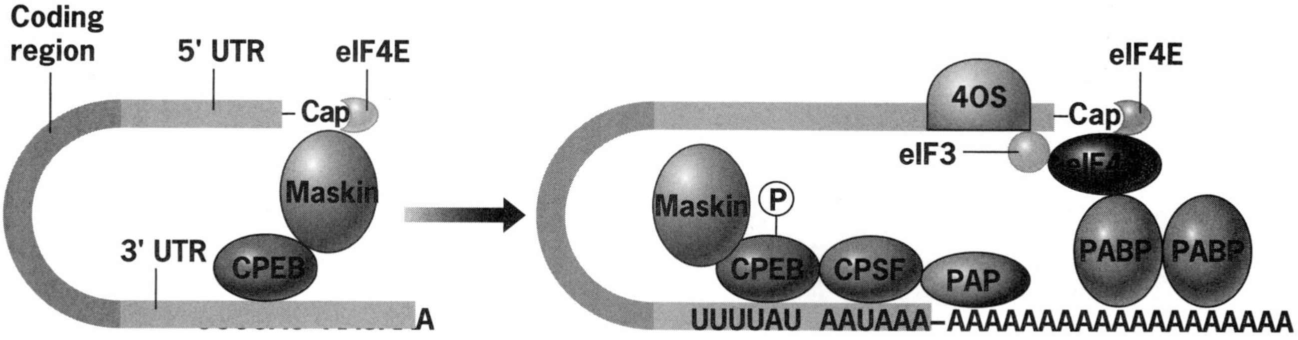

Figure 12.57 A model for the mechanism of translational activation of mRNAs following fertilization of a *Xenopus* egg.

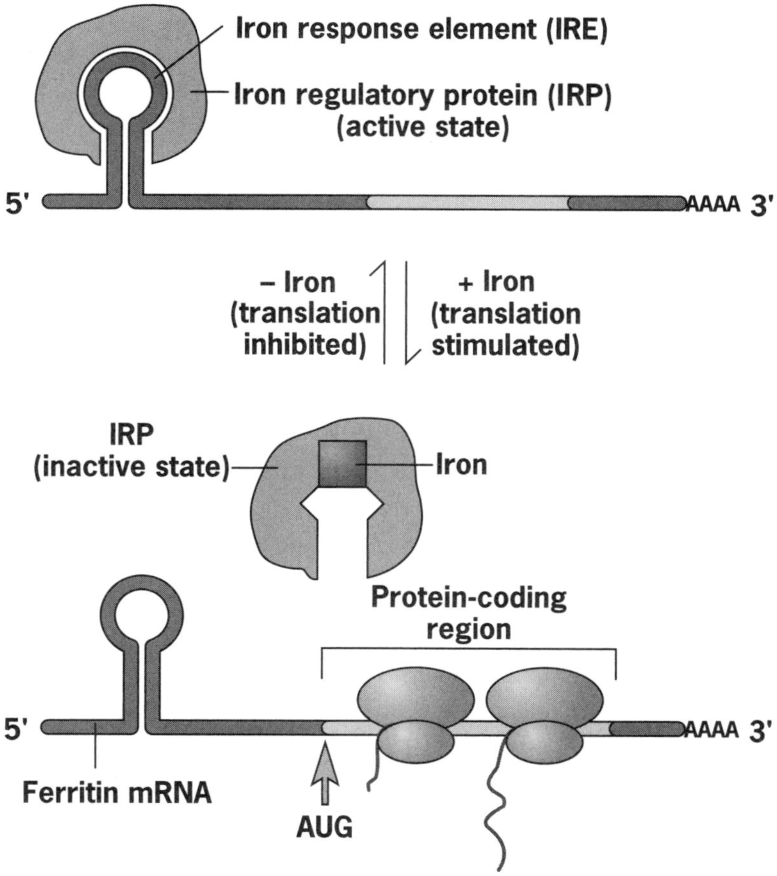

Figure 12.58 The control of ferritin mRNA translation.

Copyright © 2005 John Wiley & Sons, Inc.

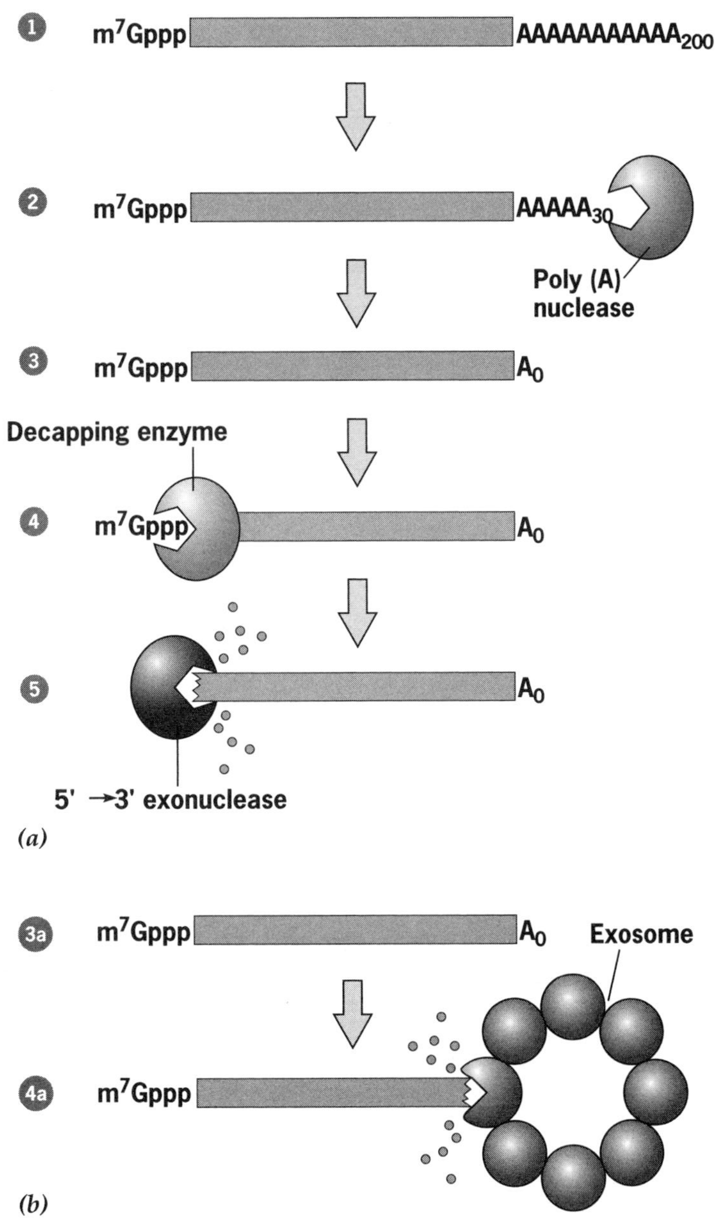

Figure 12.59 mRNA degradation in mammalian cells.

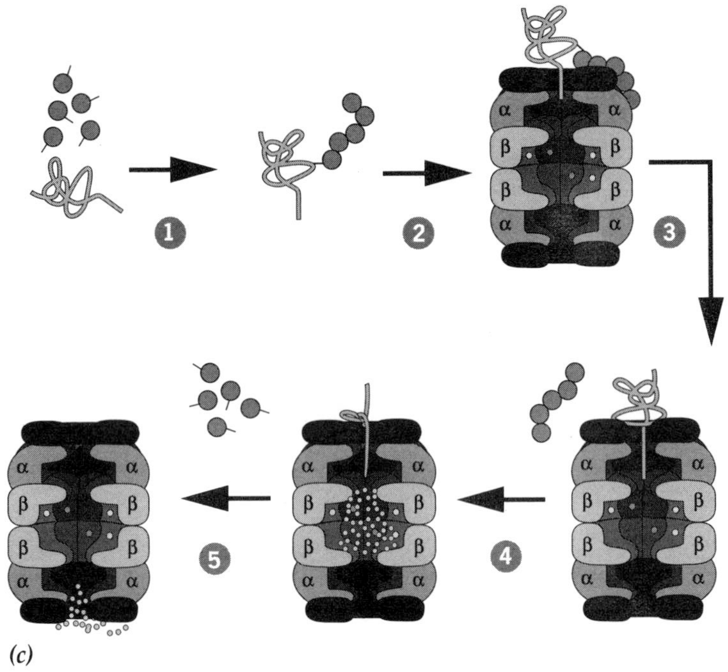

Figure 12.60c Proteasome structure and function.

Copyright © 2005 John Wiley & Sons, Inc.

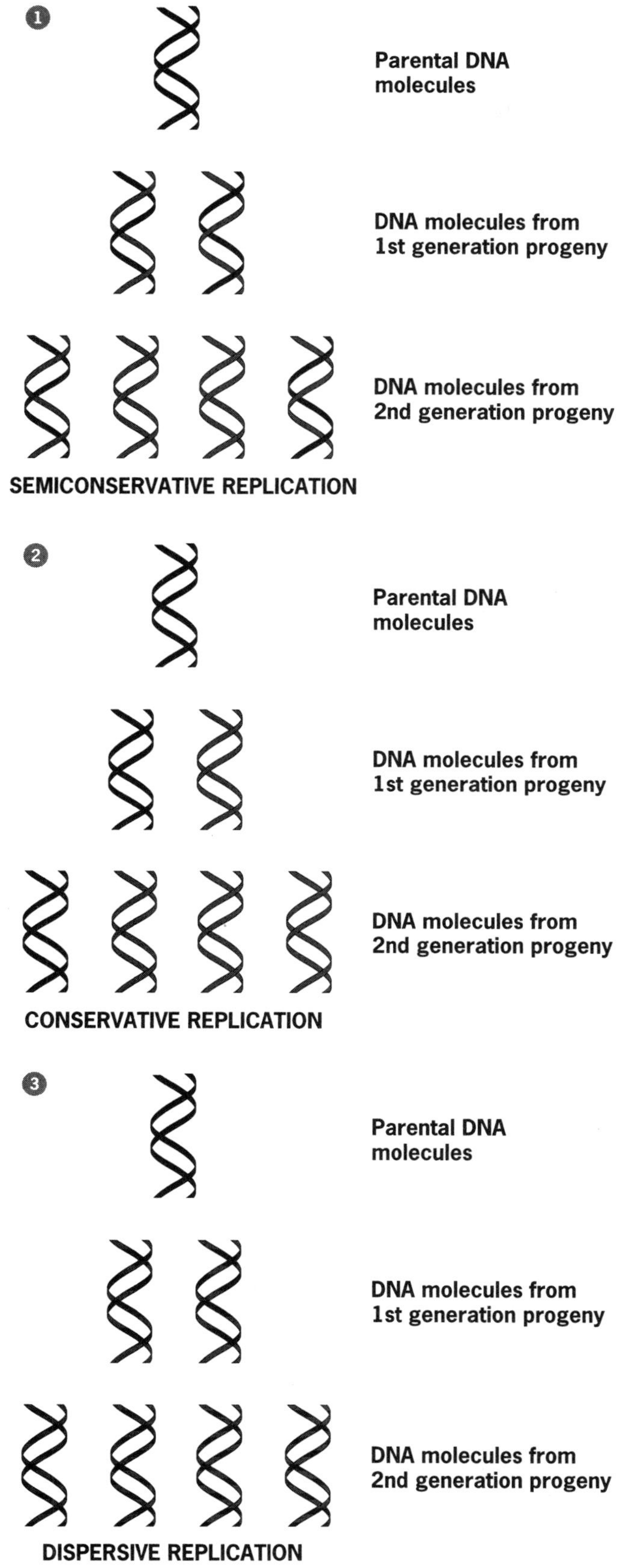

Figure 13.2 Three alternate schemes of replication.

Copyright © 2005 John Wiley & Sons, Inc.

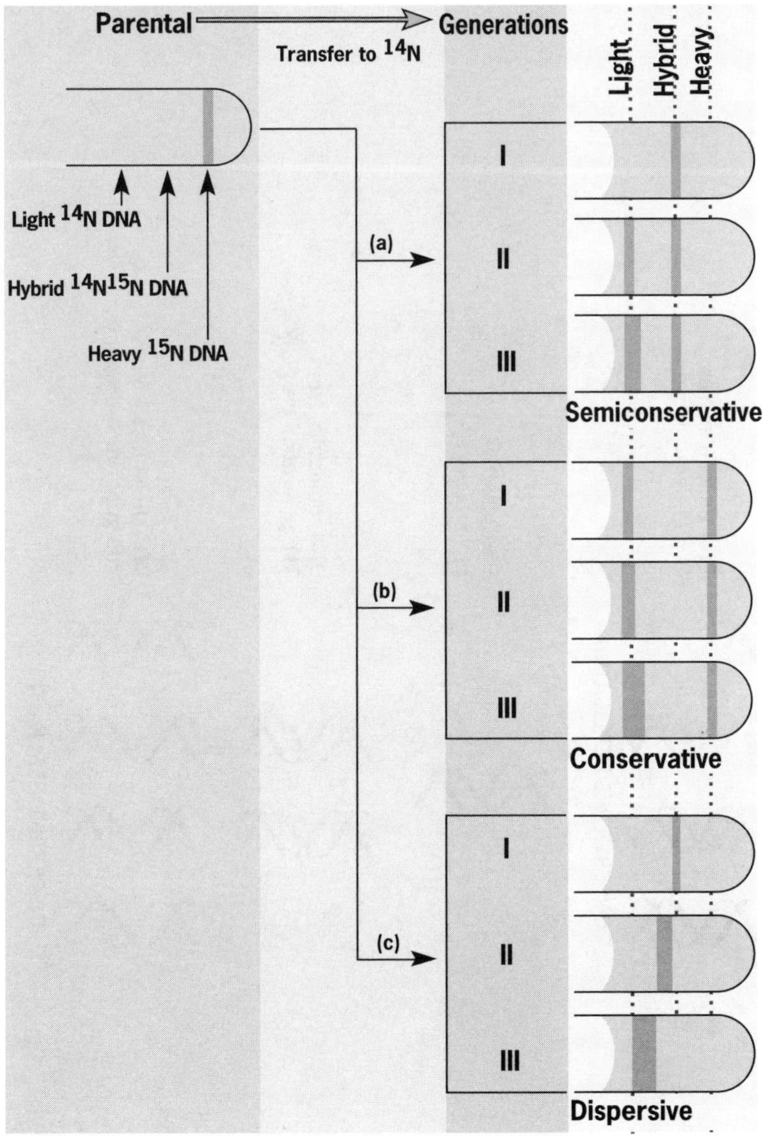

(a)

Figure 13.3a Experiment demonstrating that DNA replication in bacteria is semiconservative.

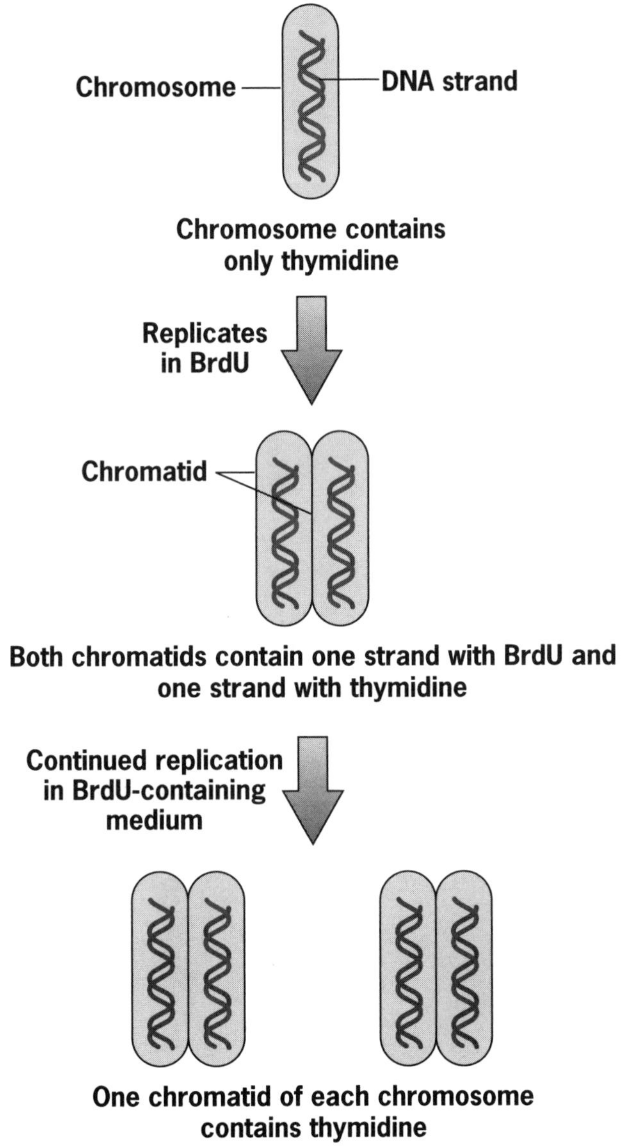

(a)

Figure 13.4a DNA replication in eukaryotic cells is semiconservative.

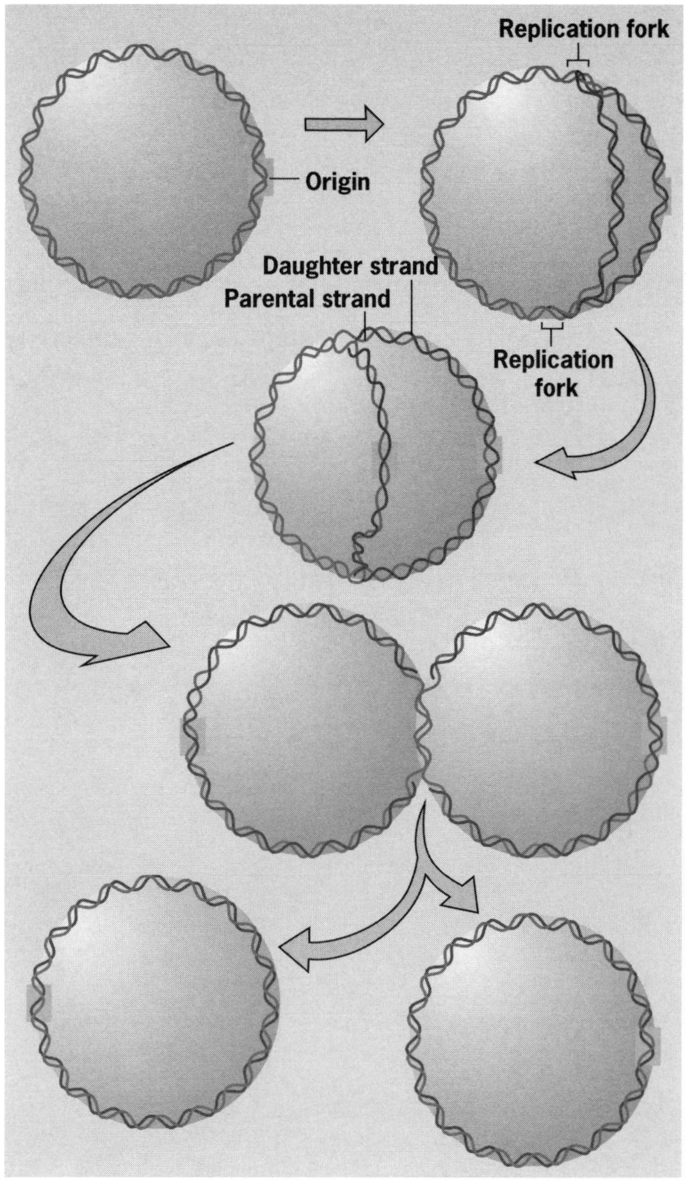

Figure 13.5 Model of a circular chromosome undergoing bidirectional, semiconservative replication.

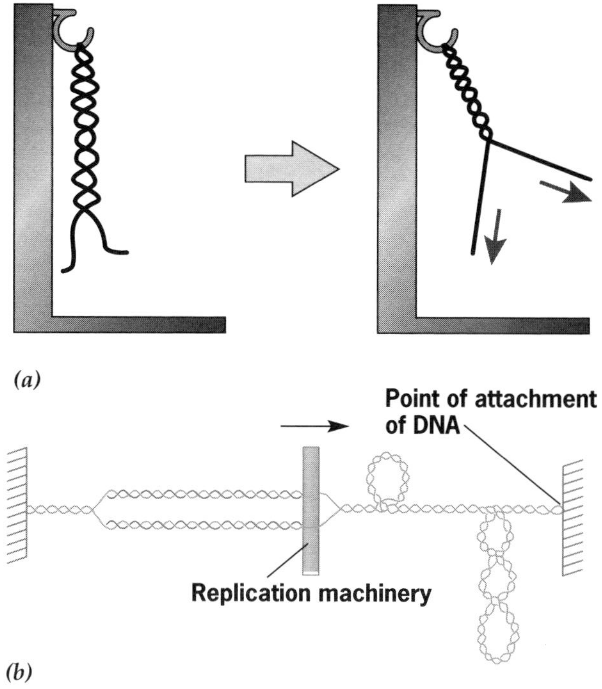

Figure 13.6 The unwinding problem.

Copyright © 2005 John Wiley & Sons, Inc.

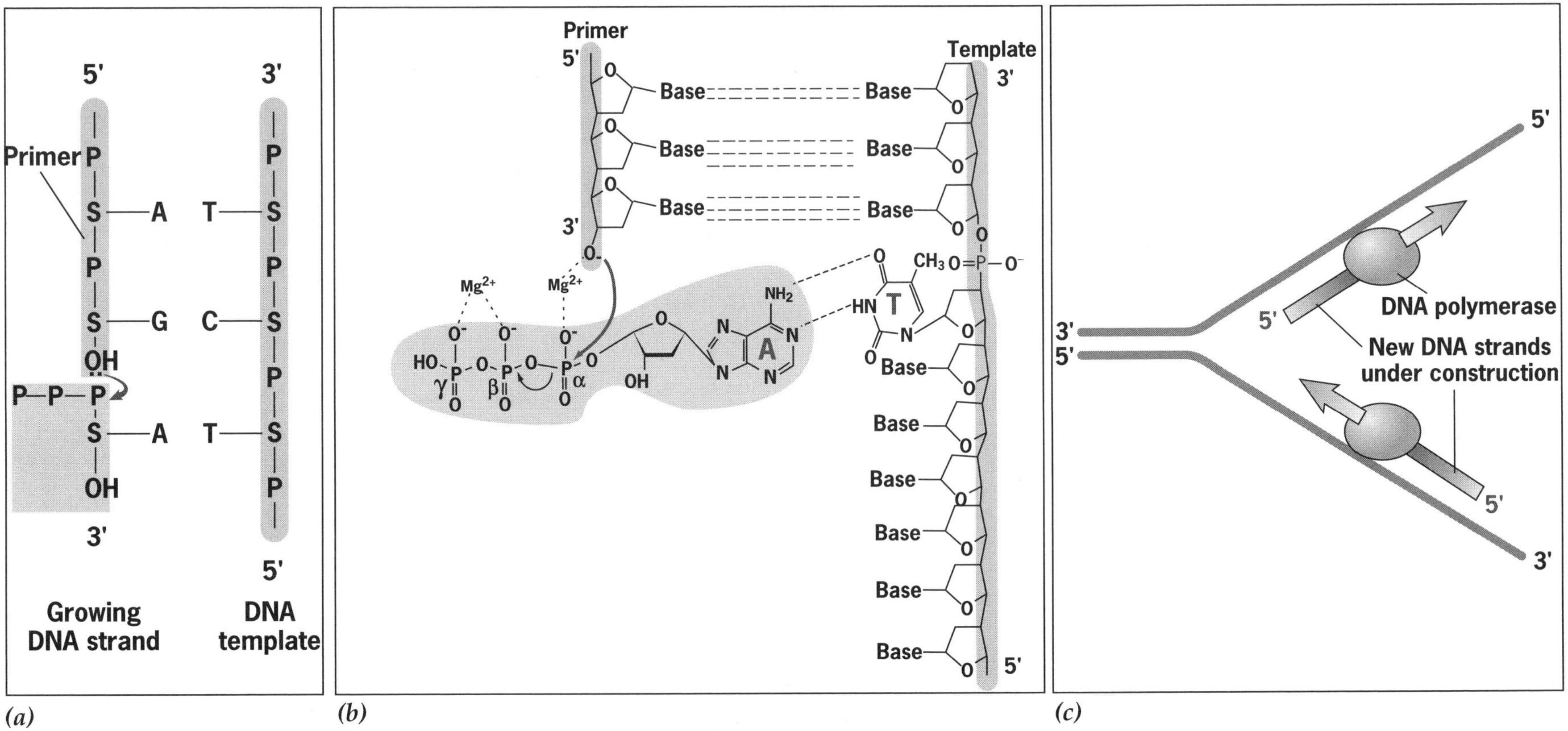

Figure 13.8 The incorporation of nucleotides onto the 3′ end of a growing strand by a DNA polymerase.

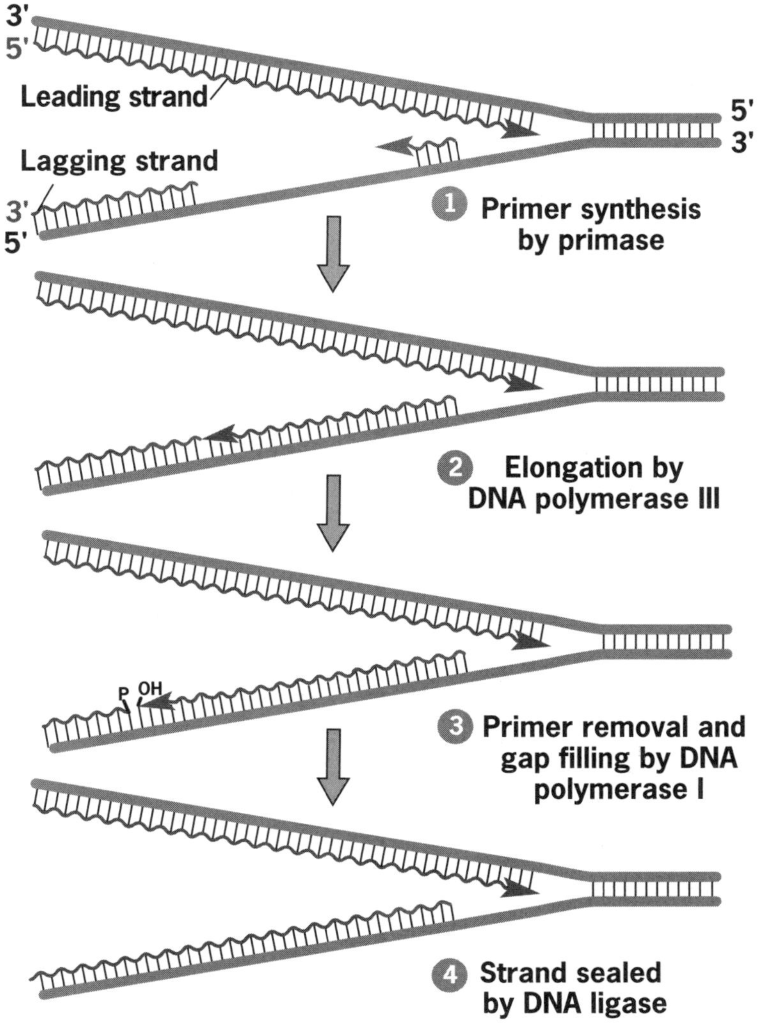

Figure 13.11 The use of short RNA fragments as removable primers in initiating synthesis of each Okazaki fragment of the lagging strand.

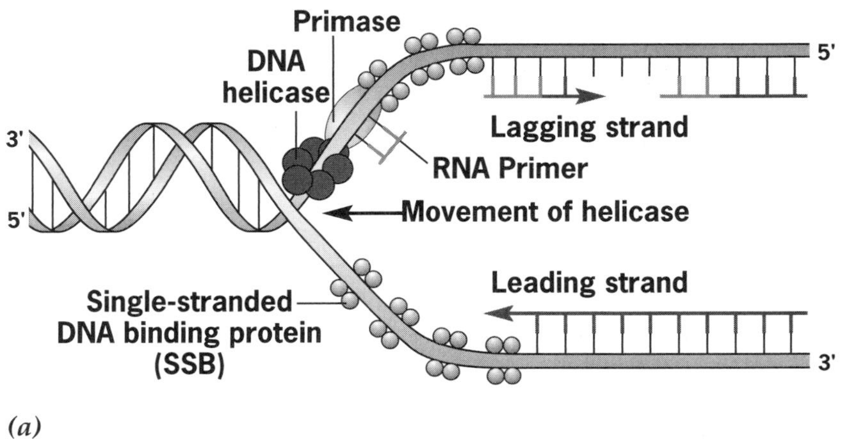

Figure 13.12a The role of the DNA helicase, single-stranded DNA-binding proteins, and primase at the replication fork.

Copyright © 2005 John Wiley & Sons, Inc.

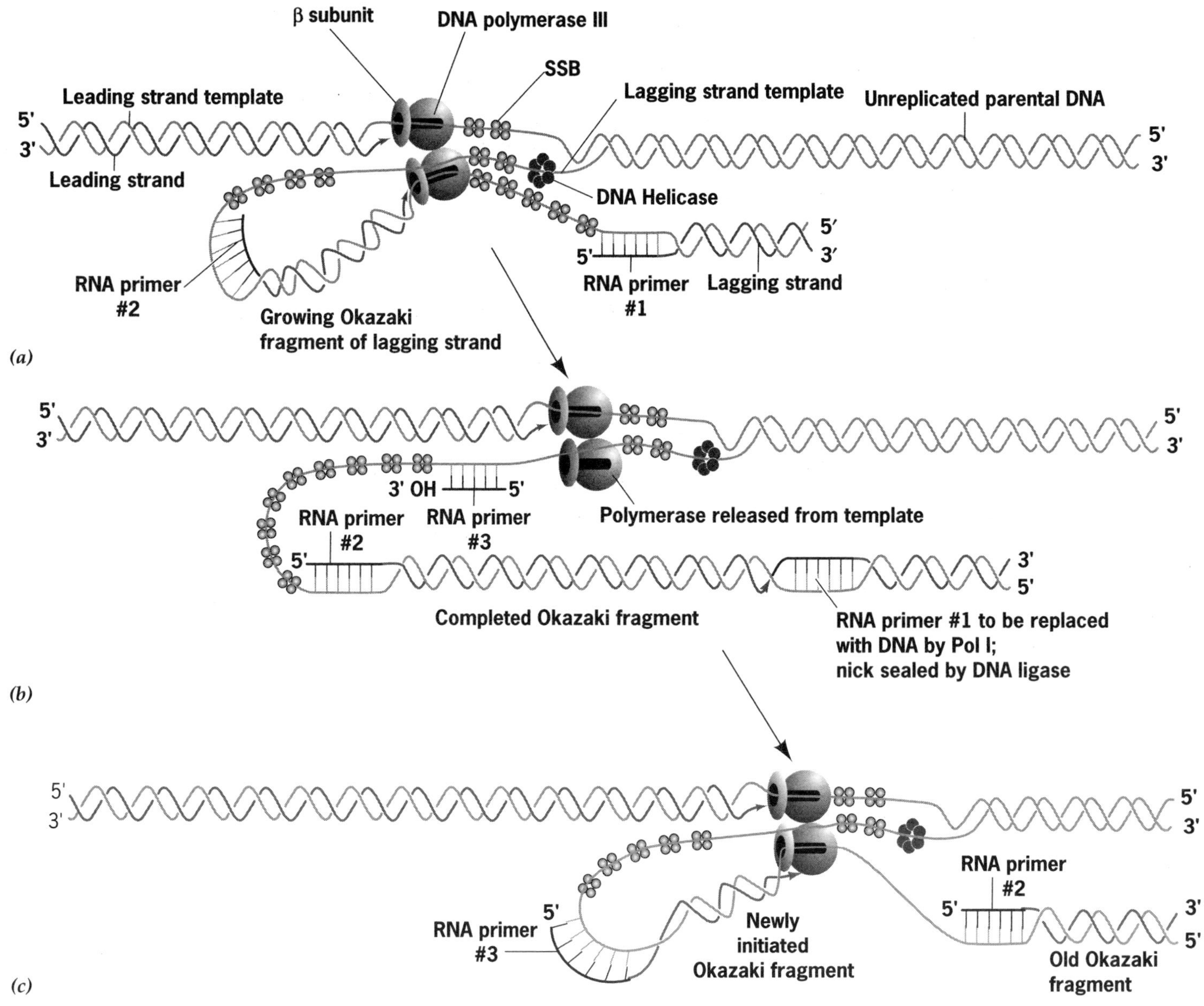

Figure 13.13 Replication of the leading and lagging strands in E. coli is accomplished by two DNA polymerases working together as part of a single complex.

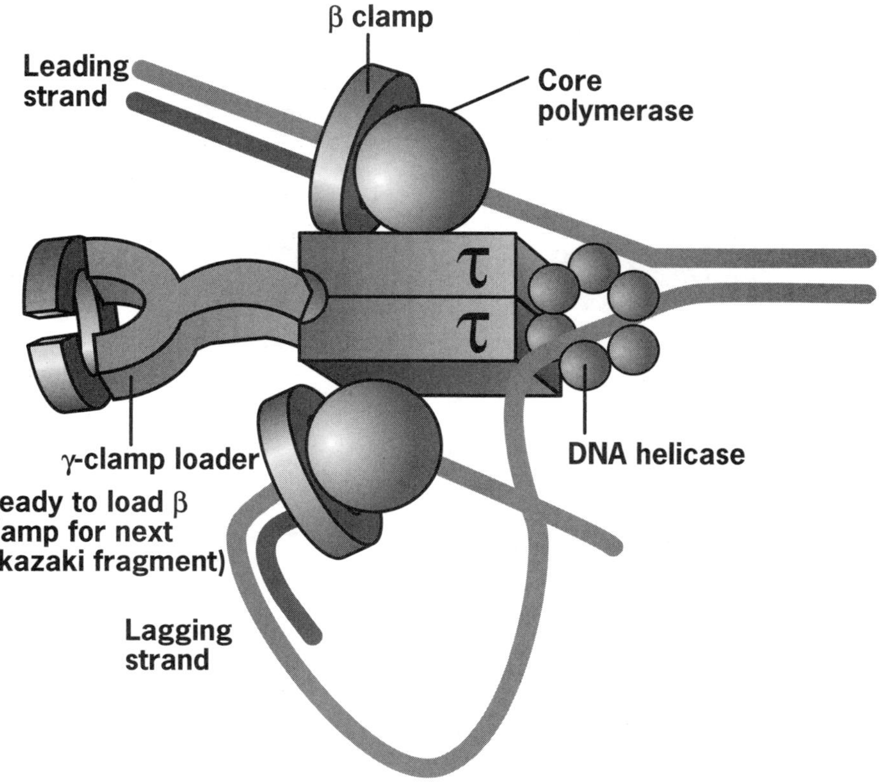

Figure 13.14 Schematic representation of DNA polymerase III holoenzyme (replisome).

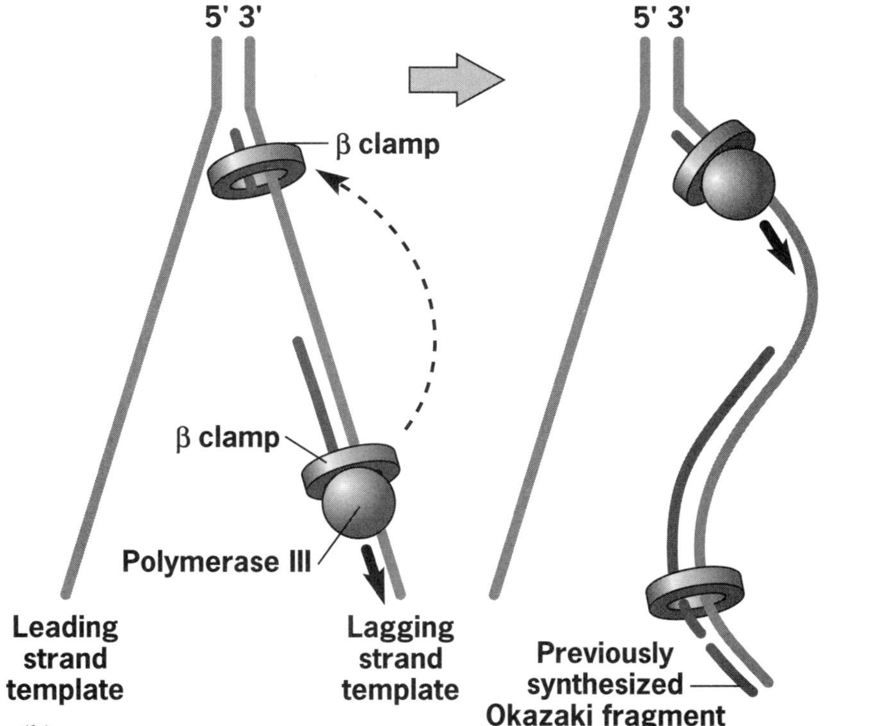

(b)

Figure 13.15b The β sliding clamp of DNA polymerase III.

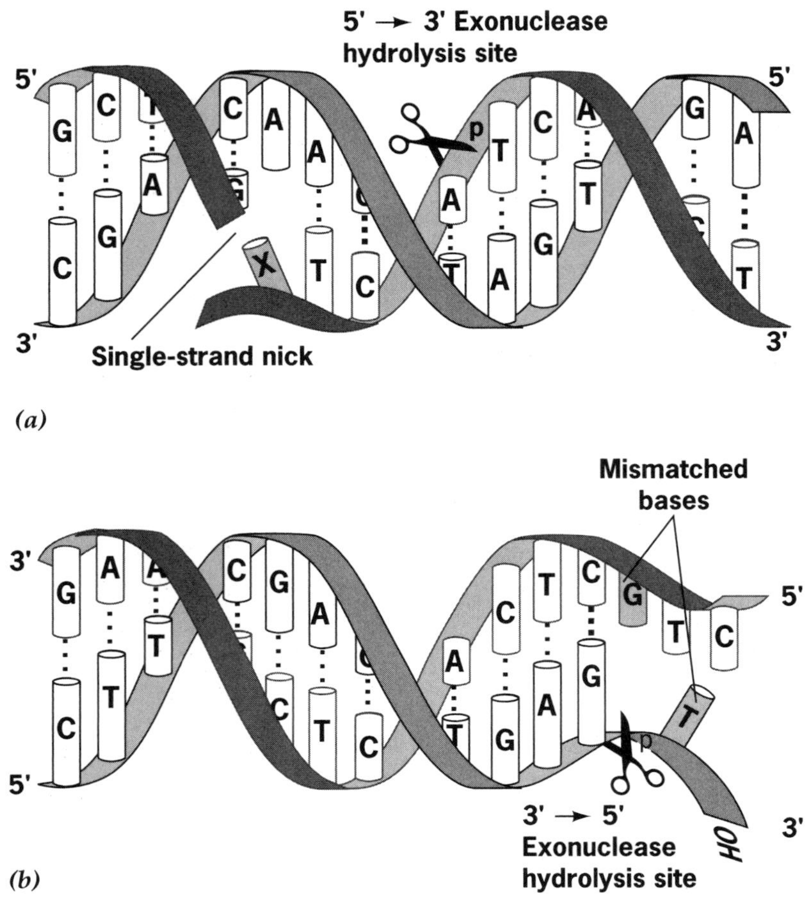

Figure 13.16 The exonuclease activities of DNA polymerase I.

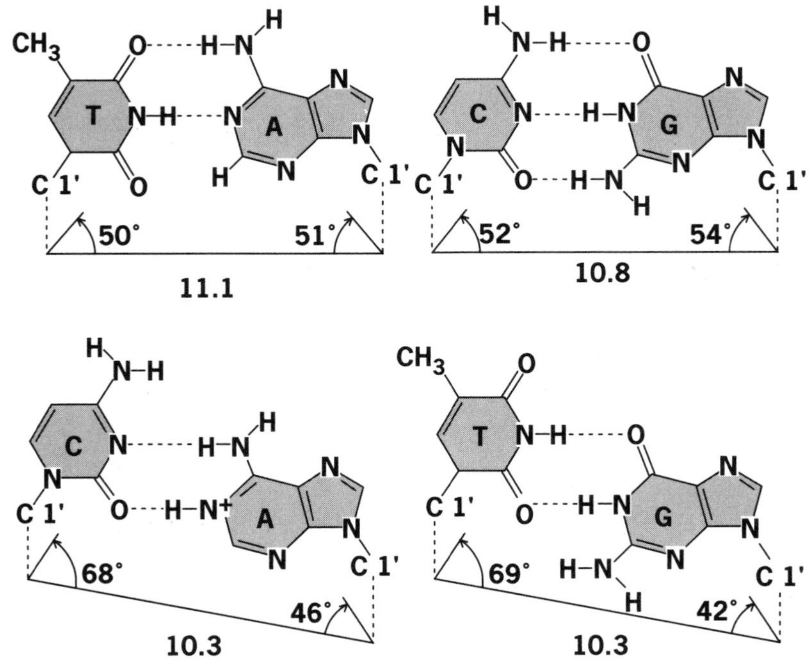

Figure 13.17 Geometry of proper and mismatched base pairs.

Copyright © 2005 John Wiley & Sons, Inc.

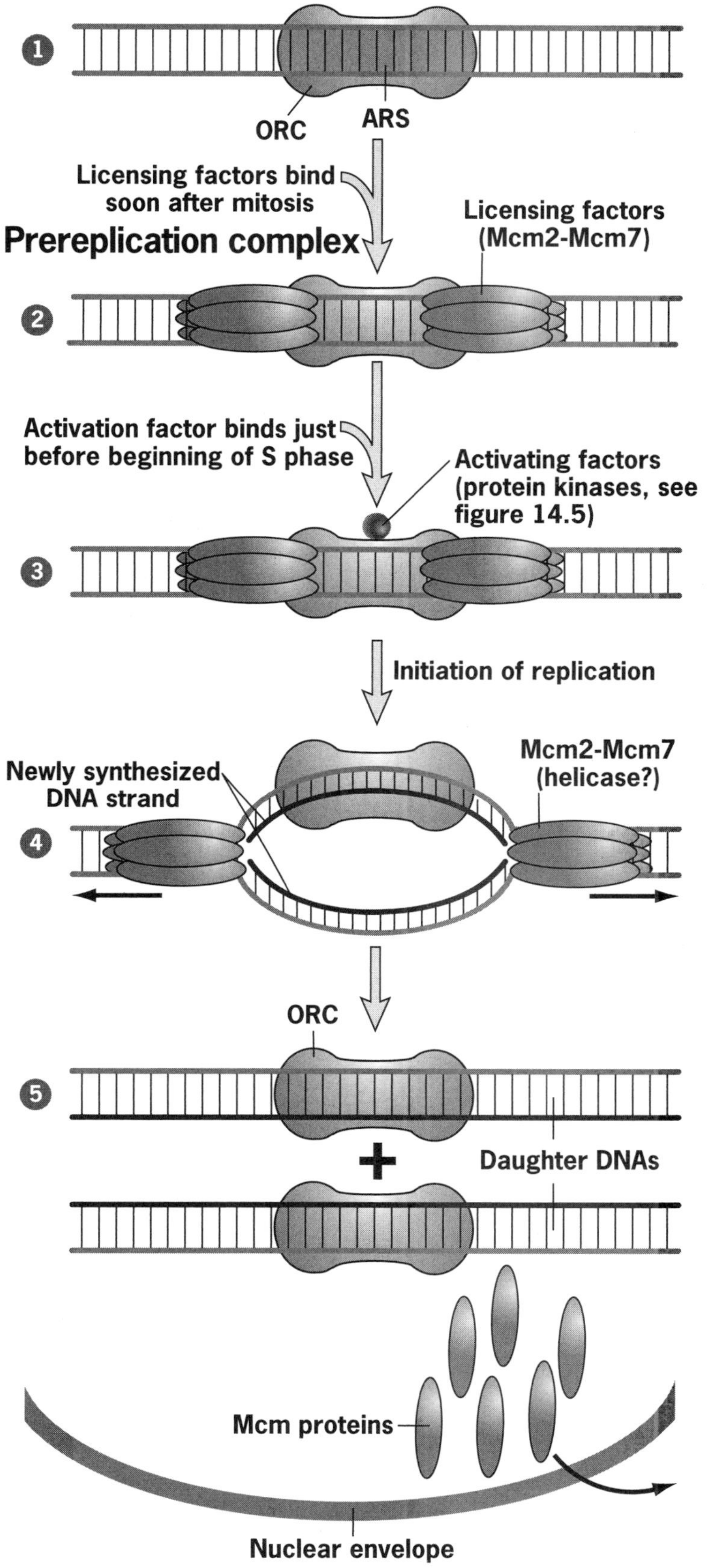

Figure 13.20 Steps leading to the replication of a yeast replicon.

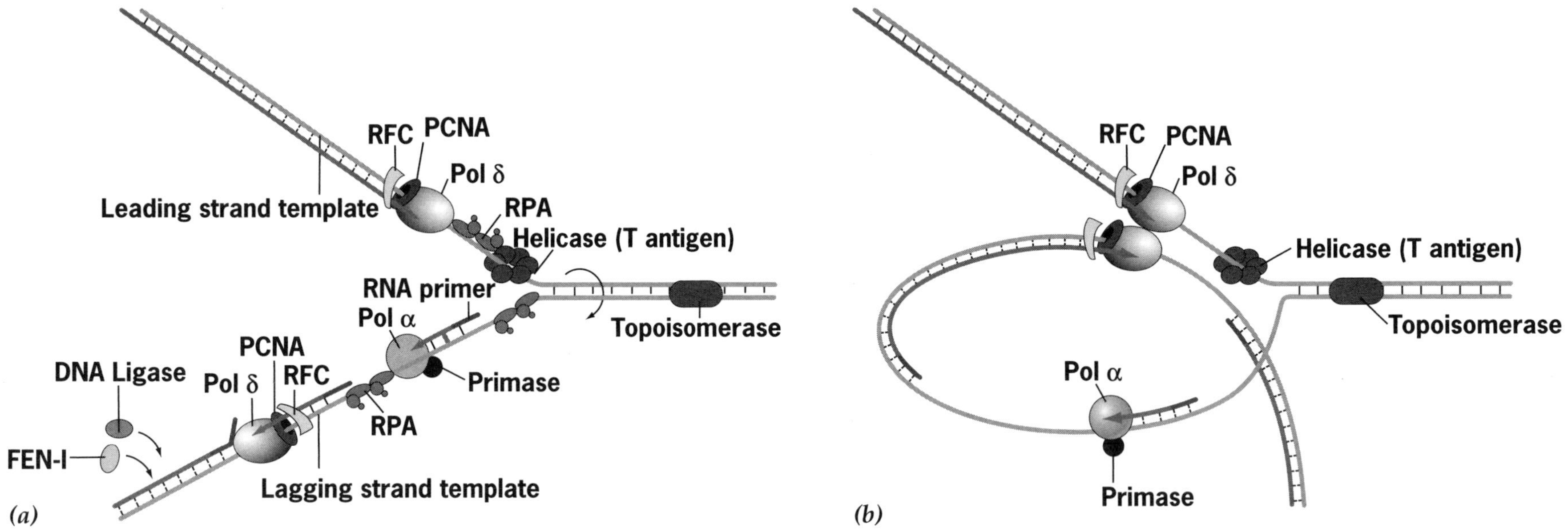

Figure 13.21 A schematic view of the major components at the eukaryotic replication fork.

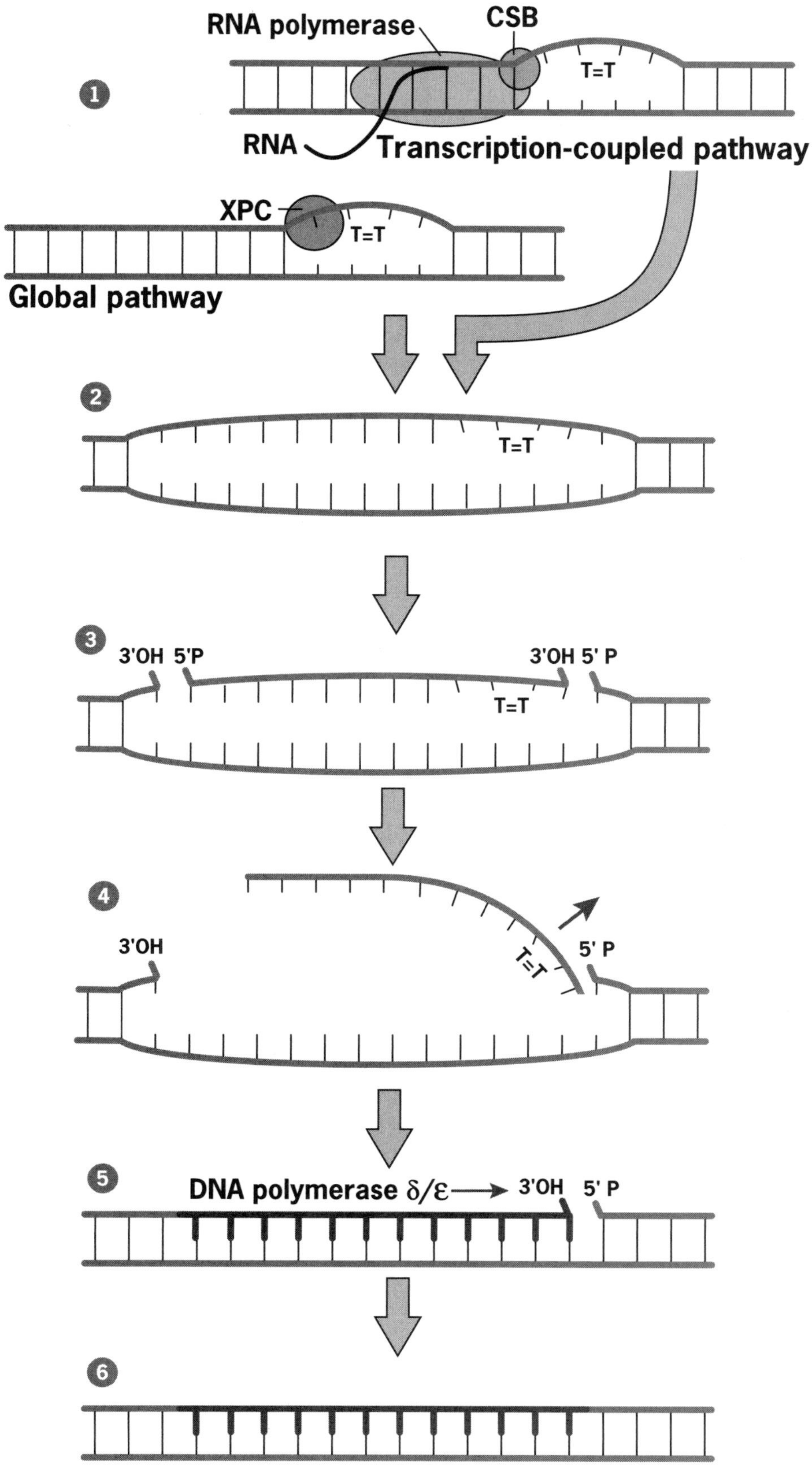

Figure 13.26 Nucleotide excision repair.

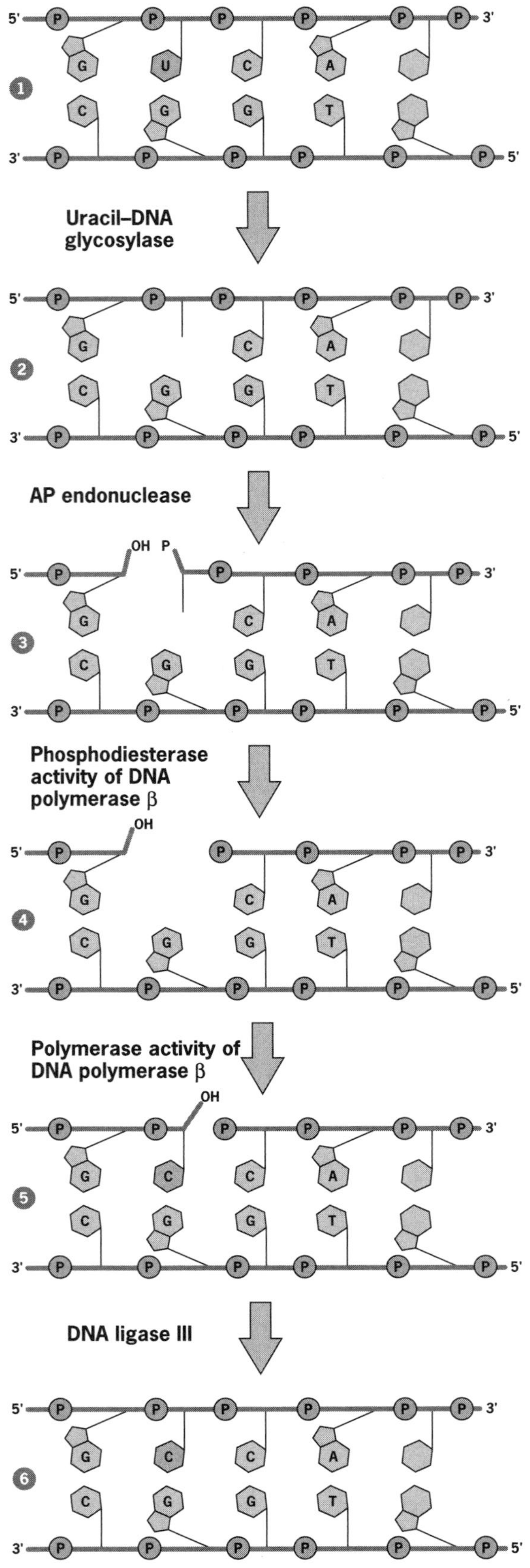

Figure 13.27 Base excision repair.

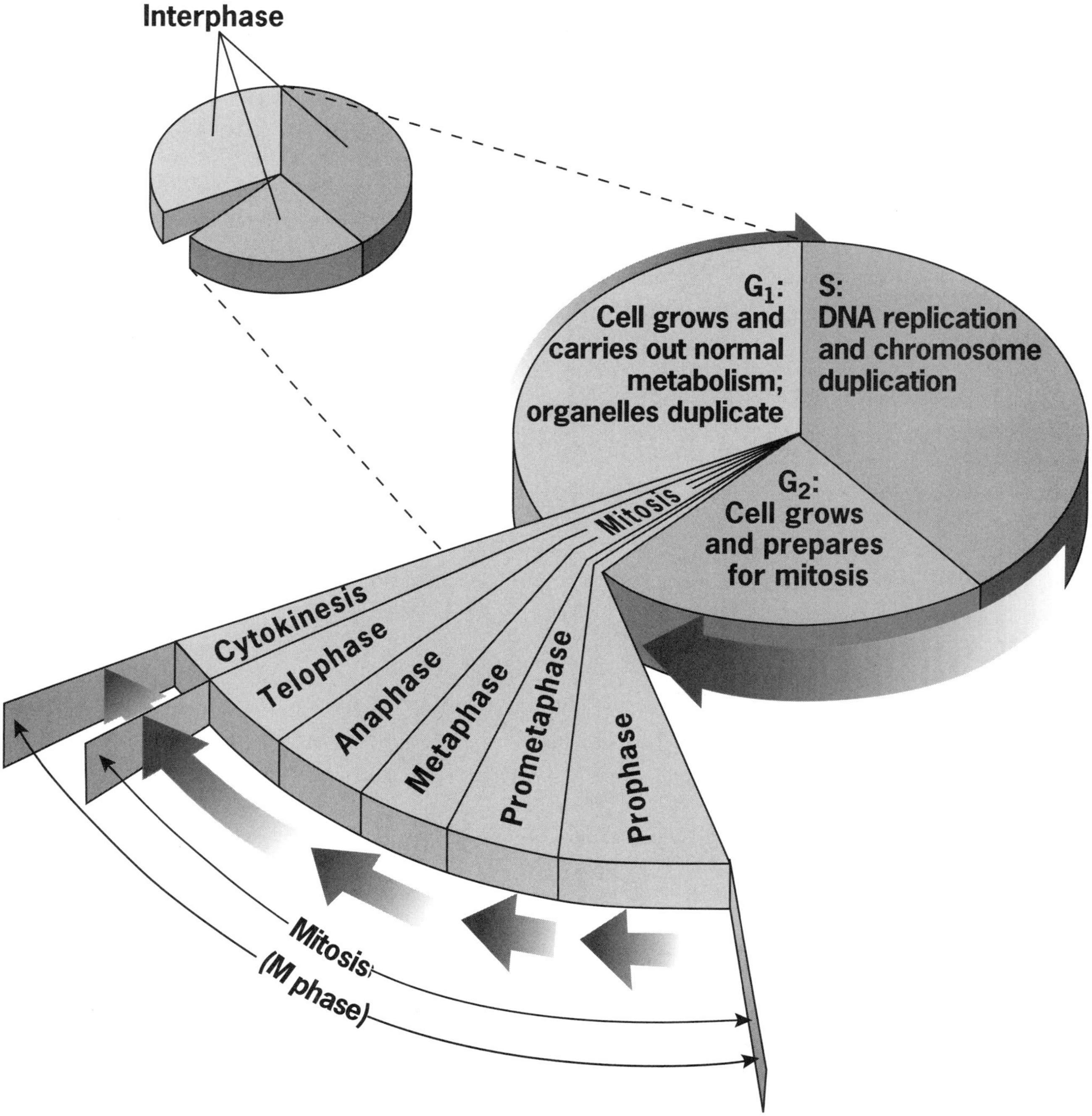

Figure 14.1 An overview of the eukaryotic cell cycle.

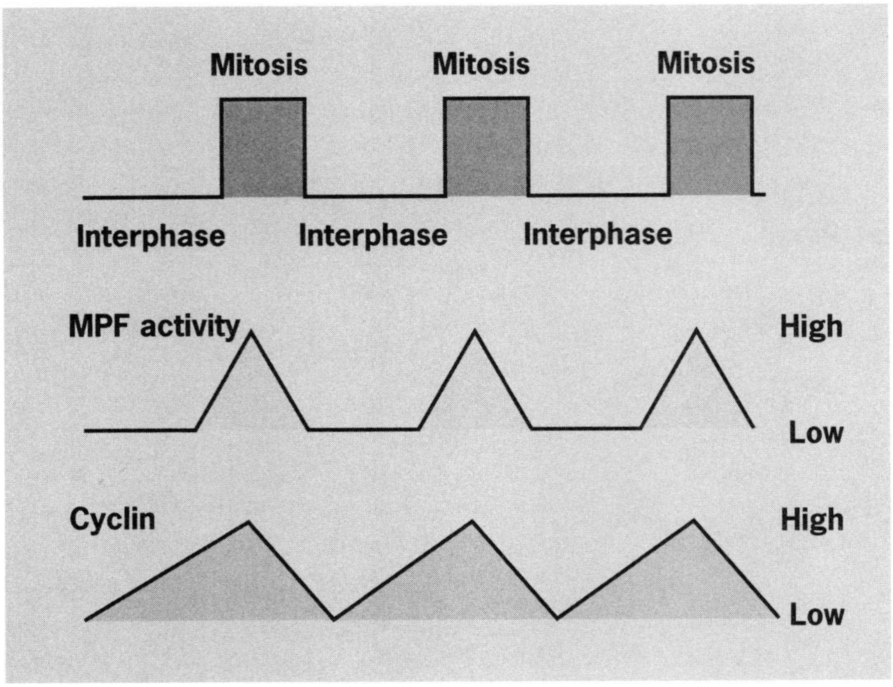

Figure 14.4 Fluctuation of cyclin and MPF levels during the cell cycle.

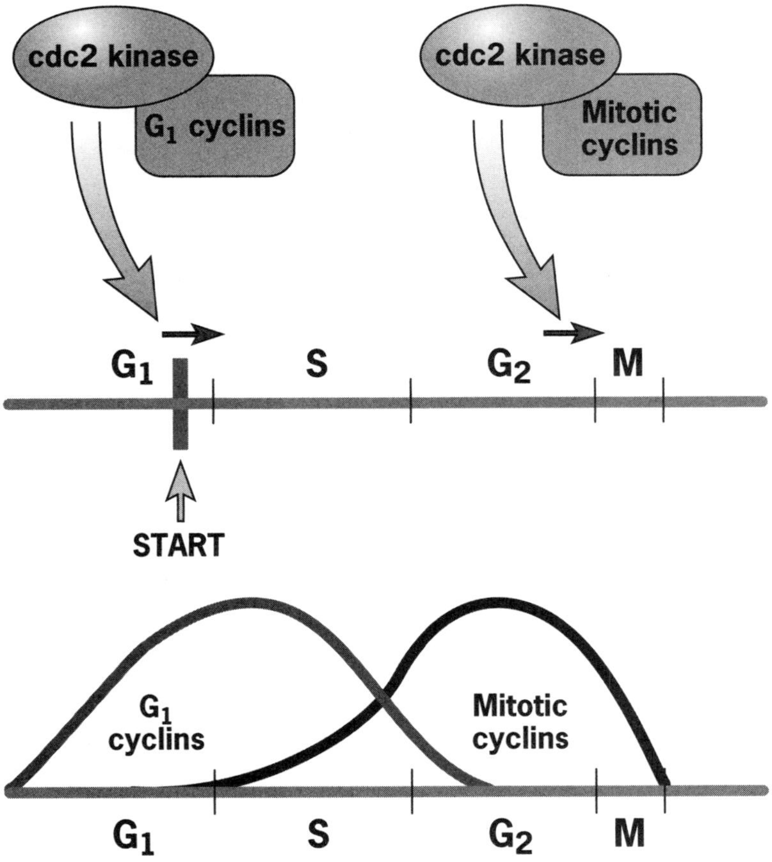

Figure 14.5 A simplified model for cell cycle regulation in fission yeast.

Copyright © 2005 John Wiley & Sons, Inc.

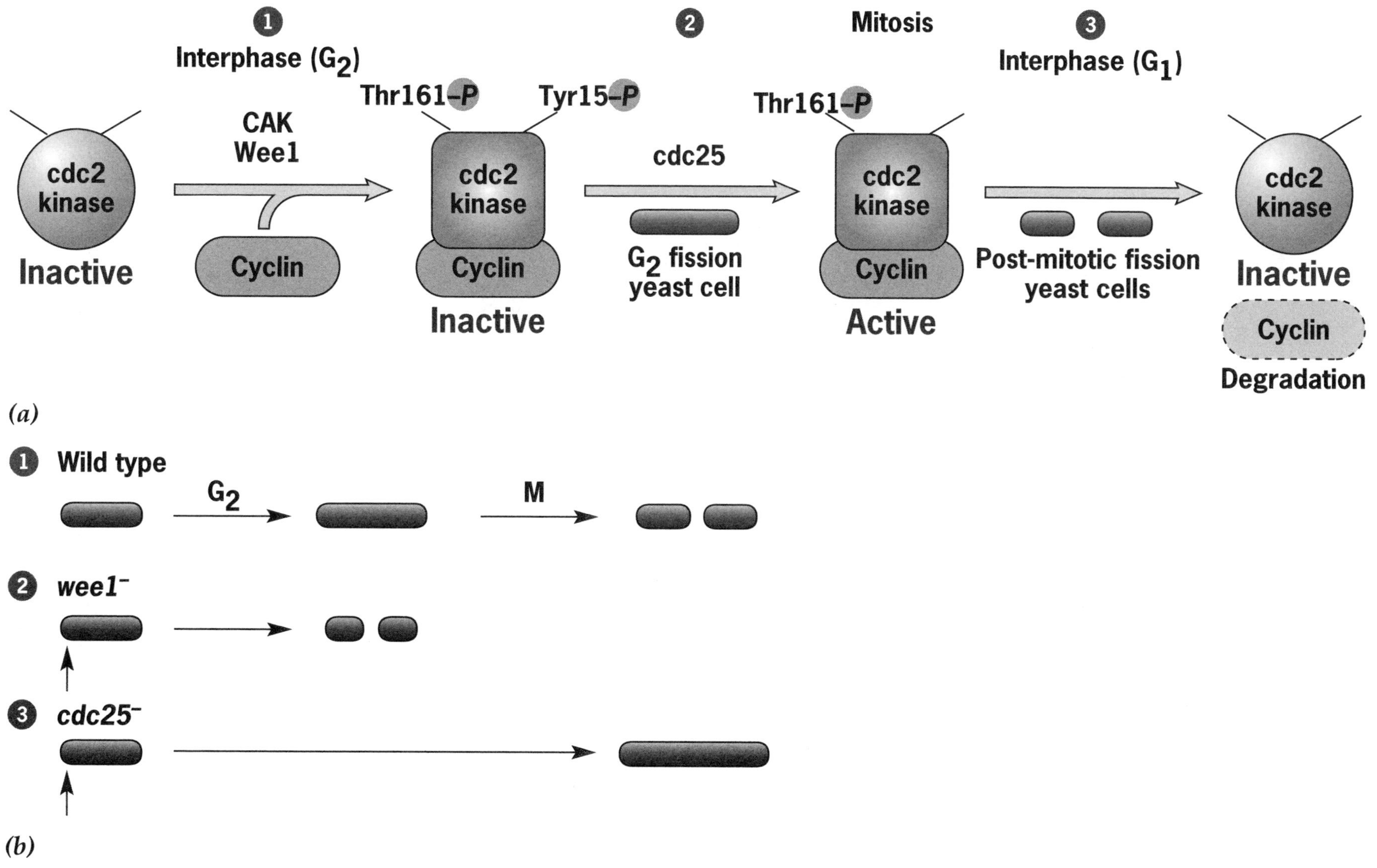

Figure 14.6 Progression through the fission yeast cell cycle requires the phosphorylation and dephosphorylation of critical cdc2 residues.

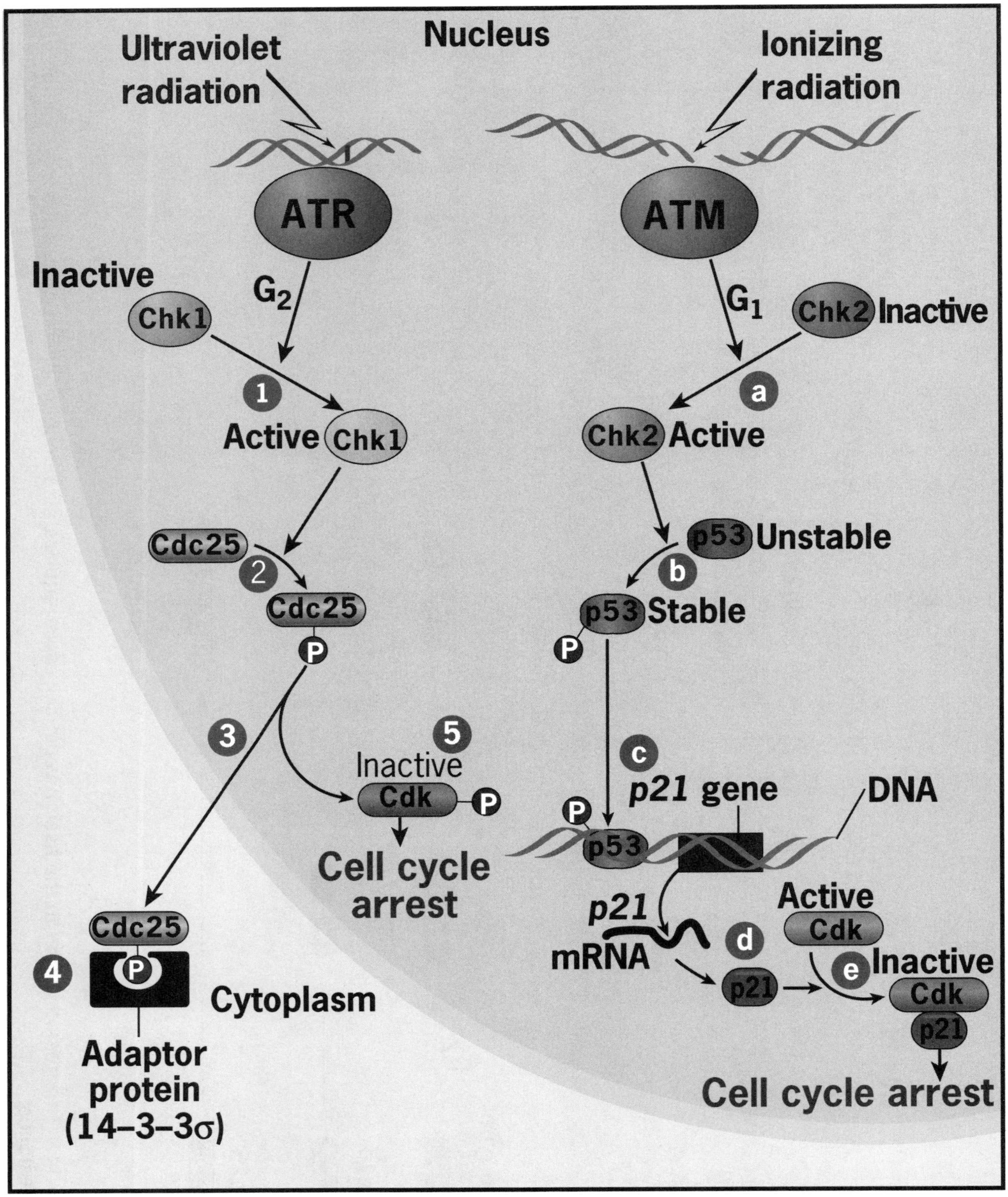

Figure 14.9 Models for the mechanism of action of two DNA-damage checkpoints.

	Prophase
1. Chromosomal material condenses to form compact mitotic chromosomes. Chromosomes are seen to be composed of two chromatids attached together at the centromere. 2. Cytoskeleton is disassembled and mitotic spindle is assembled. 3. Golgi complex and ER fragment. Nuclear envelope disperses.	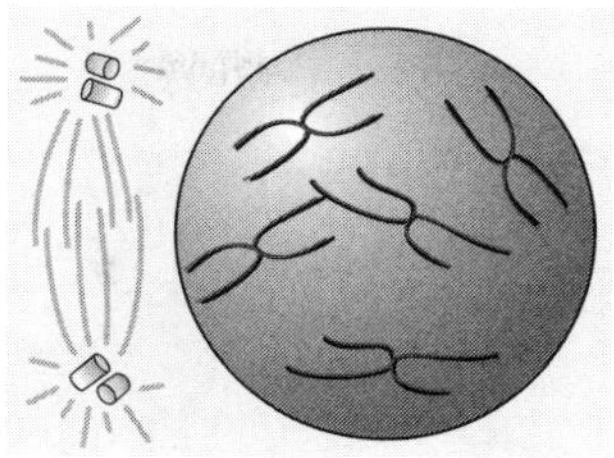
	Prometaphase
1. Chromosomal microtubules attach to kinetochores of chromosomes. 2. Chromosomes are moved to spindle equator.	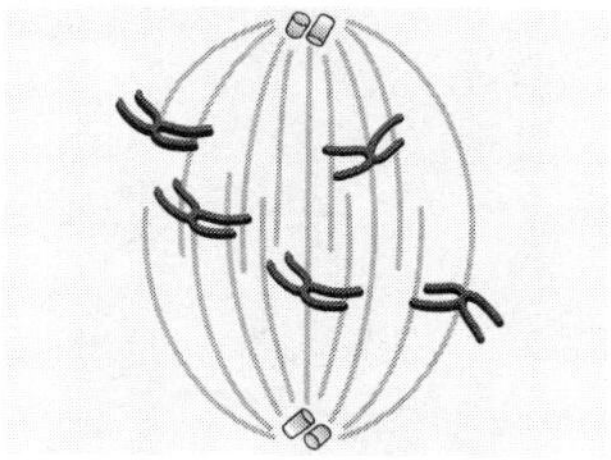
	Metaphase
1. Chromosomes are aligned along metaphase plate, attached by chromosomal microtubules to both poles.	
	Anaphase
1. Centromeres split, and chromatids separate. 2. Chromosomes move to opposite spindle poles. 3. Spindle poles move farther apart.	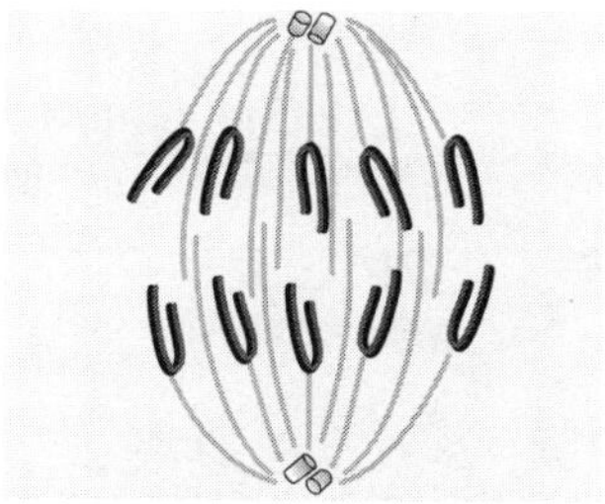
	Telophase
1. Chromosomes cluster at opposite spindle poles. 2. Chromosomes become dispersed. 3. Nuclear envelope assembles around chromosome clusters. 4. Golgi complex and ER reforms. 5. Daughter cells formed by cytokinesis.	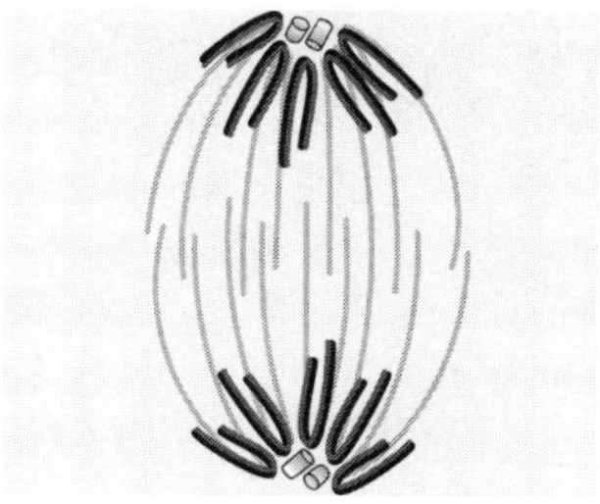

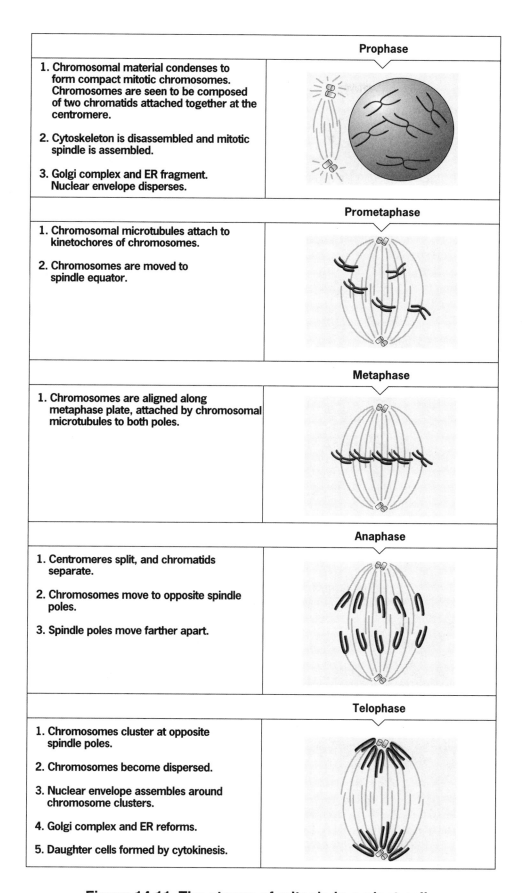

Figure 14.11 The stages of mitosis in a plant cell.

Copyright © 2005 John Wiley & Sons, Inc.

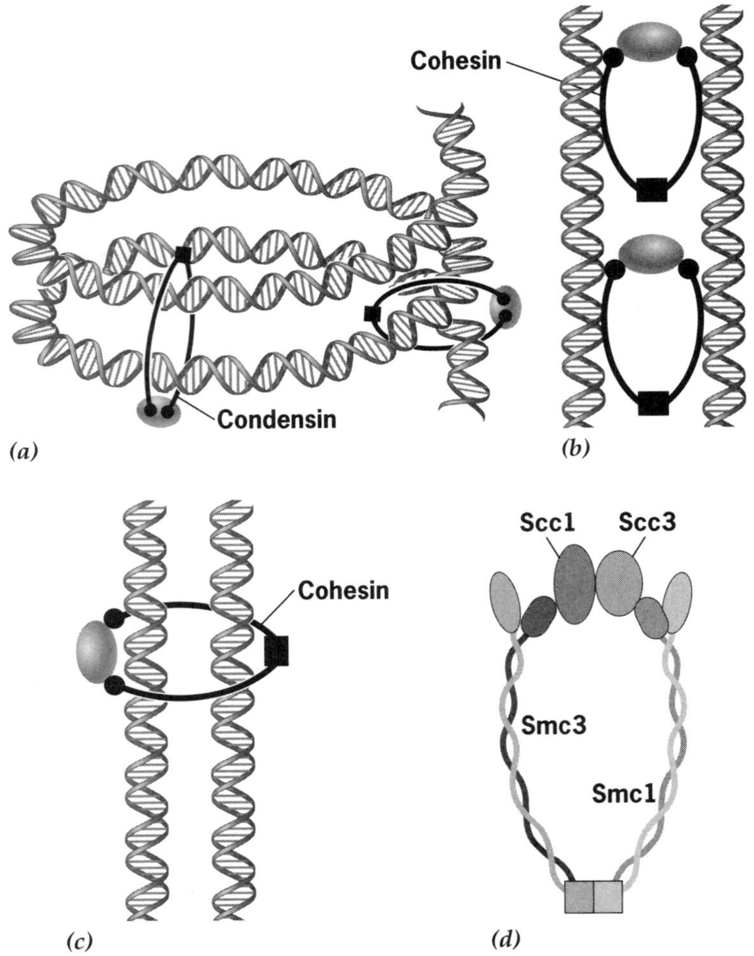

Figure 14.14 Models for the role of condensin and cohesin in the formation of mitotic chromosomes.

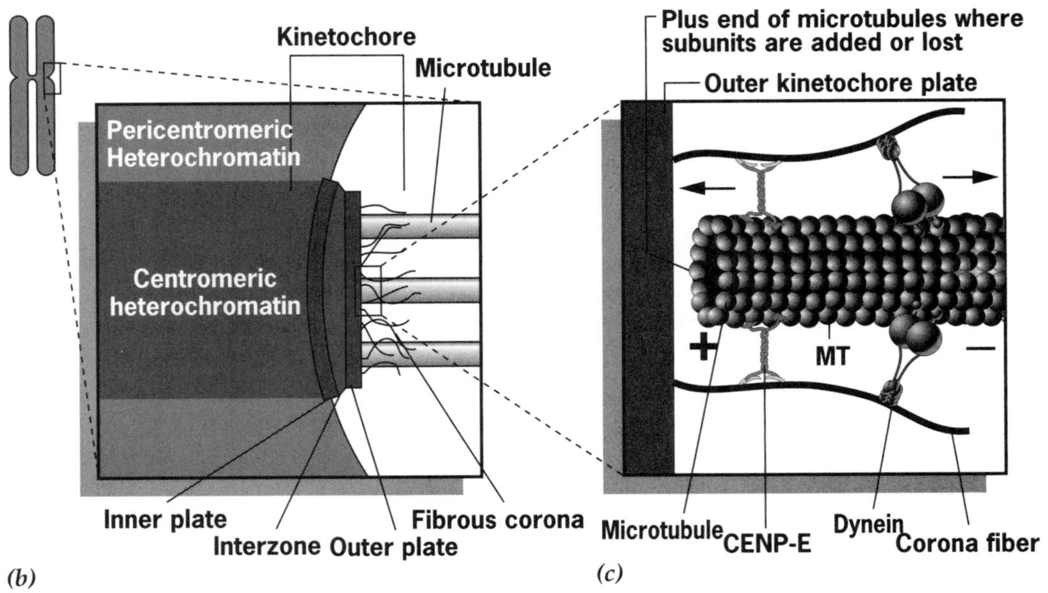

Figure 14.16b,c The kinetochore.

Copyright © 2005 John Wiley & Sons, Inc.

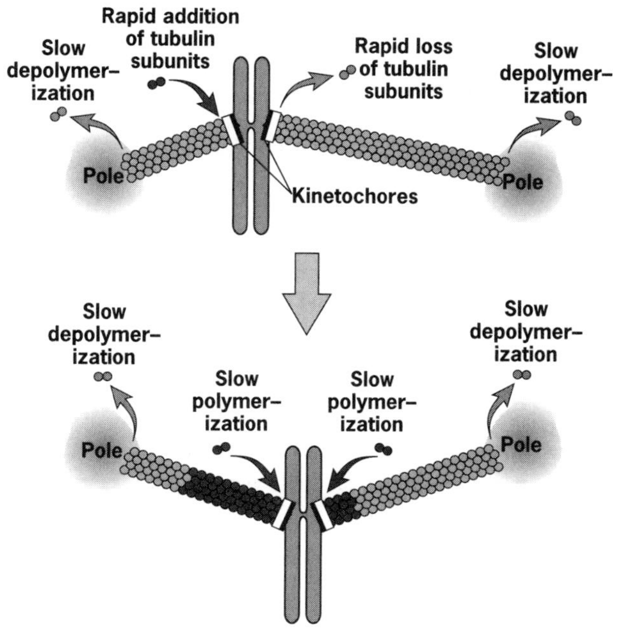

Figure 14.23 Microtubule behavior during formation of the metaphase plate.

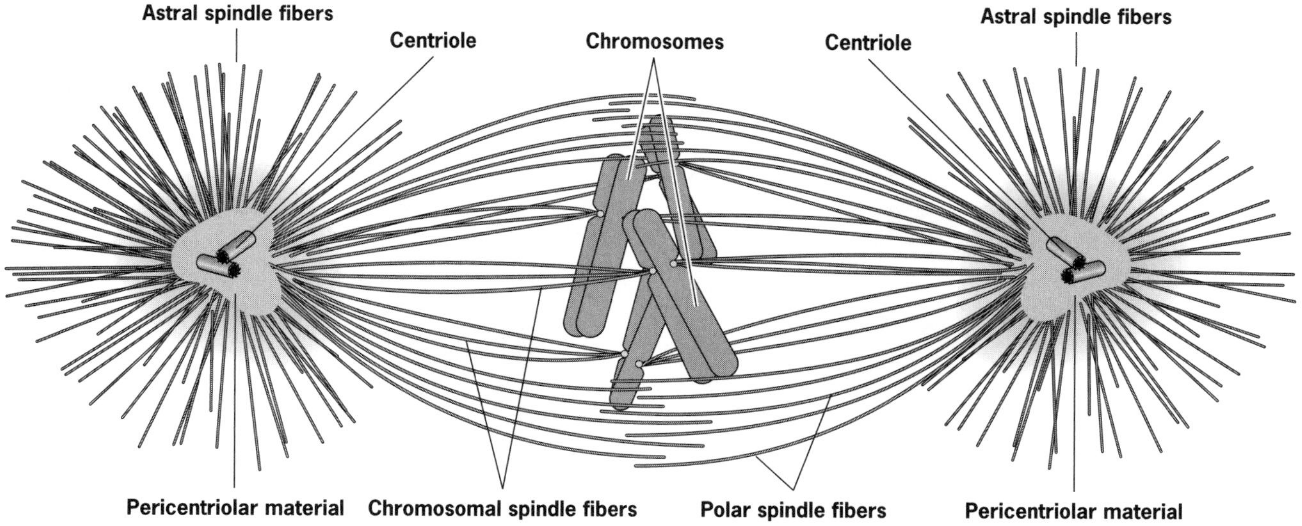

Figure 14.25 The mitotic spindle of an animal cell.

Copyright © 2005 John Wiley & Sons, Inc.

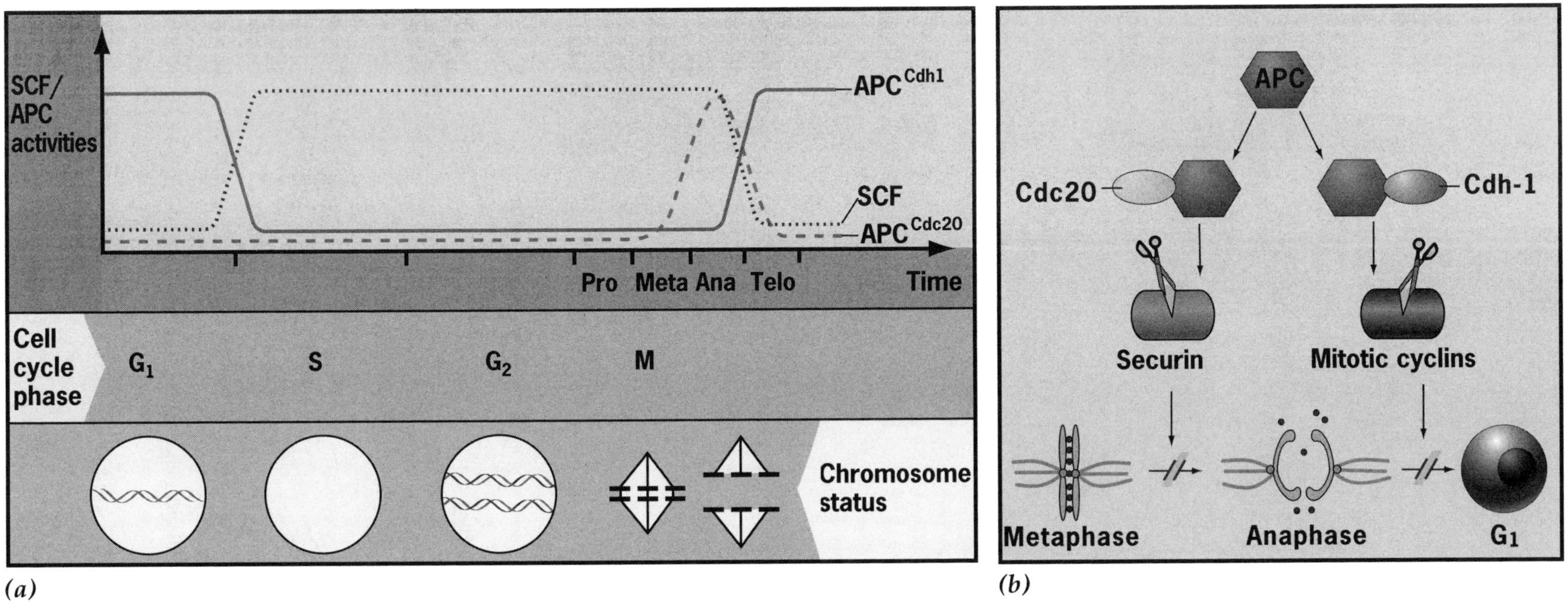

Figure 14.27 SCF and APC activities during the cell cycle.

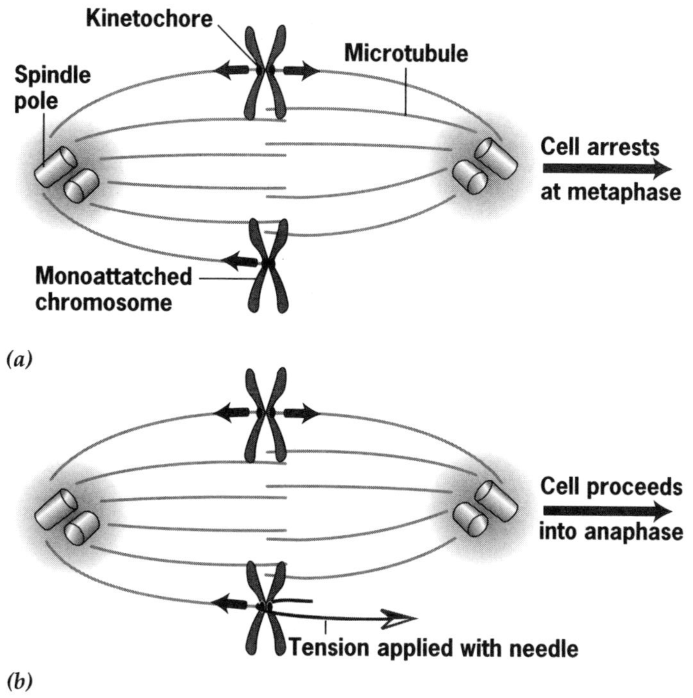

Figure 14.31 Experimental demonstration of the importance of mechanical tension in metaphase checkpoint control.

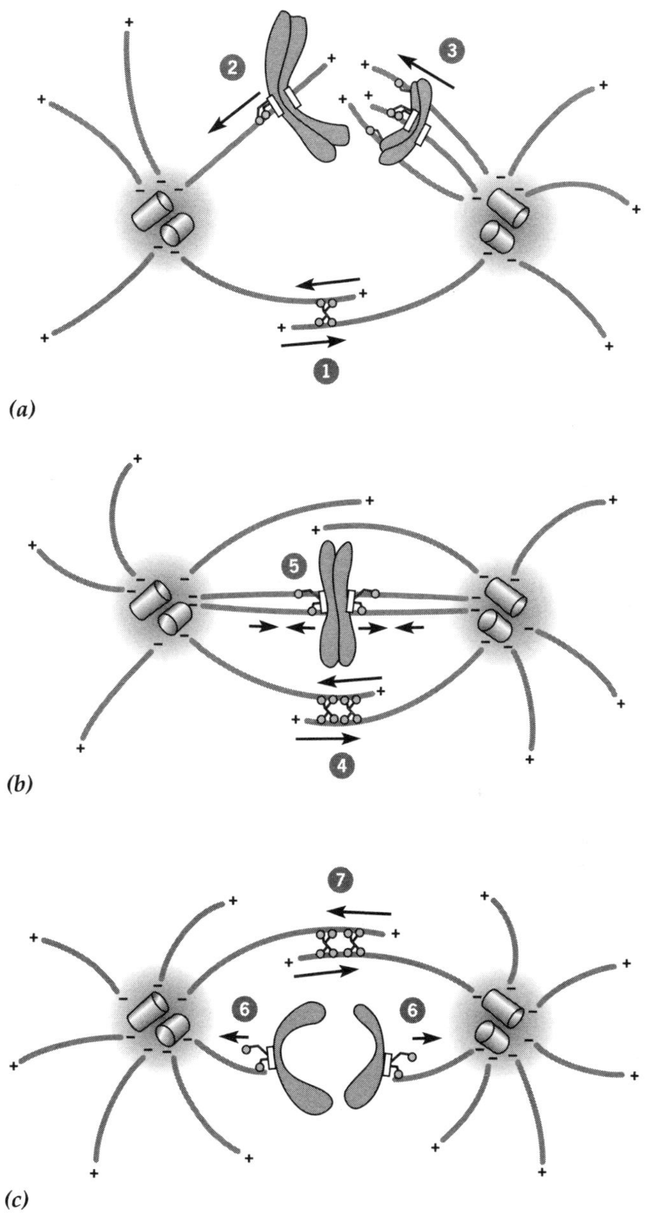

Figure 14.33 Proposed activity of motor proteins during mitosis.

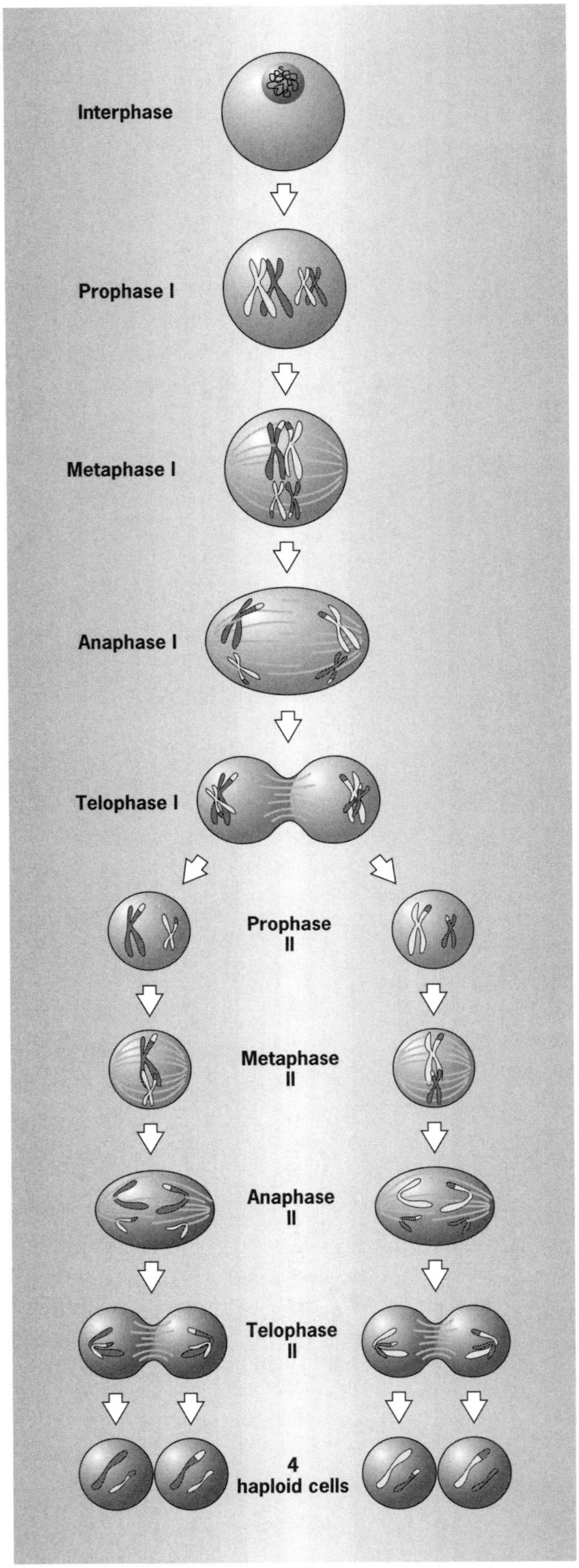

Figure 14.39 The stages of meiosis.

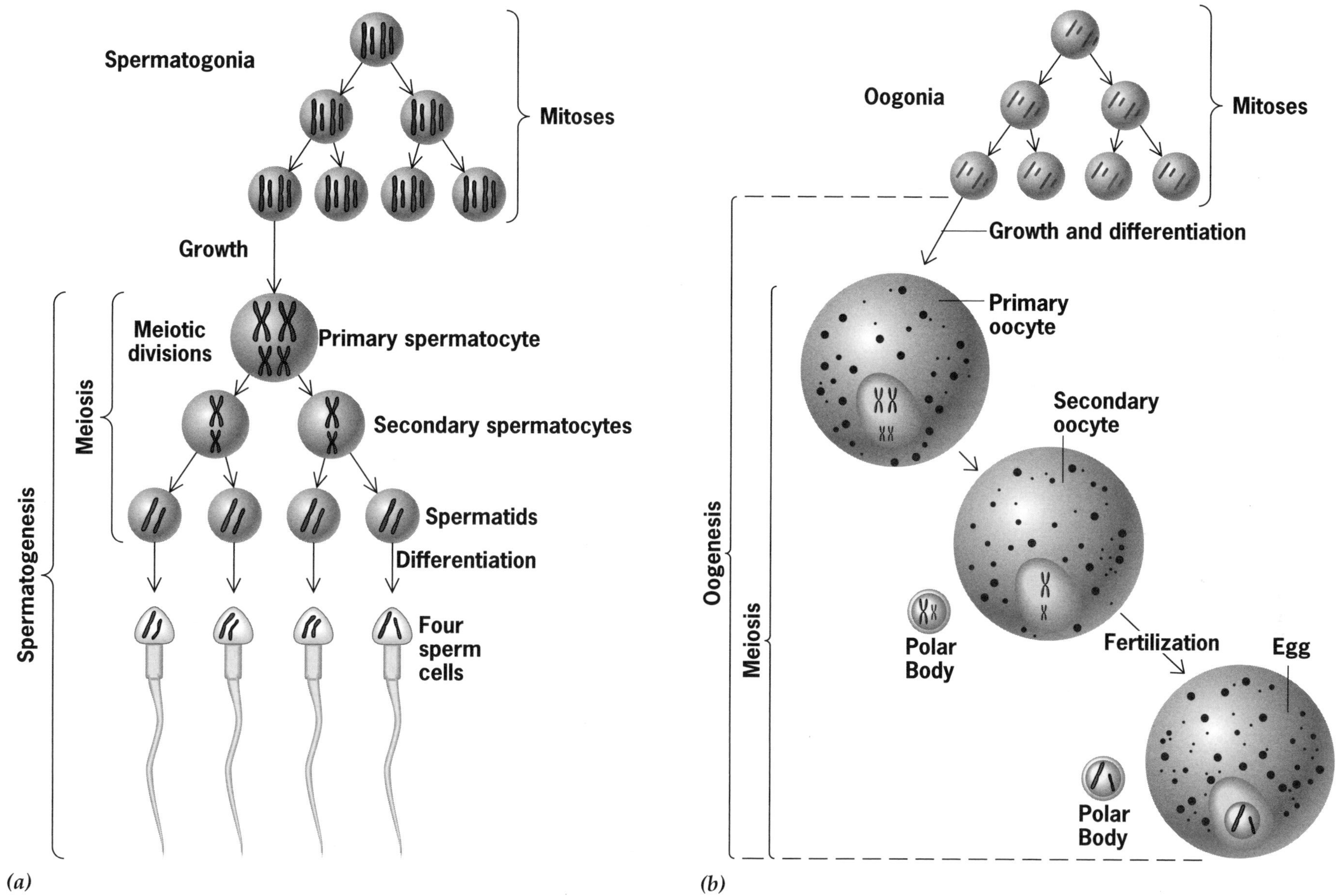

Figure 14.41 The stages of gametogenesis in vertebrates: a comparison between the formation of sperm and eggs.

Copyright © 2005 John Wiley & Sons, Inc.

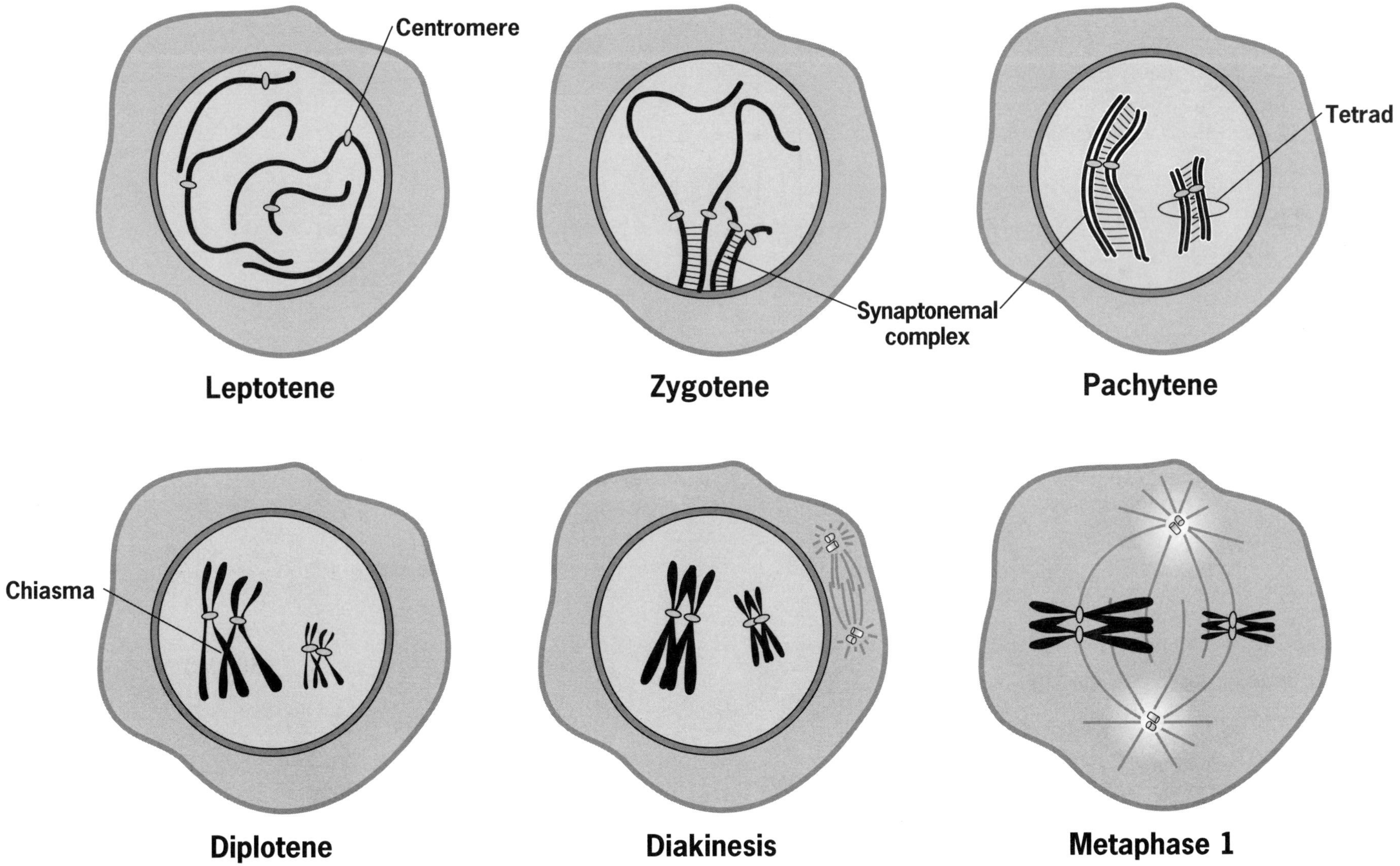

Figure 14.42 The stages of prophase I.

Copyright © 2005 John Wiley & Sons, Inc.

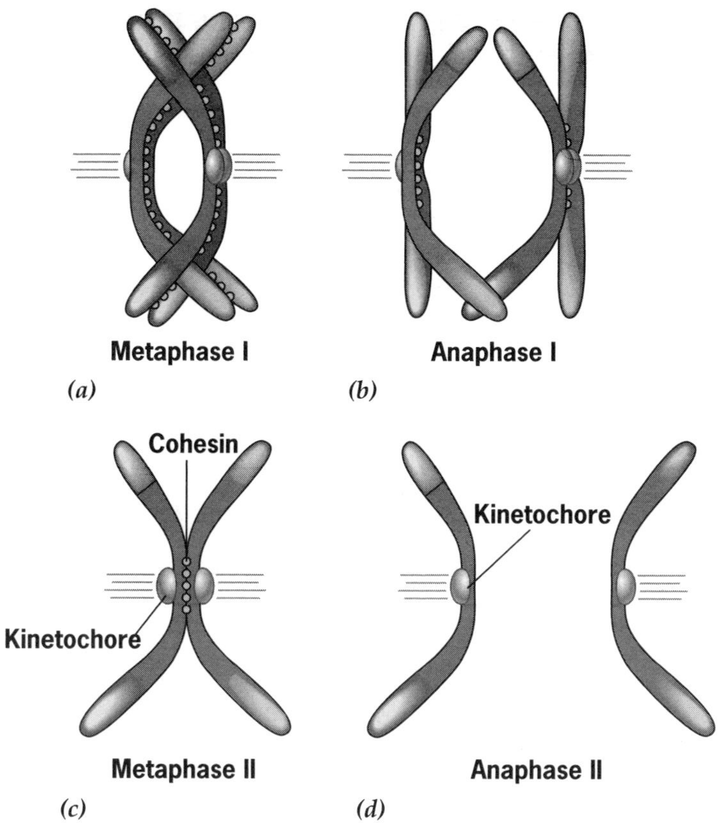

Figure 14.46 Separation of homologous chromosomes during meiosis I and separation of chromatids during meiosis II.

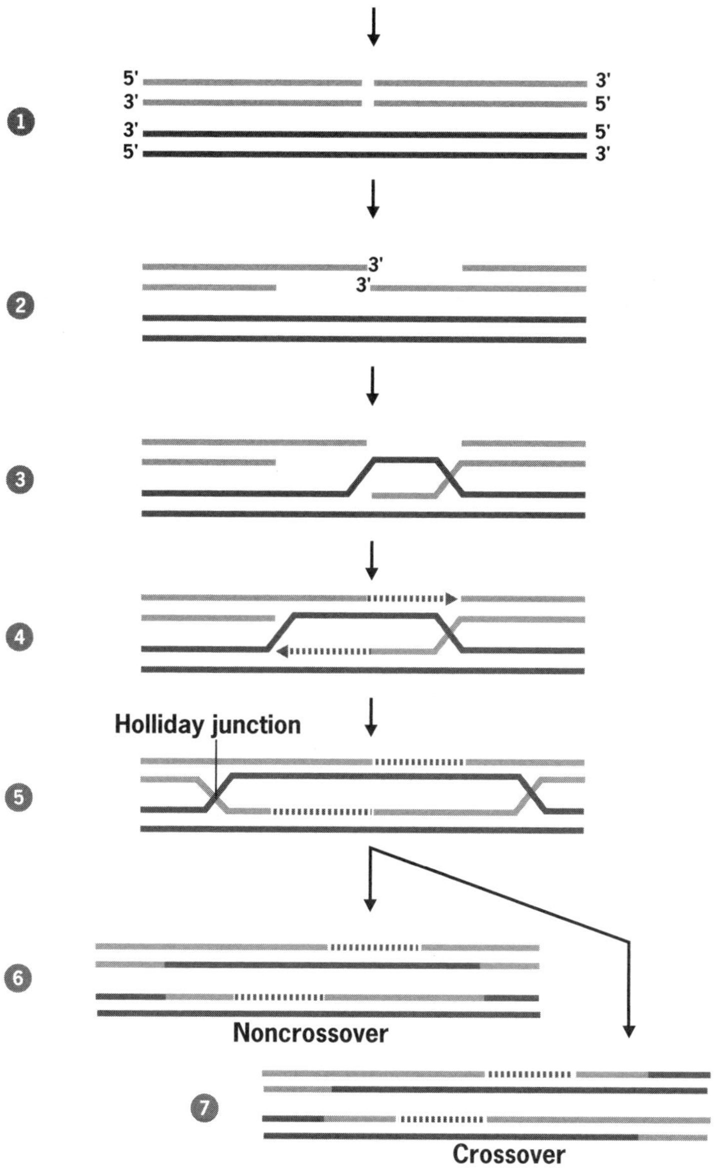

Figure 14.47 A proposed mechanism for genetic recombination initiated by double-strand breaks.

Figure 14.46 & 14.47

Copyright © 2005 John Wiley & Sons, Inc.

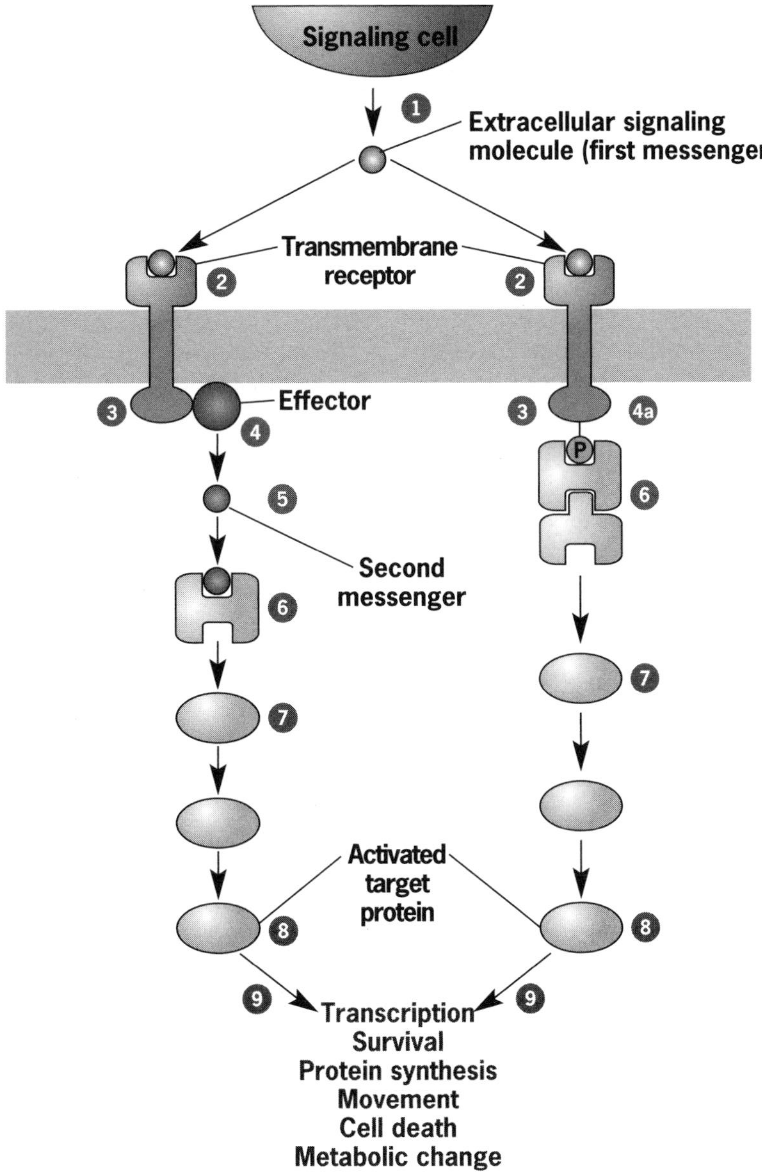

Figure 15.1 An overview of the signaling pathways by which extra-cellular messenger molecules can elicit intracellular responses.

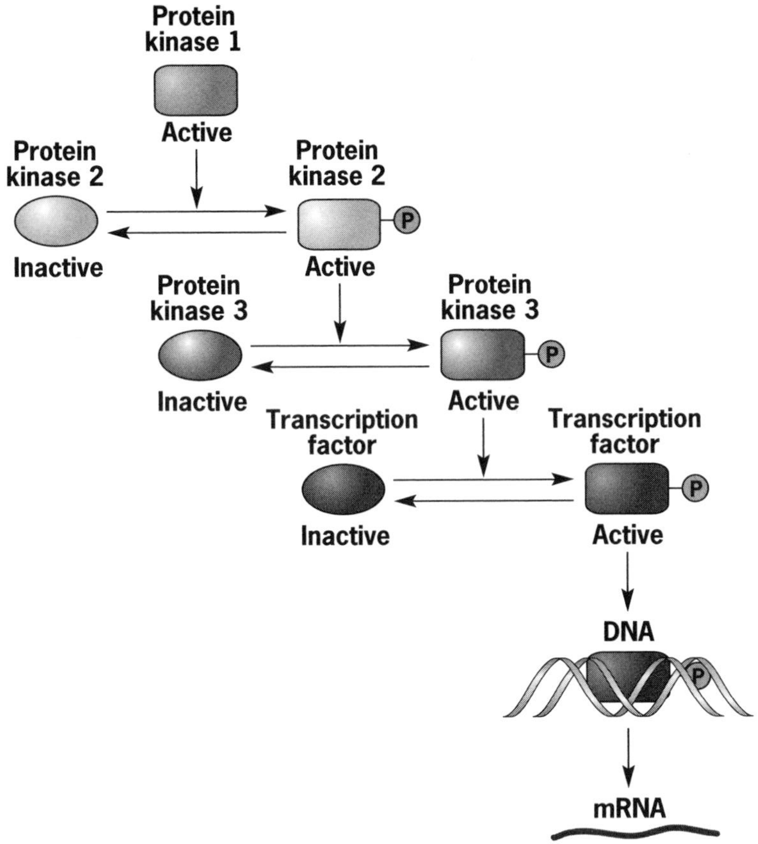

Figure 15.2 Signal transduction pathway consisting of protein kinases and protein phosphatases whose catalytic actions change the conformations, and thus the activities, of the proteins they modify.

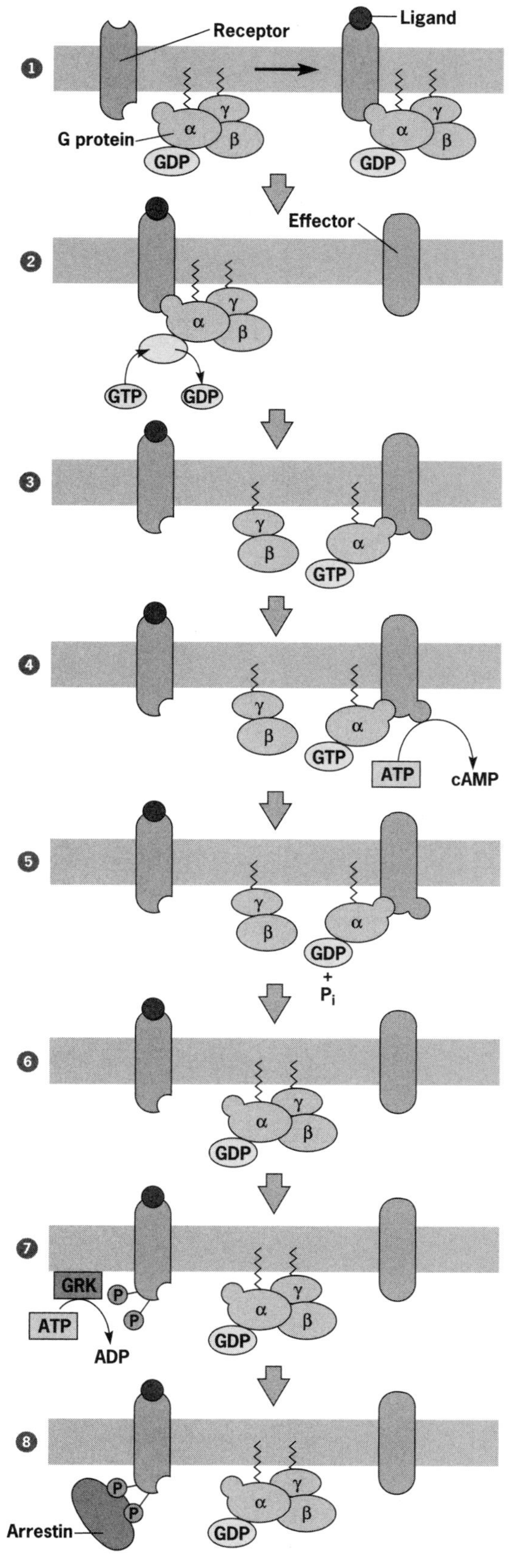

Figure 15.4 The mechanism of receptor-mediated activation (or inhibition) of effectors by means of heterotrimeric G proteins.

Copyright © 2005 John Wiley & Sons, Inc.

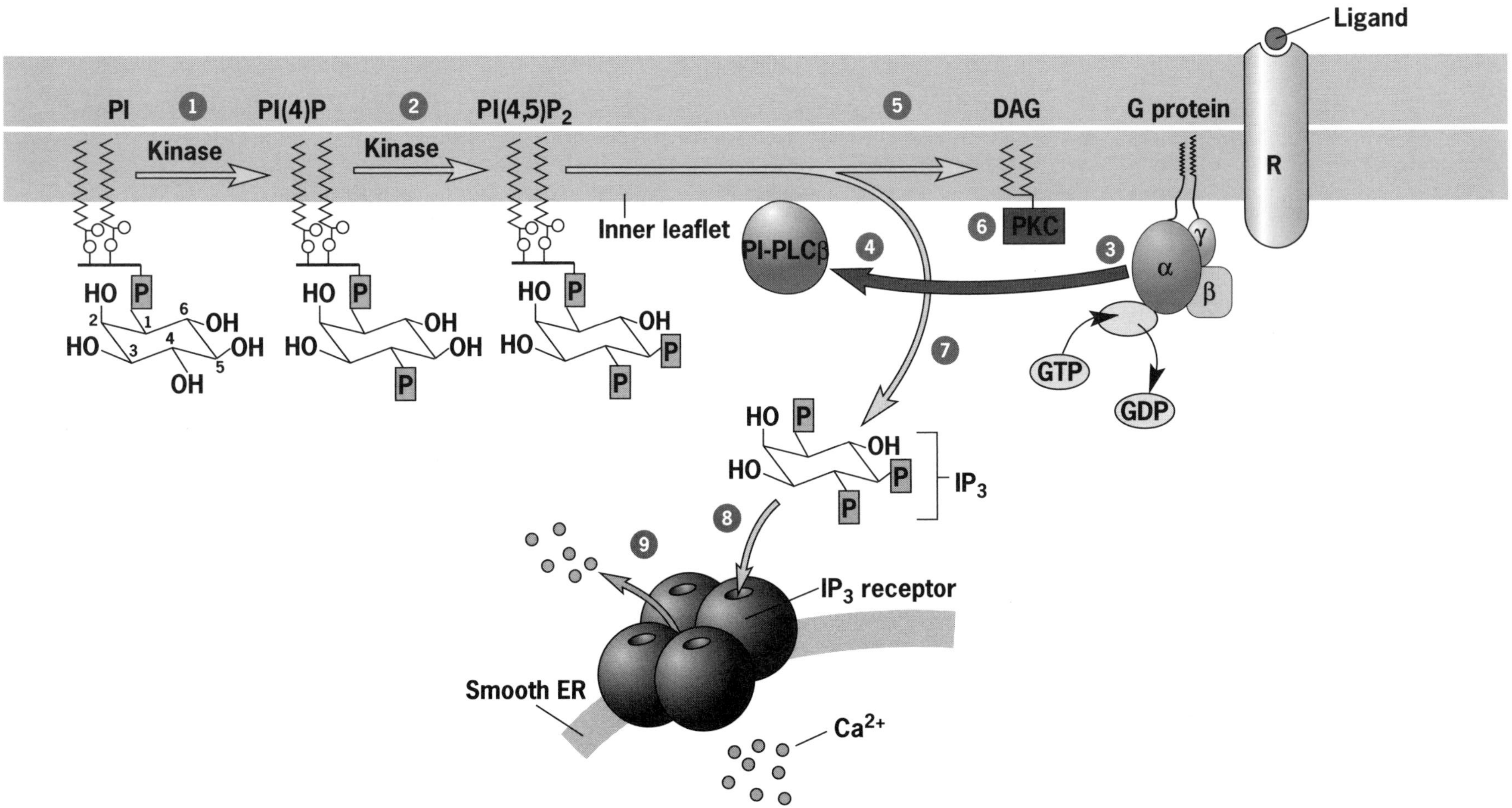

Figure 15.7 The generation of second messengers as a result of ligand-induced breakdown of phosphoinositides (PI) in the lipid bilayer.

Copyright © 2005 John Wiley & Sons, Inc.

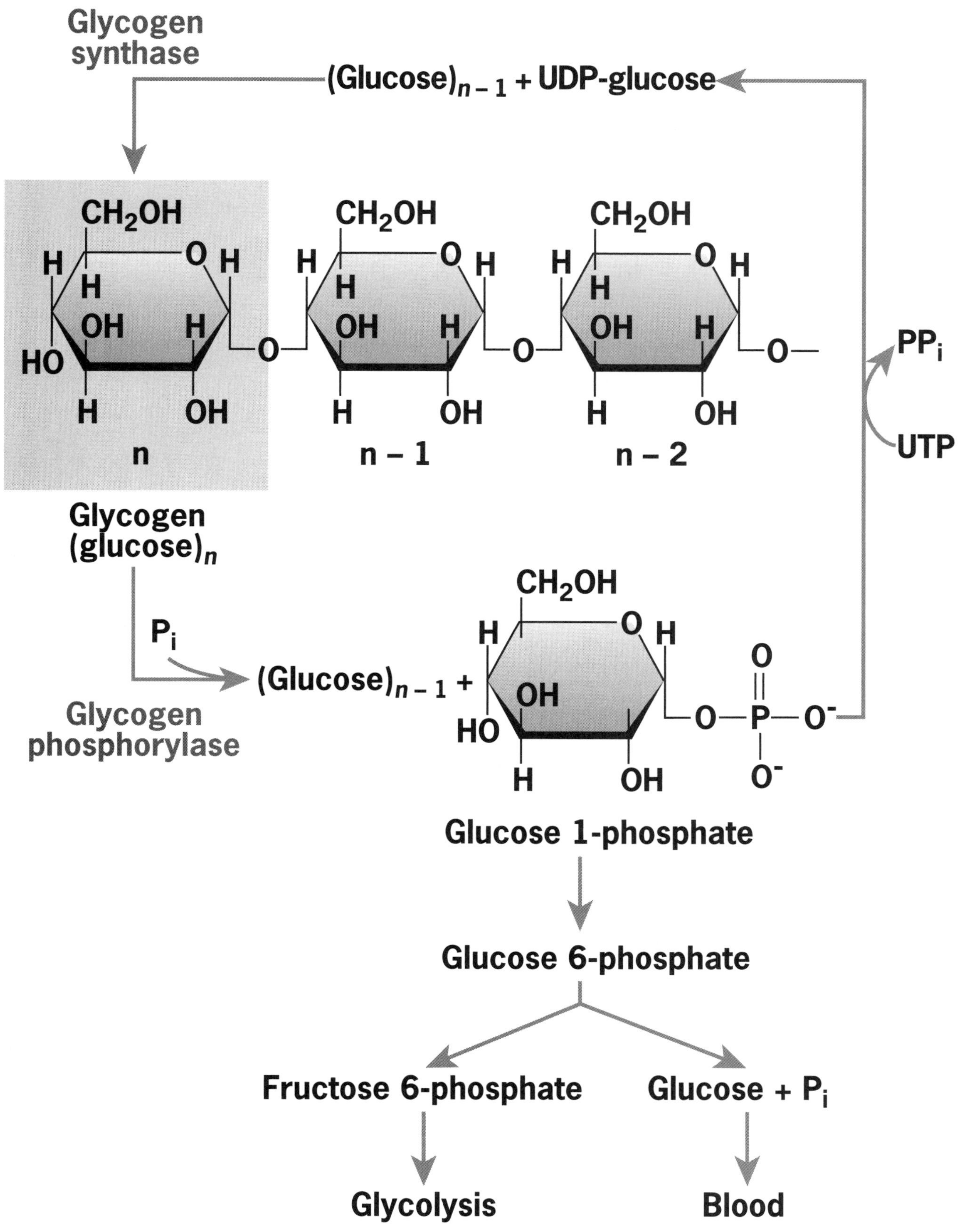

Figure 15.9 The reactions that lead to glucose storage or mobilization.

Copyright © 2005 John Wiley & Sons, Inc.

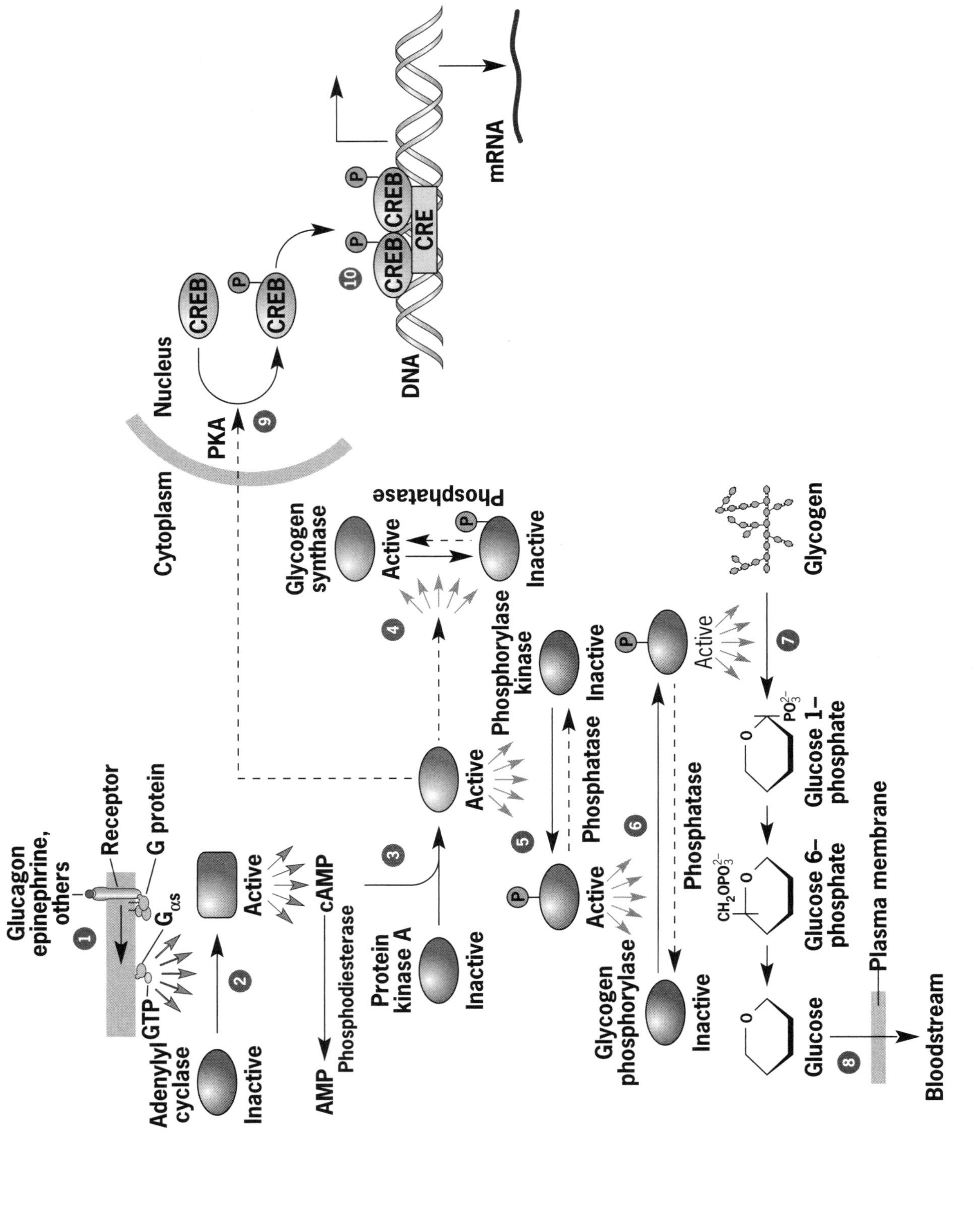

Figure 15.11 The response by a liver cell to glucagon or epinephrine.

Copyright © 2005 John Wiley & Sons, Inc.

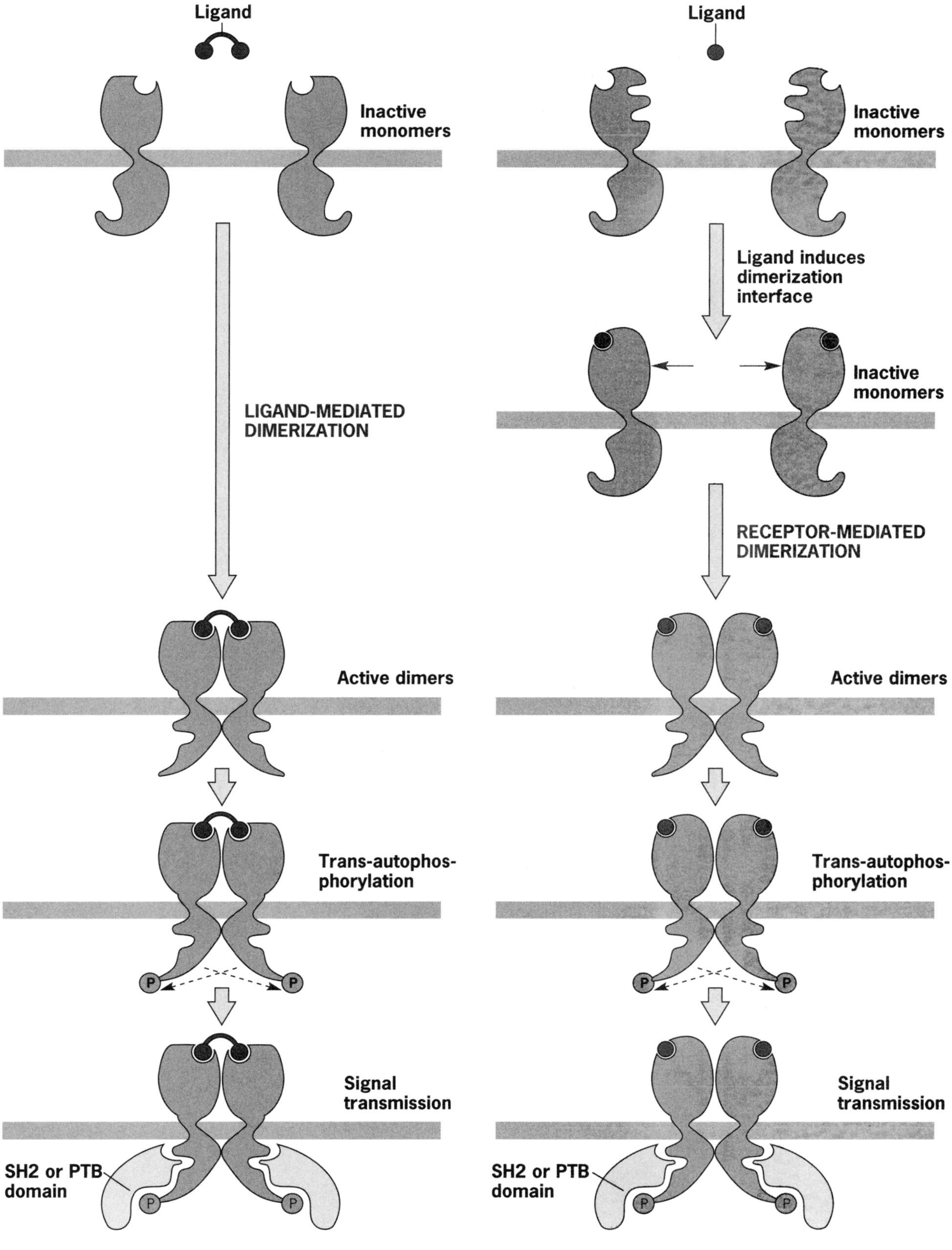

(a) *(b)*

Figure 15.13 Steps in the activation of a receptor protein-tyrosine kinase (RTK).

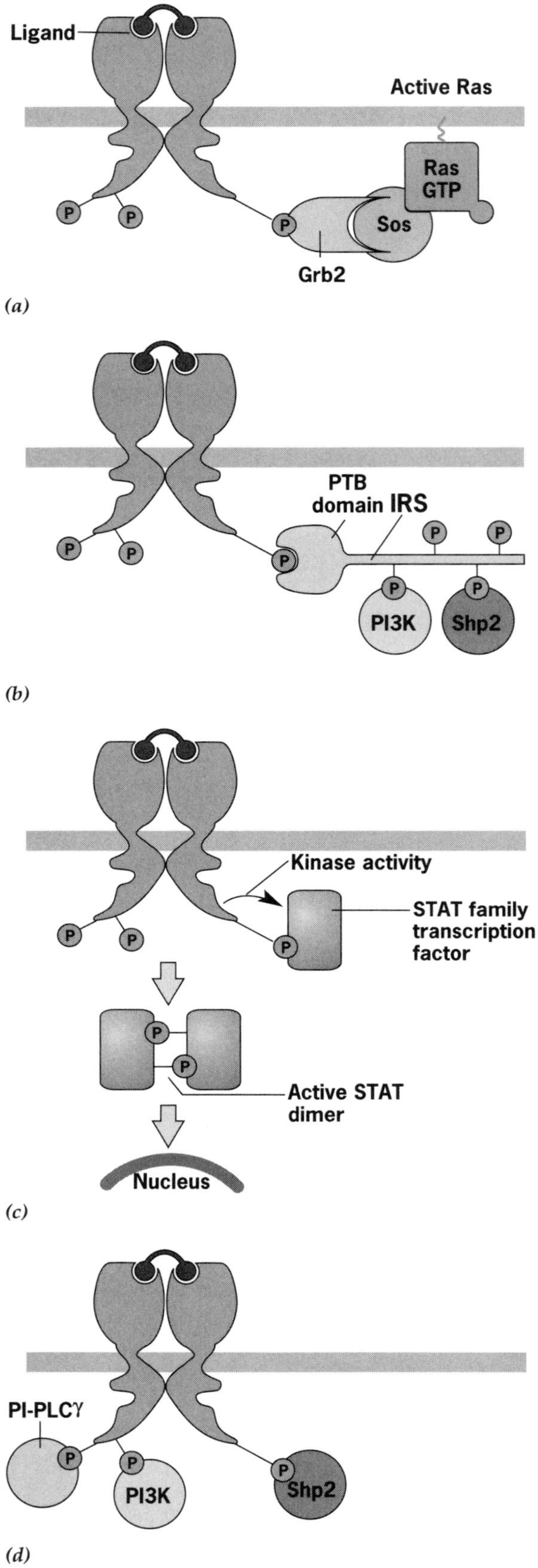

Figure 15.15 A diversity of signaling proteins.

Copyright © 2005 John Wiley & Sons, Inc.

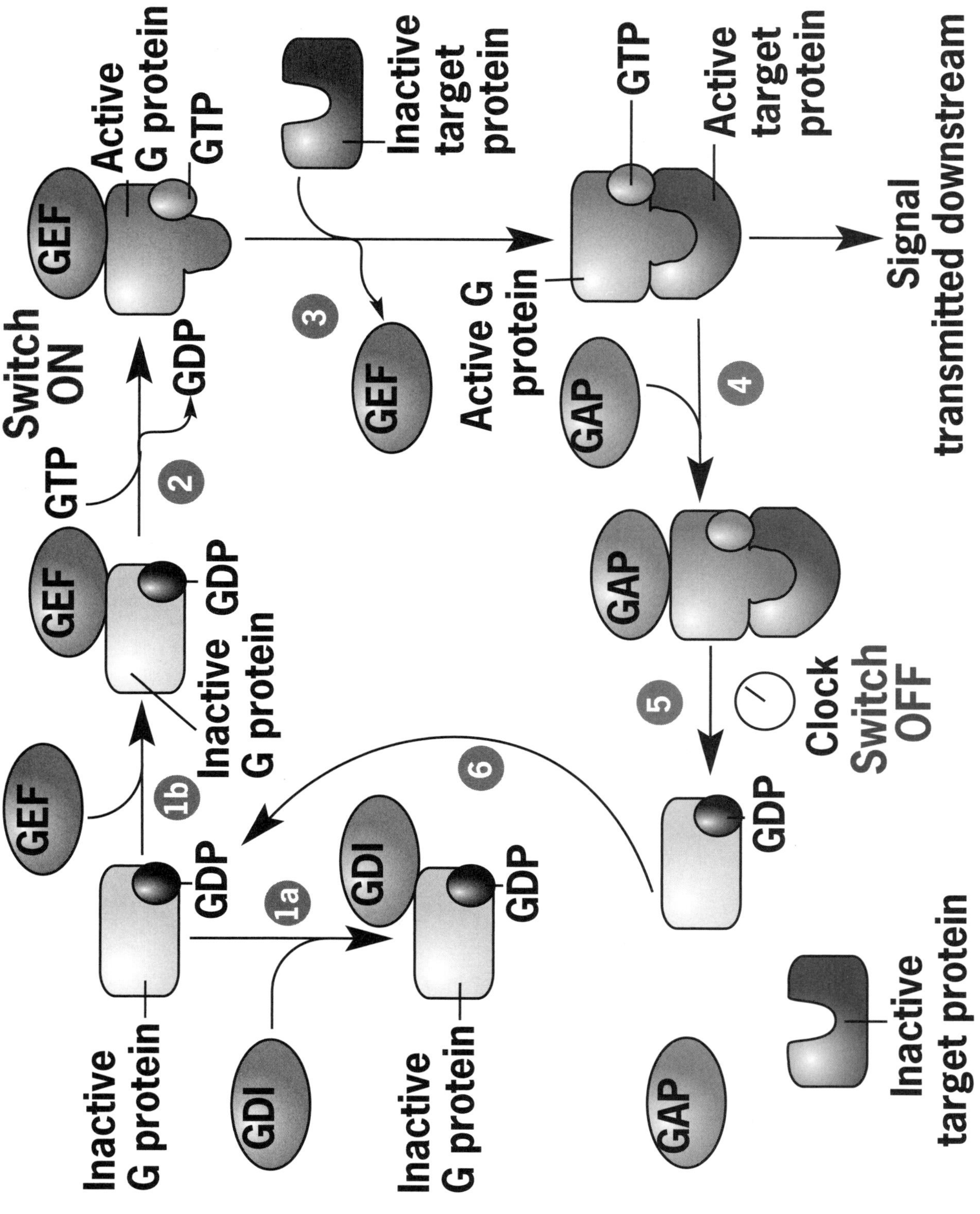

Figure 15.17b The structure of a G protein and the G protein cycle.

Copyright © 2005 John Wiley & Sons, Inc.

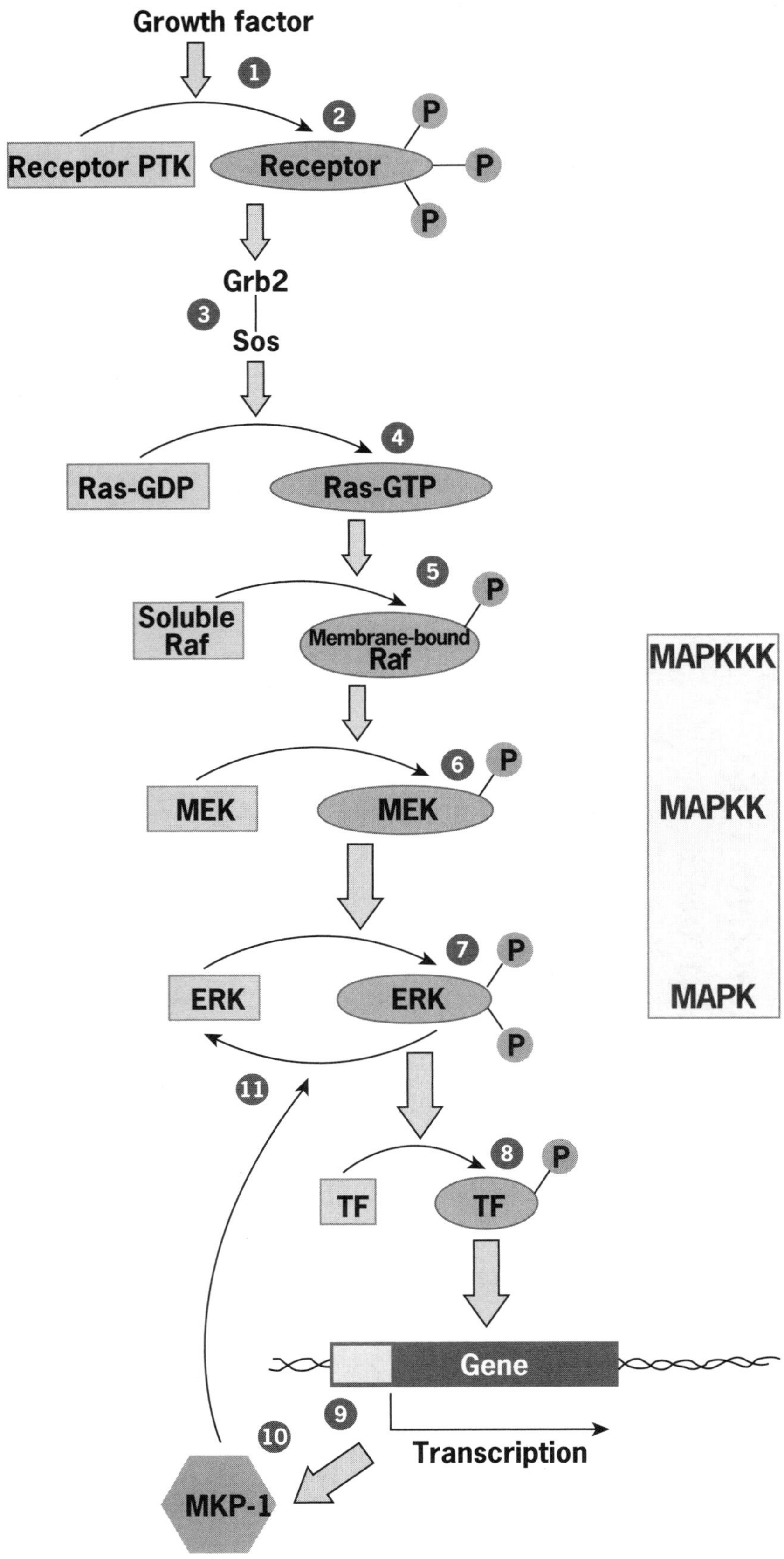

Figure 15.18 The steps of a generalized MAP kinase cascade.

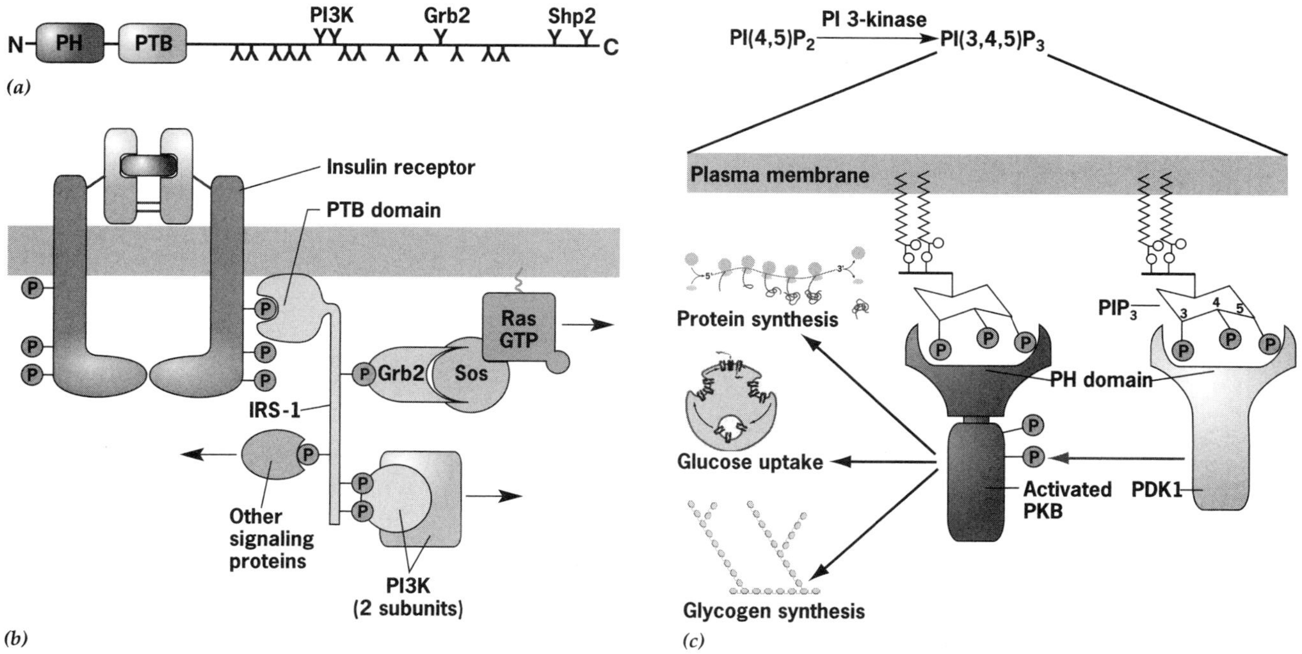

Figure 15.21 The role of tyrosine-phosphorylated IRS in activating a variety of signaling pathways.

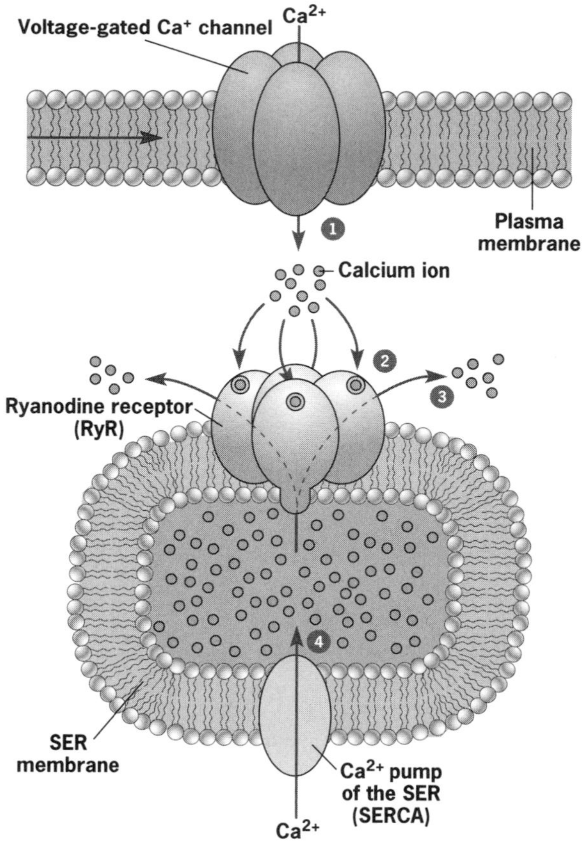

Figure 15.24 Calcium-induced calcium release.

Copyright © 2005 John Wiley & Sons, Inc.

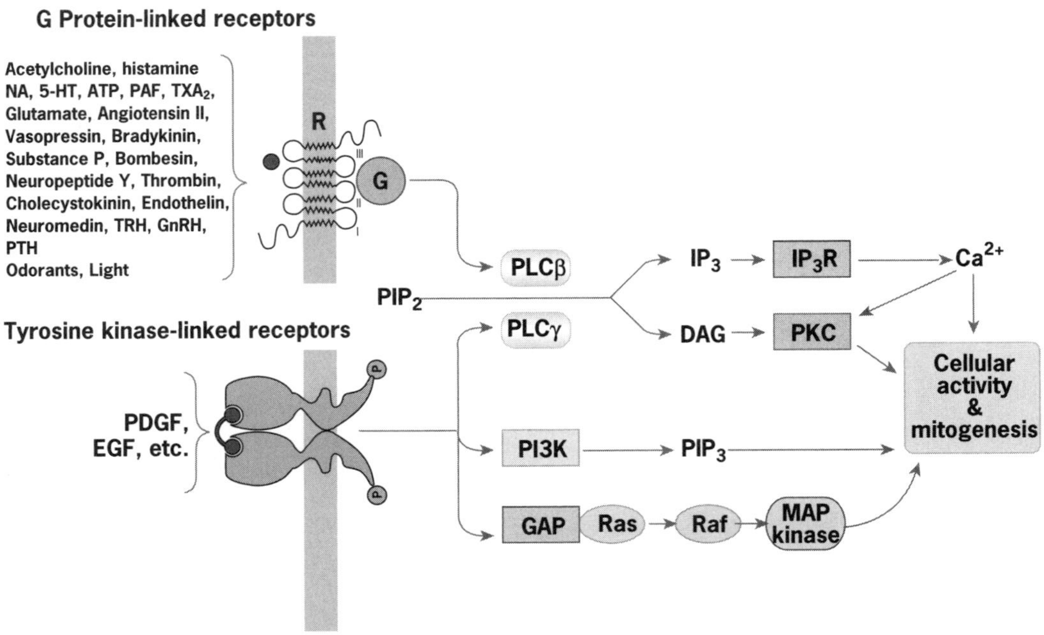

Figure 15.28 Examples of convergence, divergence, and crosstalk among various signal-transduction pathways.

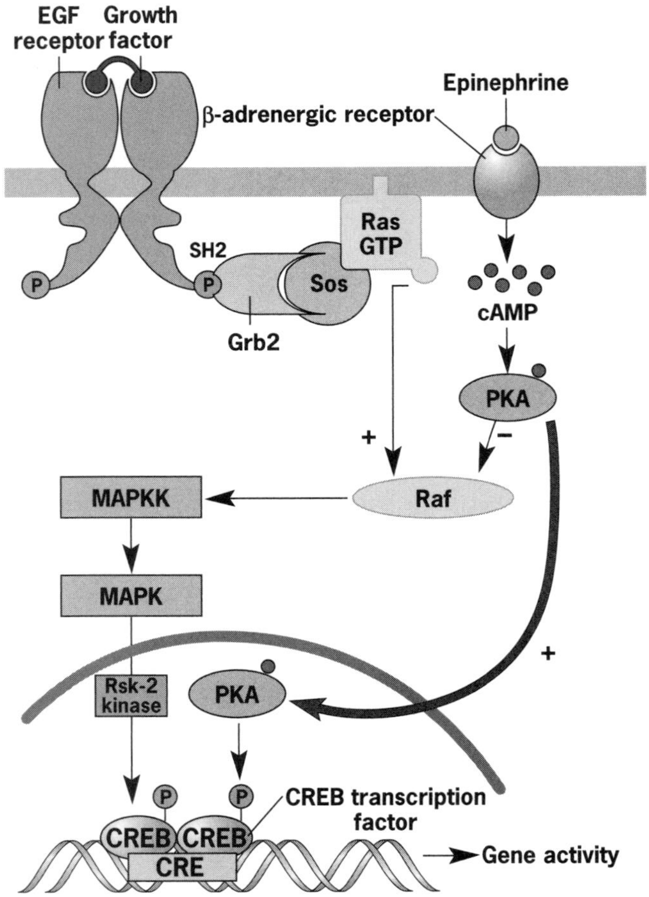

Figure 15.30 An example of crosstalk between two major signaling pathways.

Copyright © 2005 John Wiley & Sons, Inc.

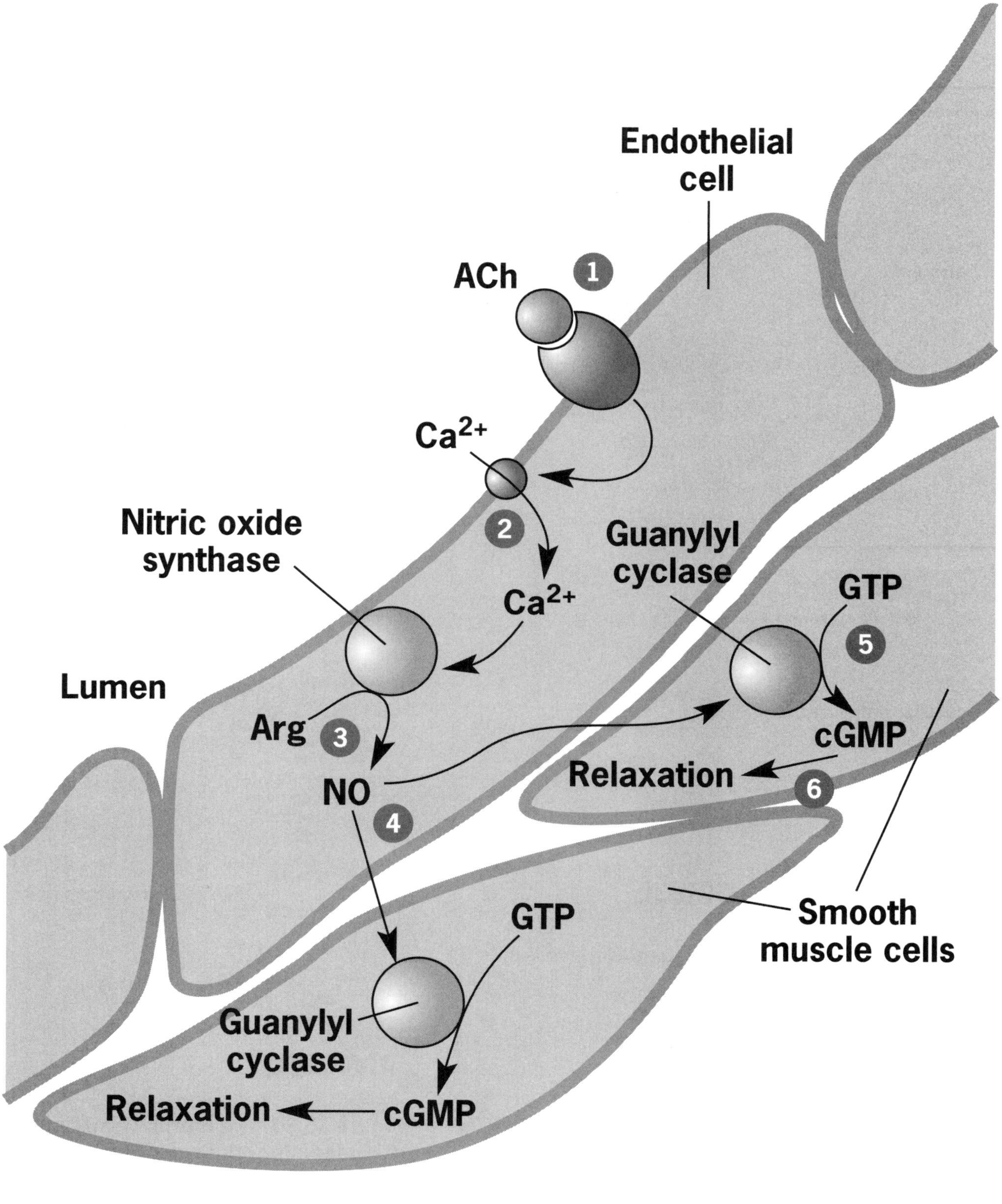

Figure 15.31 A signal transduction pathway that operates by means of NO and cyclic GMP that leads to the dilation of blood vessels.

Copyright © 2005 John Wiley & Sons, Inc.

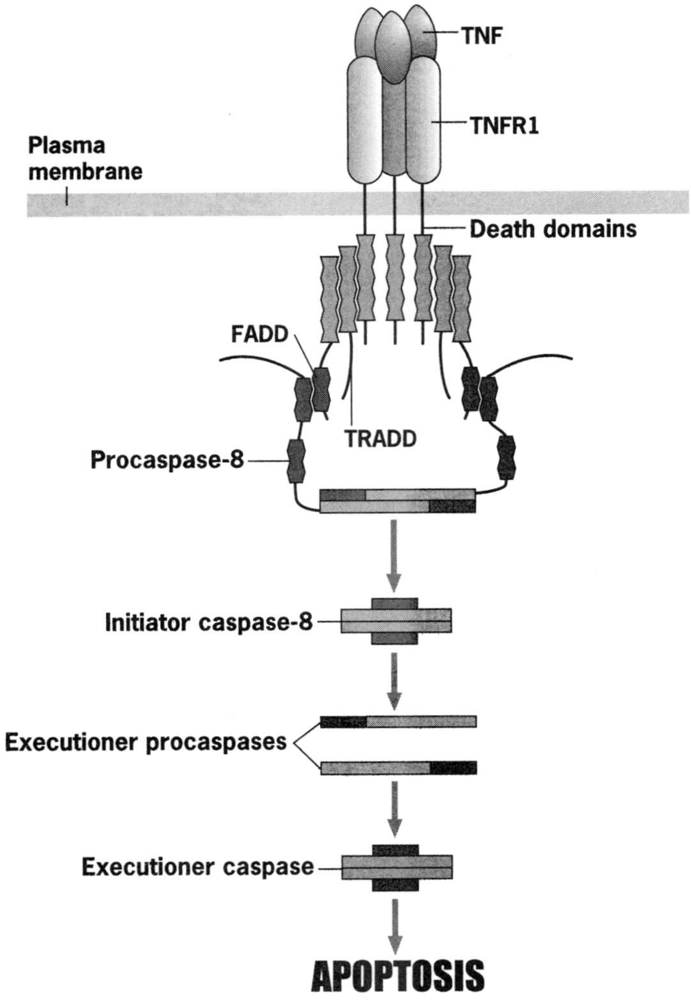

Figure 15.33 The extrinsic (receptor-mediated) pathway of apoptosis.

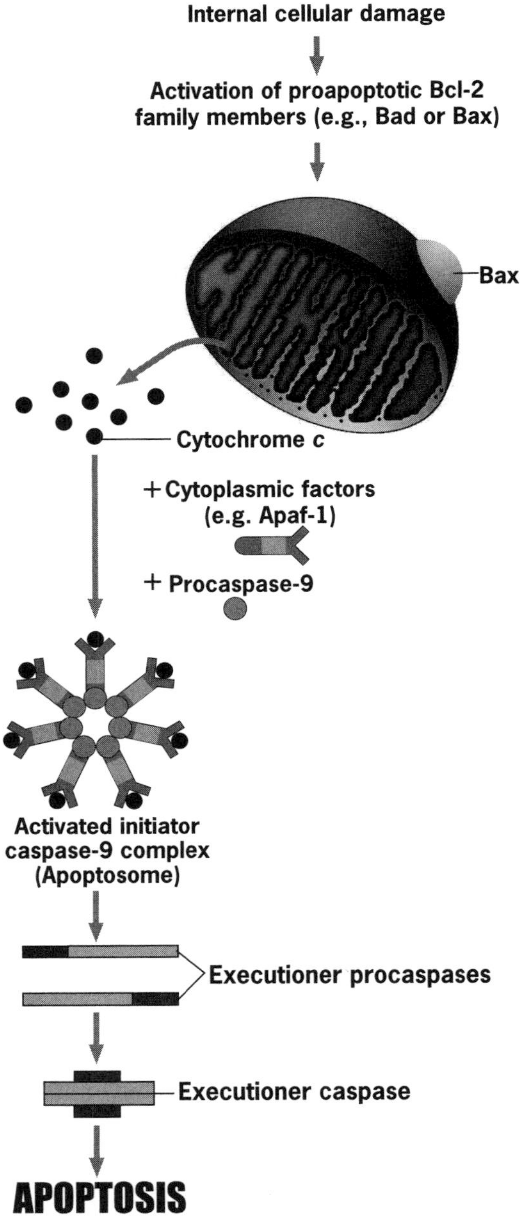

Figure 15.34 The intrinsic (mitochondria-mediated) pathway of apoptosis.

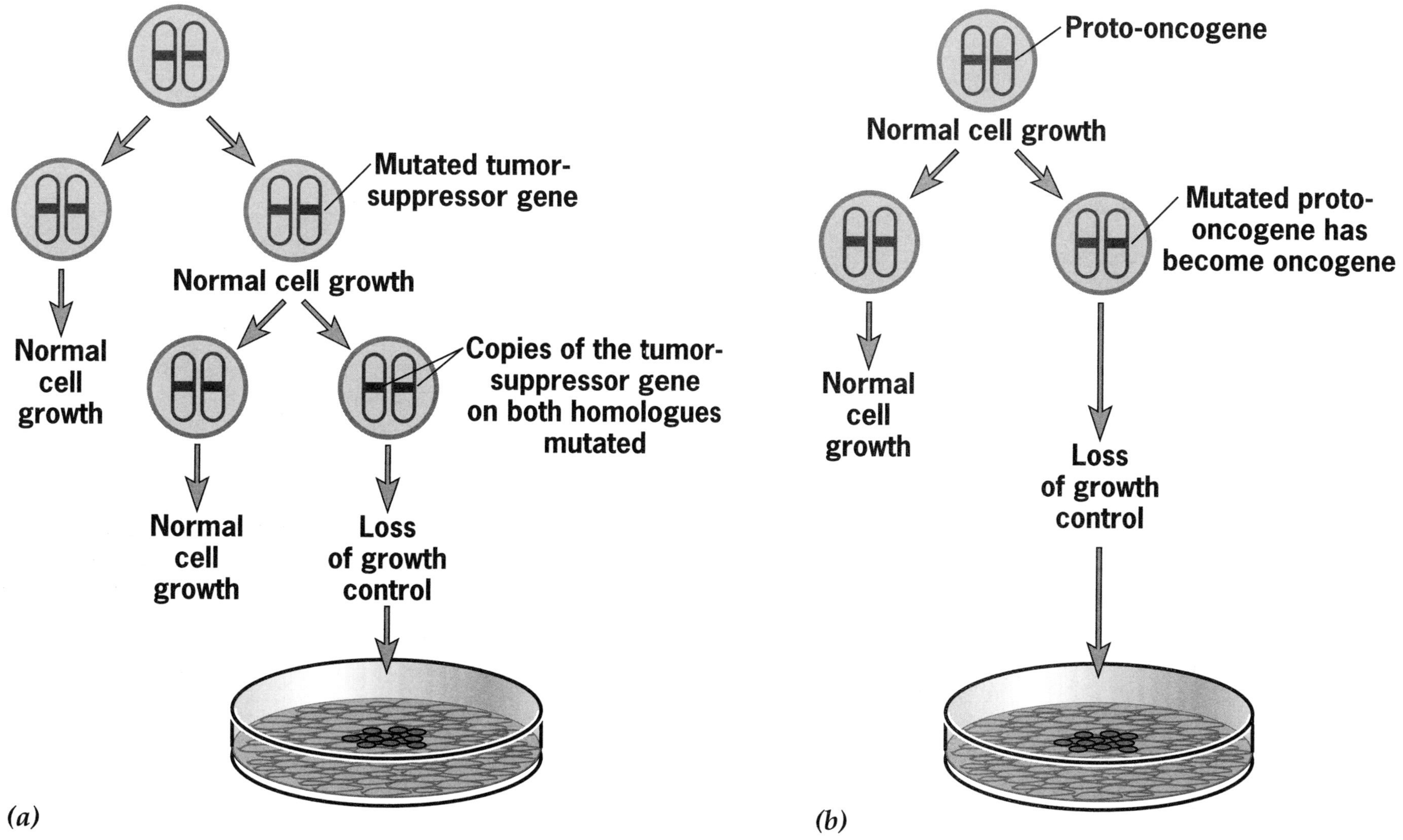

Figure 16.11 Contrasting effects of mutations in tumor-suppressor genes (a) and oncogenes (b).

Copyright © 2005 John Wiley & Sons, Inc.

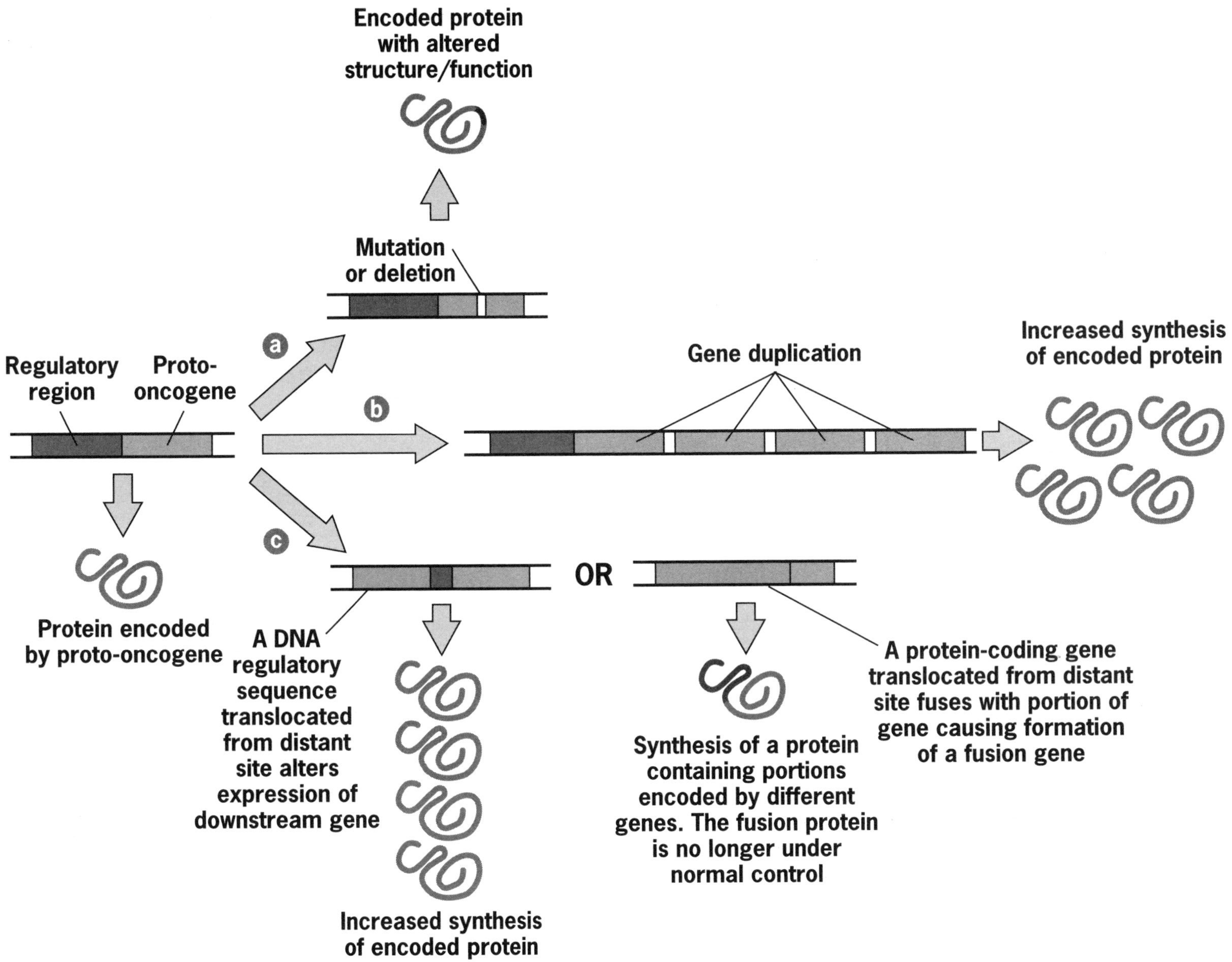

Figure 16.12 Activation of a proto-oncogene to an oncogene.

Copyright © 2005 John Wiley & Sons, Inc.

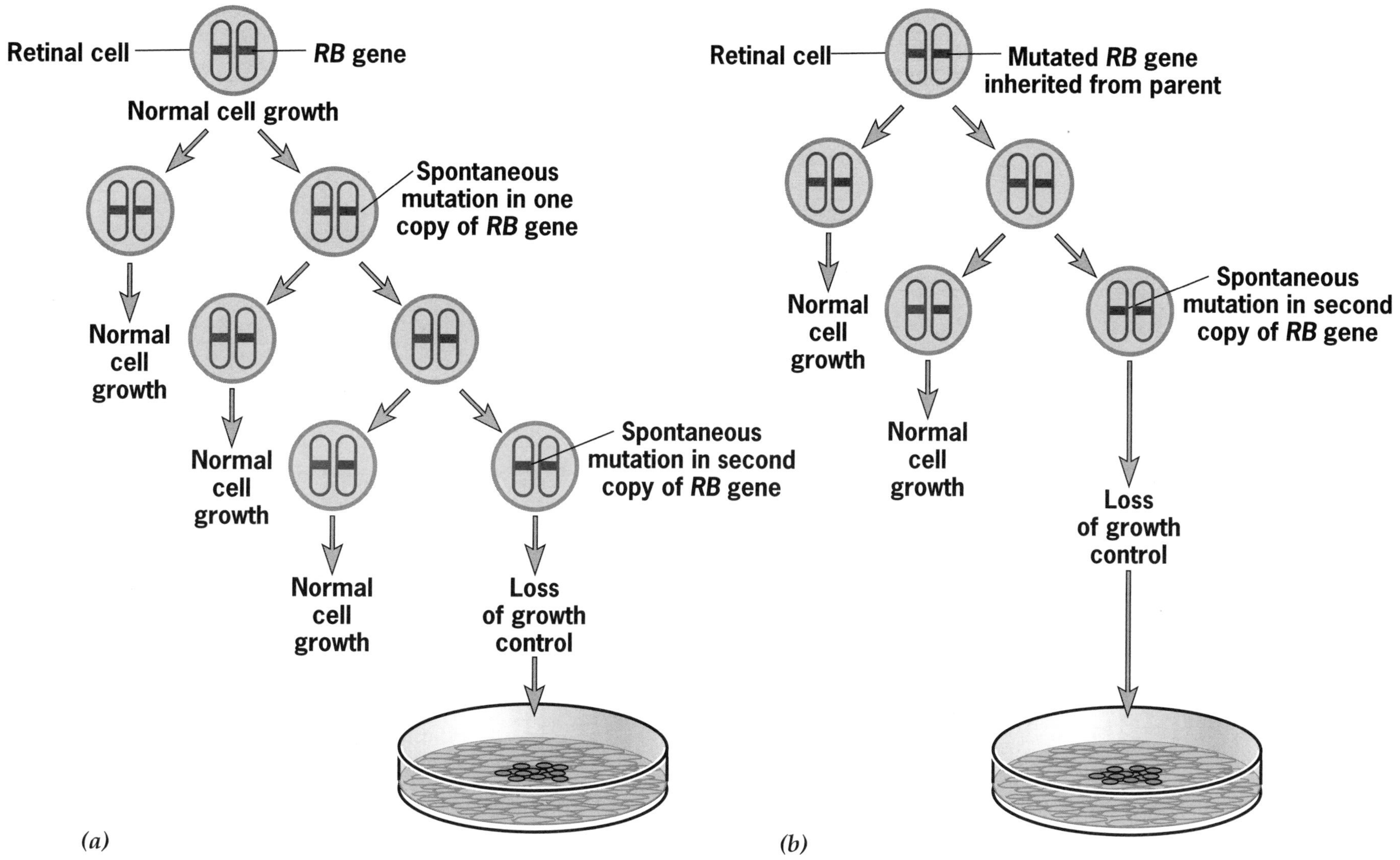

Figure 16.13 Mutations in the *RB* gene that can lead to retinoblastoma.

Copyright © 2005 John Wiley & Sons, Inc.

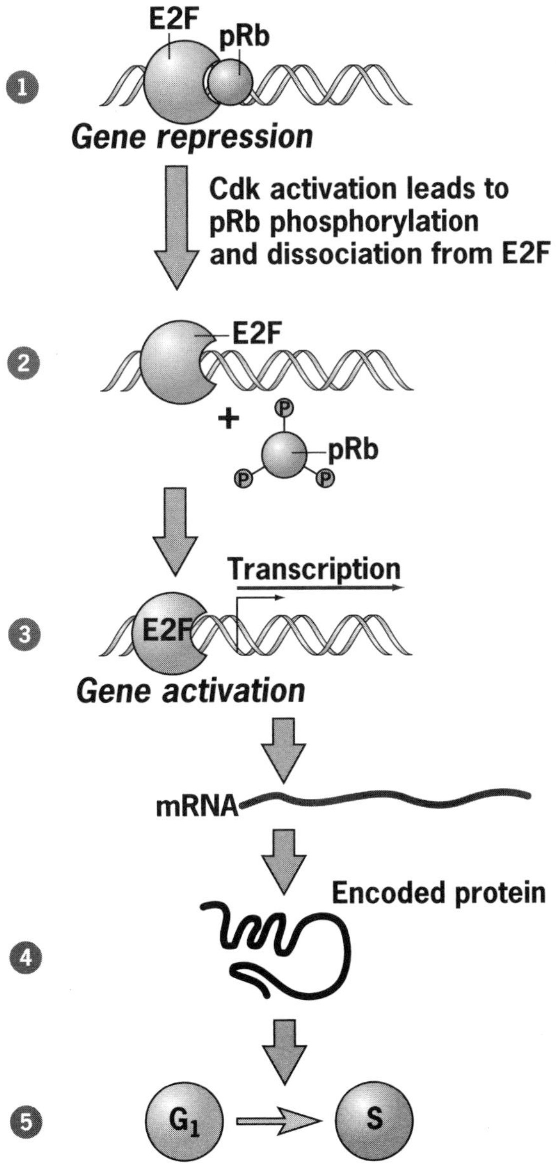

Figure 16.14 The role of pRb in controlling transcription of genes required for progression of the cell cycle.

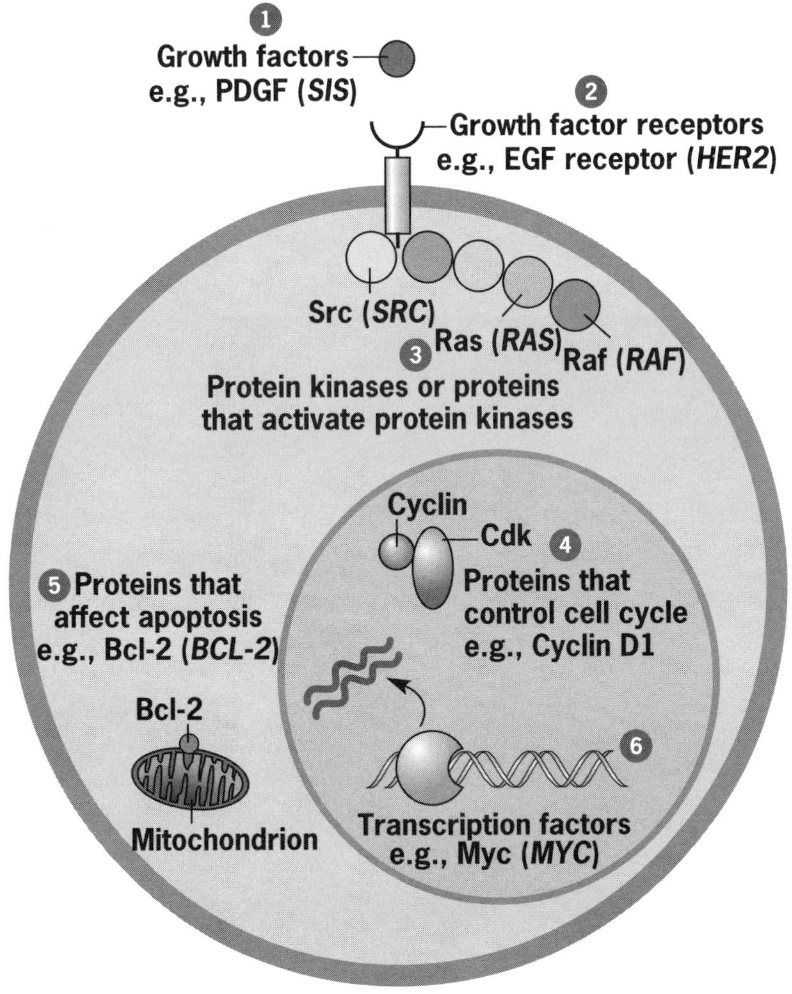

Figure 16.20 A schematic diagram summarizing the types of proteins encoded by proto-oncogenes.

Copyright © 2005 John Wiley & Sons, Inc.

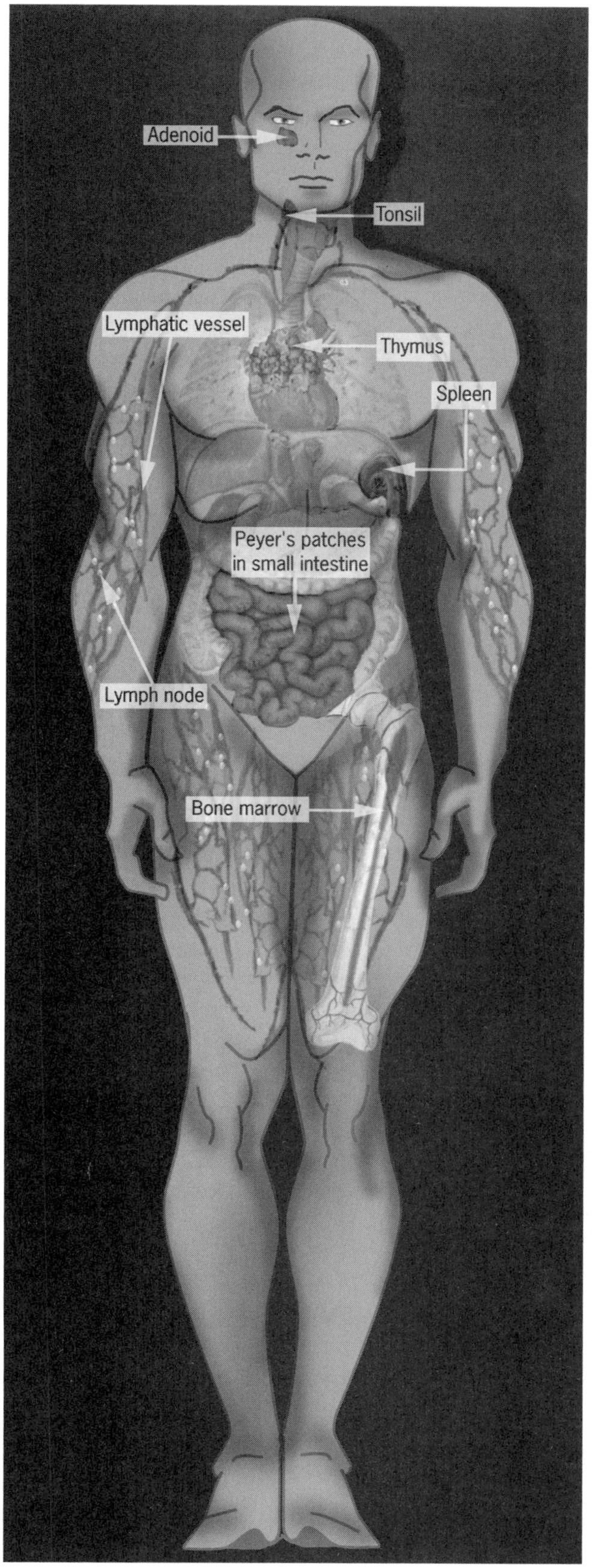

Figure 17.1 The human immune system.

Copyright © 2005 John Wiley & Sons, Inc.

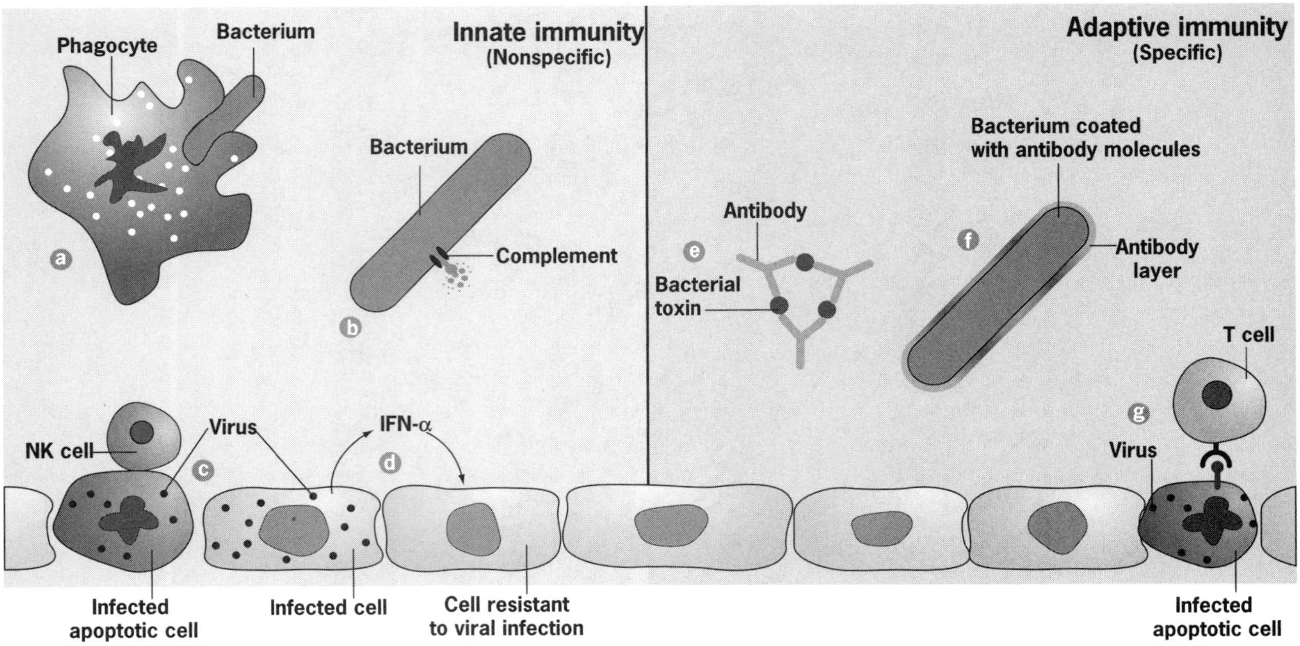

Figure 17.2 An overview of some of the mechanisms by which the immune system rids the body of invading pathogens.

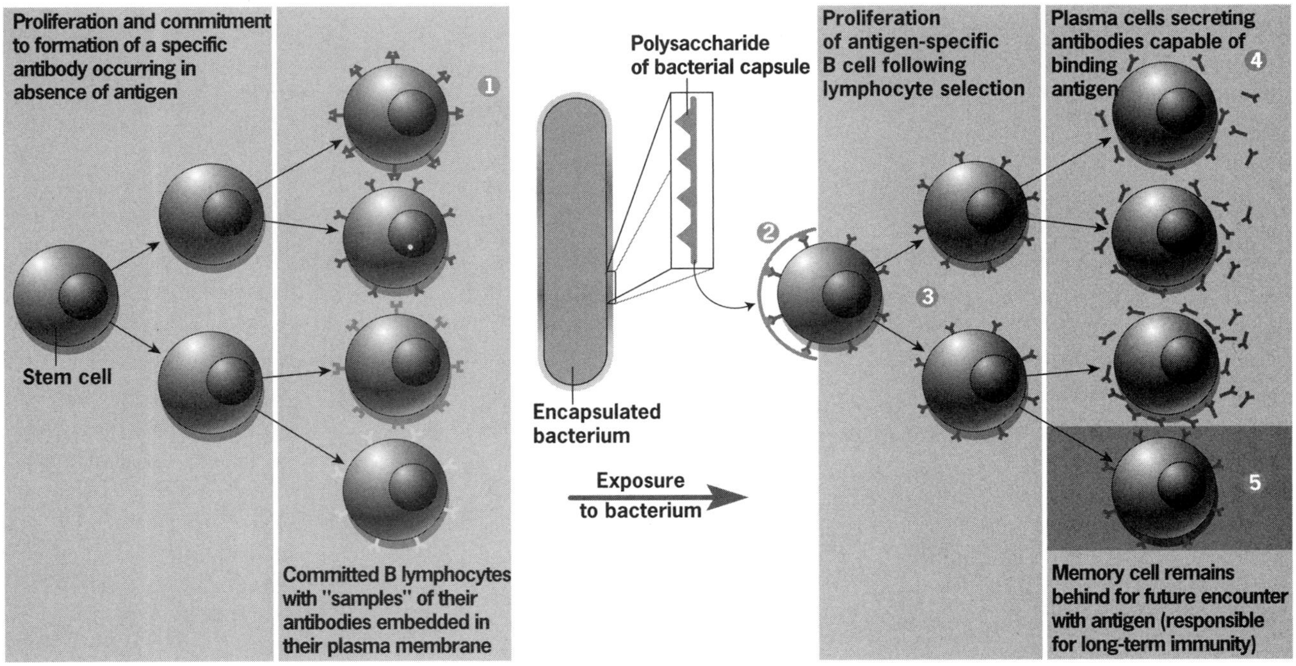

Figure 17.5 The clonal selection of B cells by a thymus-independent antigen.

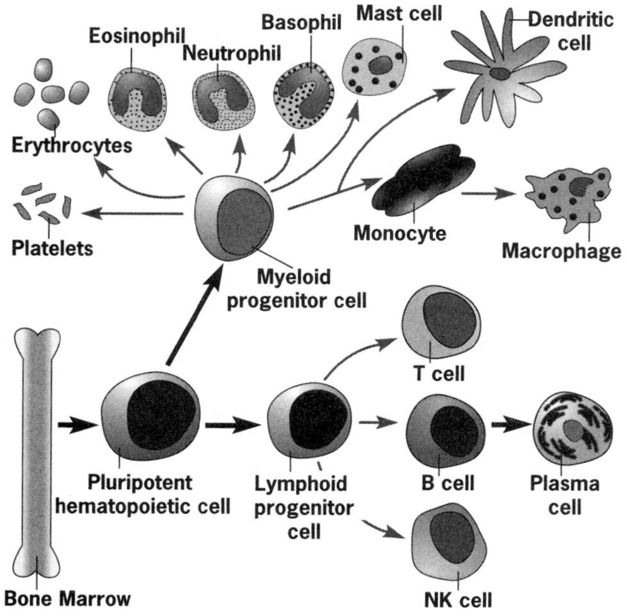

Figure 17.4 Pathways of differentiation of a pluripotent hematopoietic stem cell of the bone marrow.

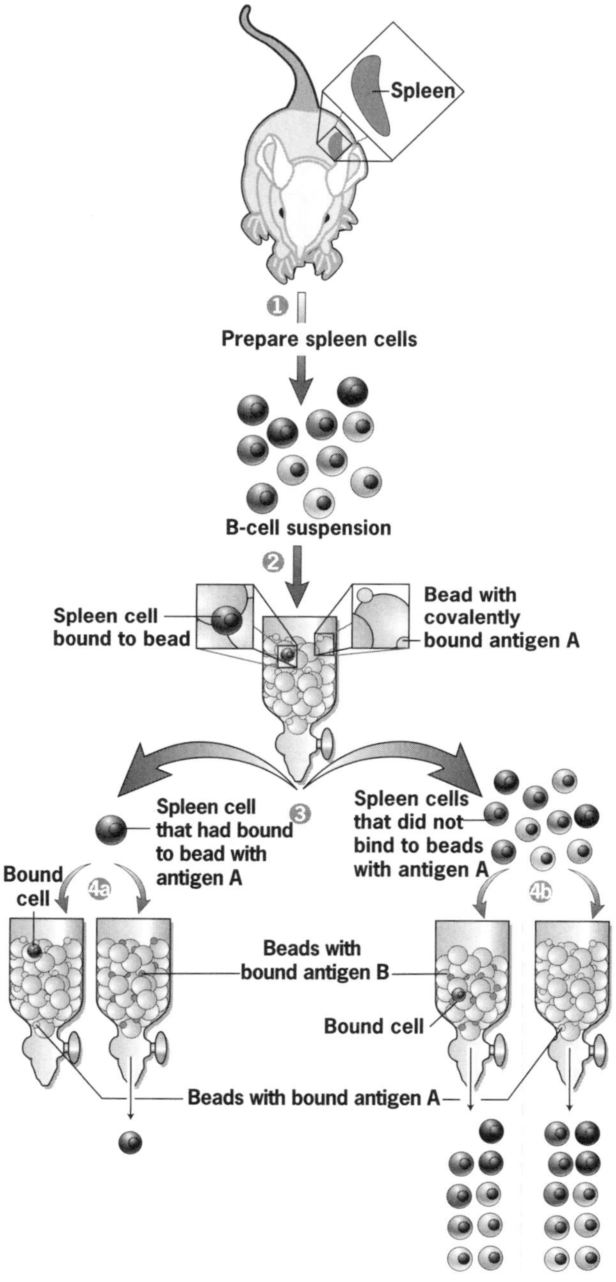

Figure 17.6 Experimental demonstration that different B cells each contain a different membrane-bound antibody and that these antibodies are produced in the absence of antigen.

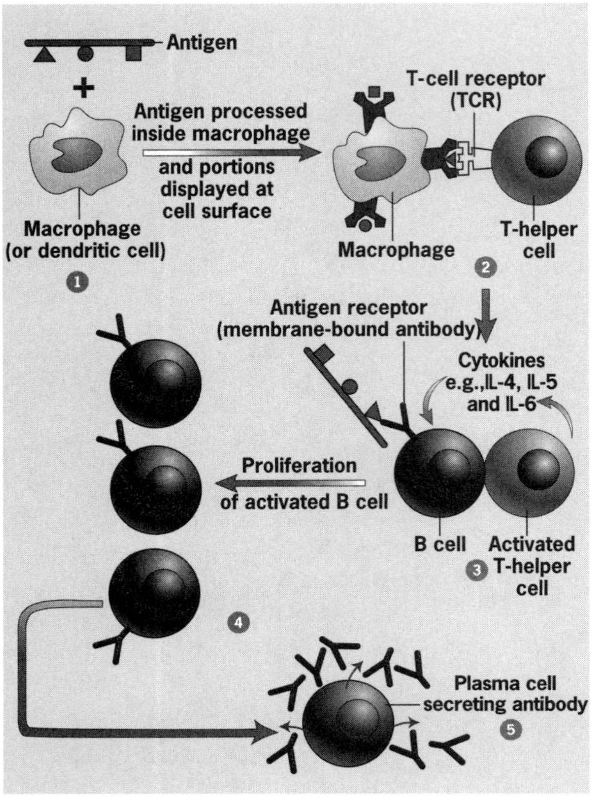

Figure 17.9 Highly simplified, schematic drawing showing the role of T_H cells in antibody formation.

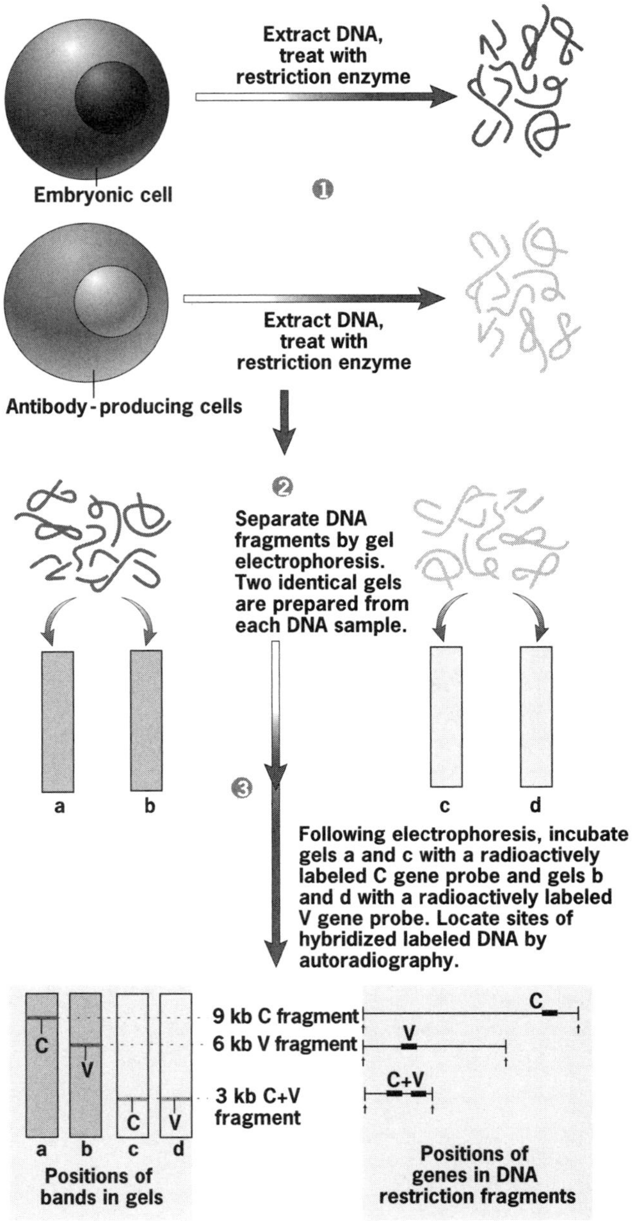

Figure 17.14 Experimental demonstration that genes encoding antibody light chains are formed by DNA rearrangement.

Copyright © 2005 John Wiley & Sons, Inc.

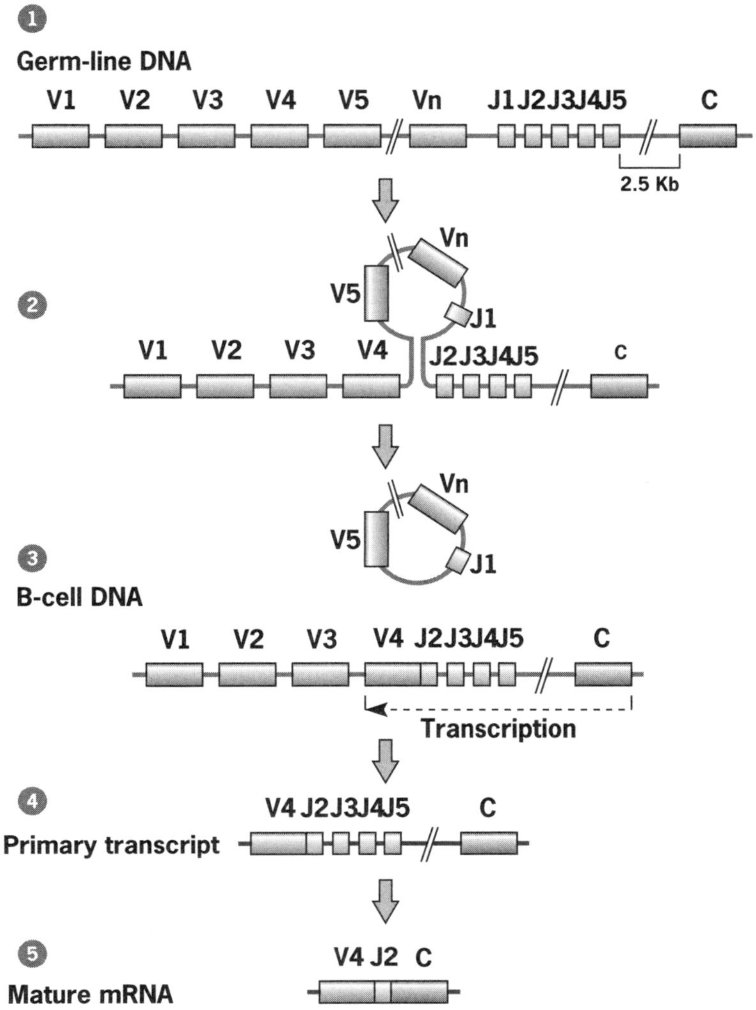

Figure 17.15 DNA rearrangements that lead to the formation of a functional gene that encodes an immunoglobulin κ chain.

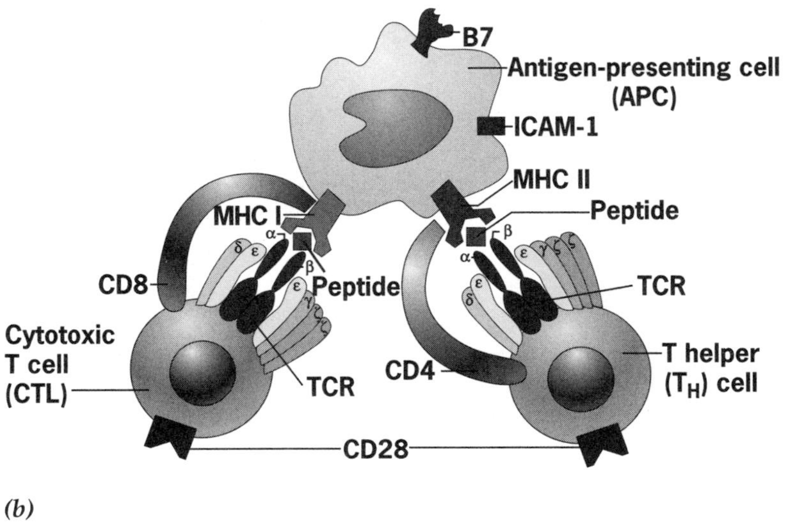

Figure 17.19 Interaction between a macrophage and a T cell during antigen presentation.

Copyright © 2005 John Wiley & Sons, Inc.

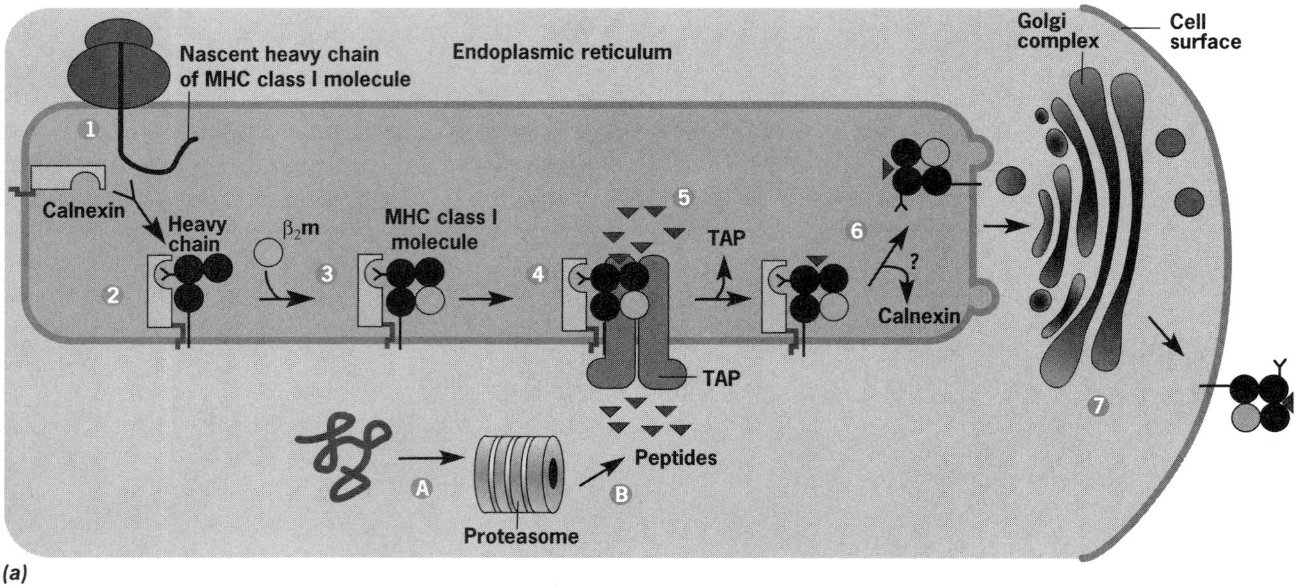

(a)

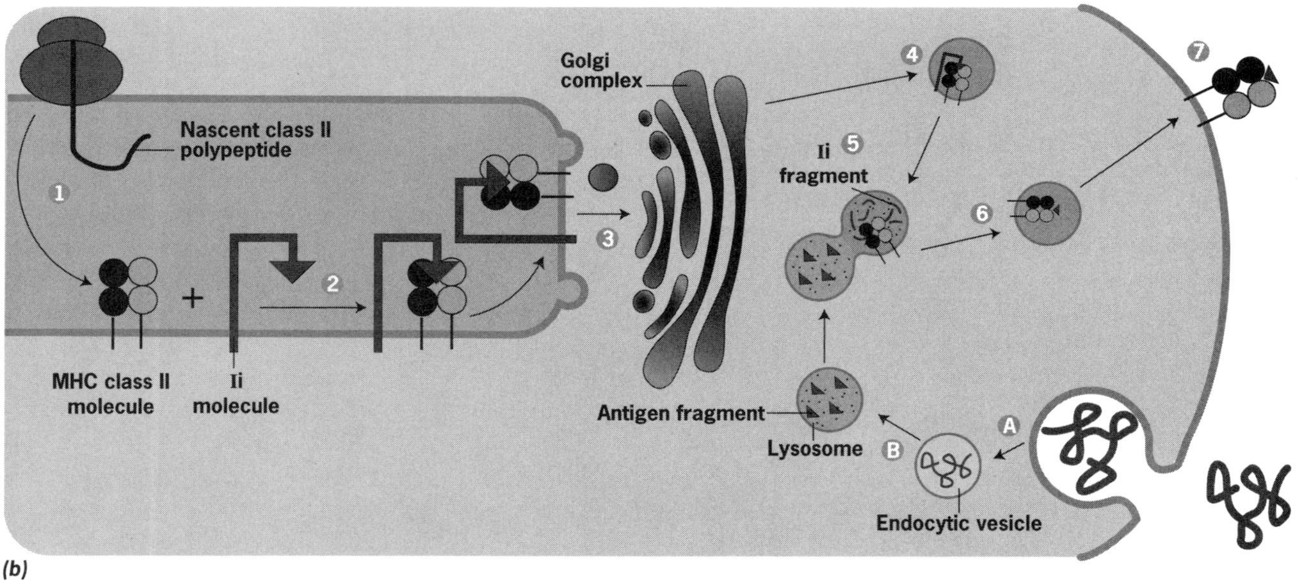

(b)

Figure 17.20 Processing pathways for antigens that become associated with MHC class I and class II molecules.

Copyright © 2005 John Wiley & Sons, Inc.

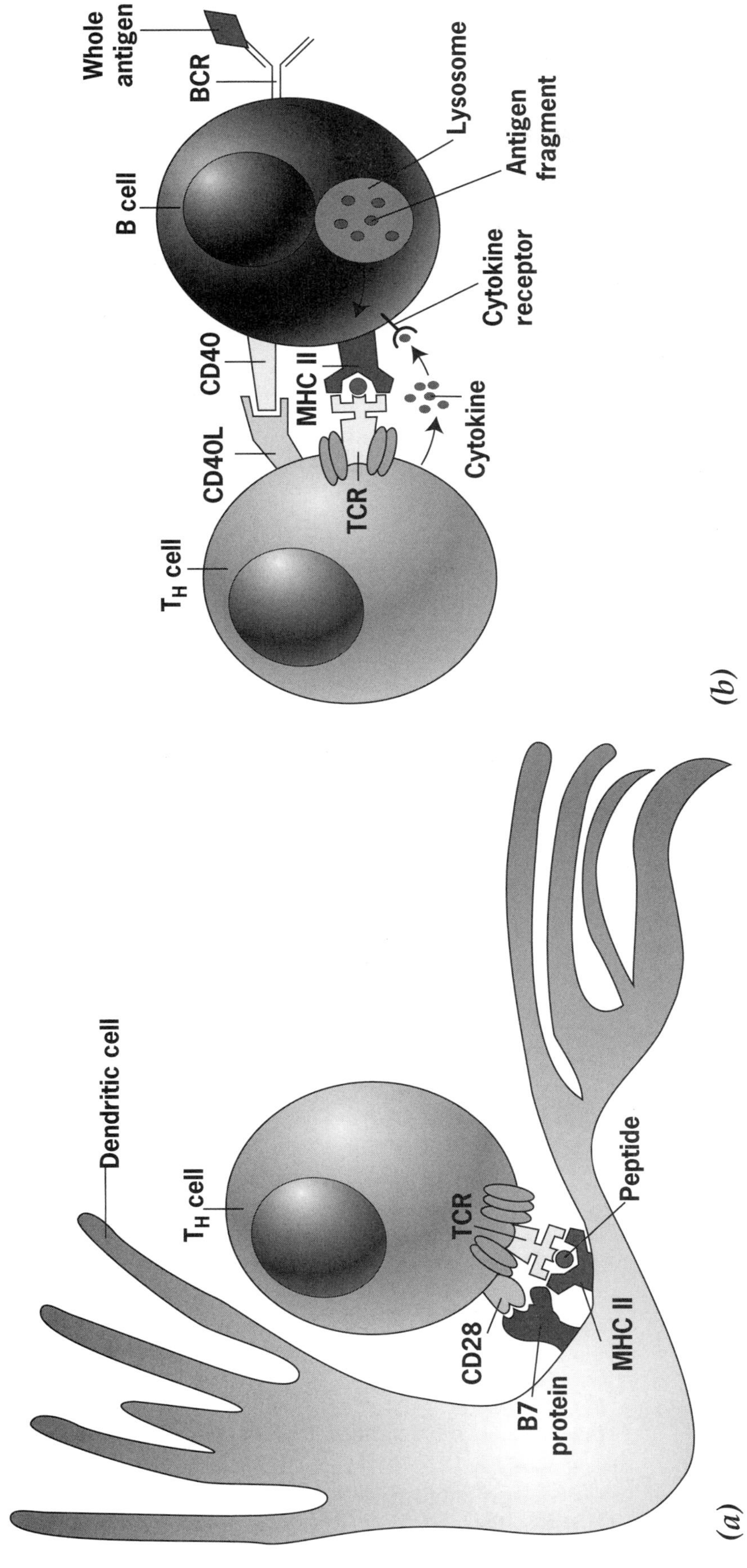

Figure 17.24 Lymphocyte activation.

Copyright © 2005 John Wiley & Sons, Inc.

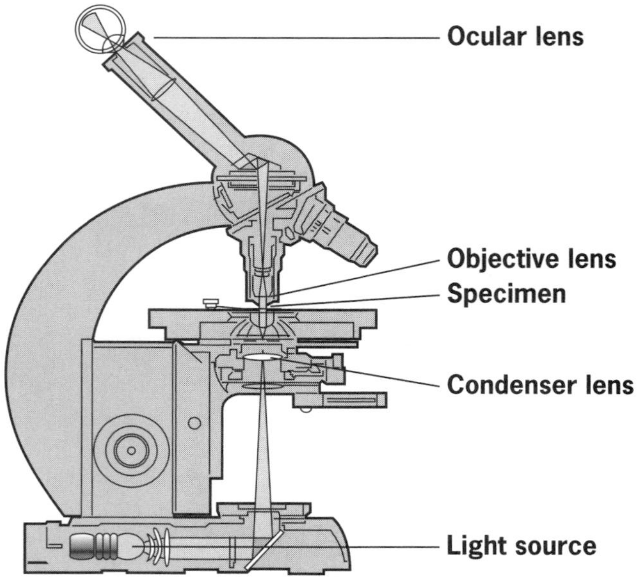

Figure 18.1 Sectional diagram through a compound light microscope.

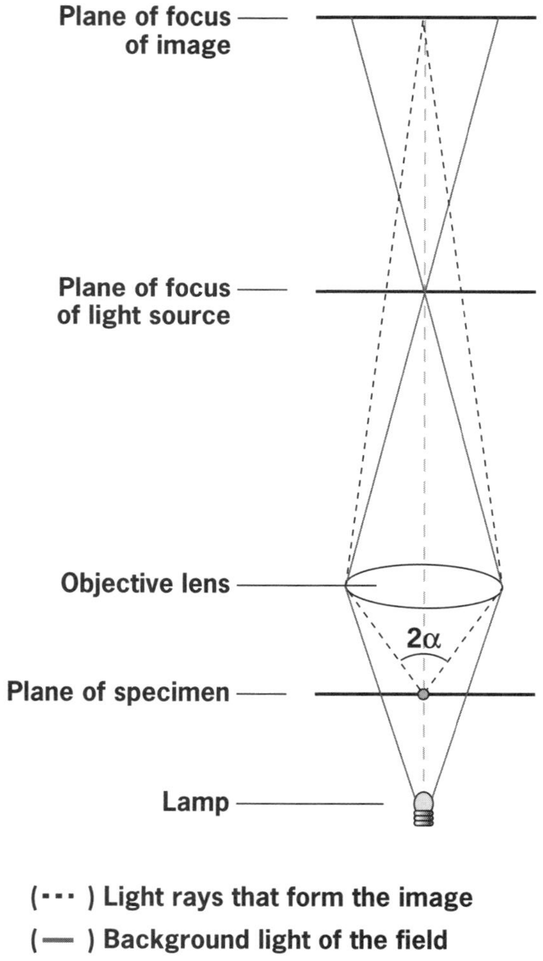

Figure 18.2 The paths taken by light rays that form the image of the specimen and those that form the background light of the field.

Copyright © 2005 John Wiley & Sons, Inc.

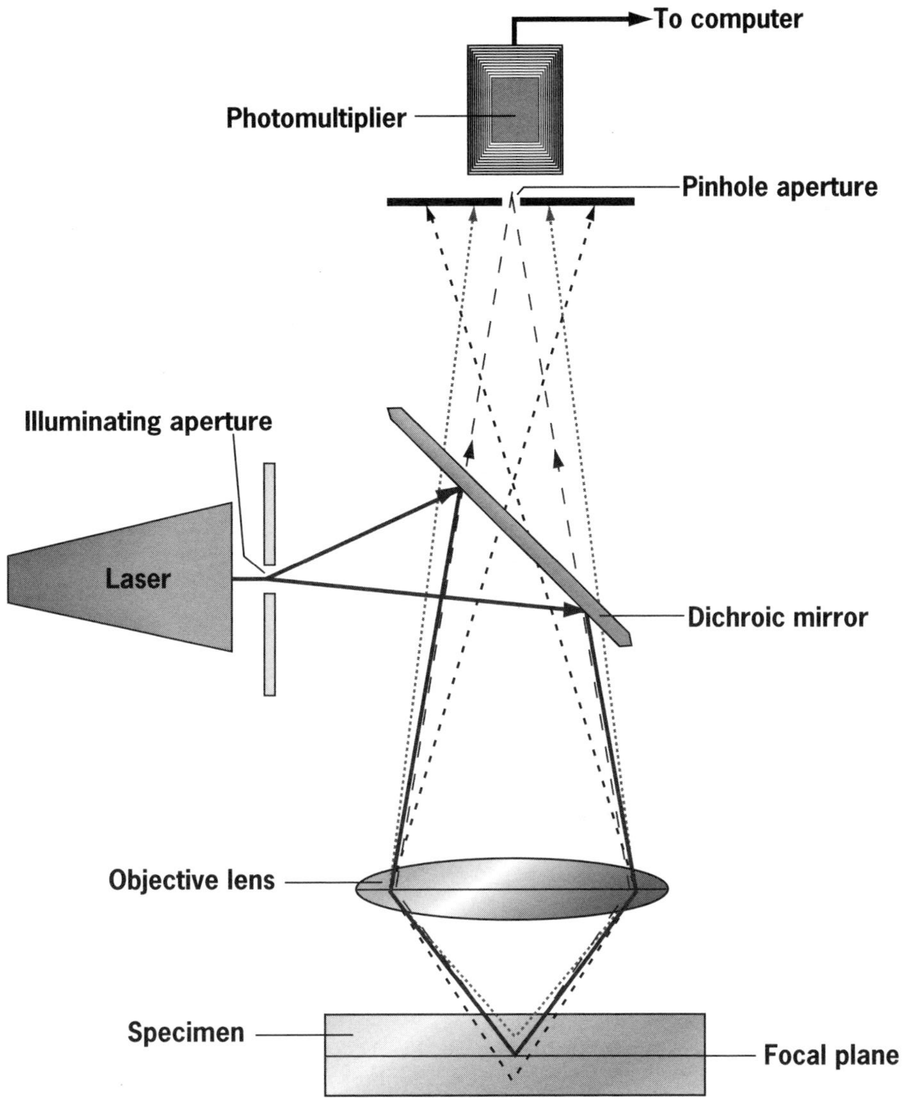

Figure 18.9 The light paths in a fluorescence confocal scanning light microscope.

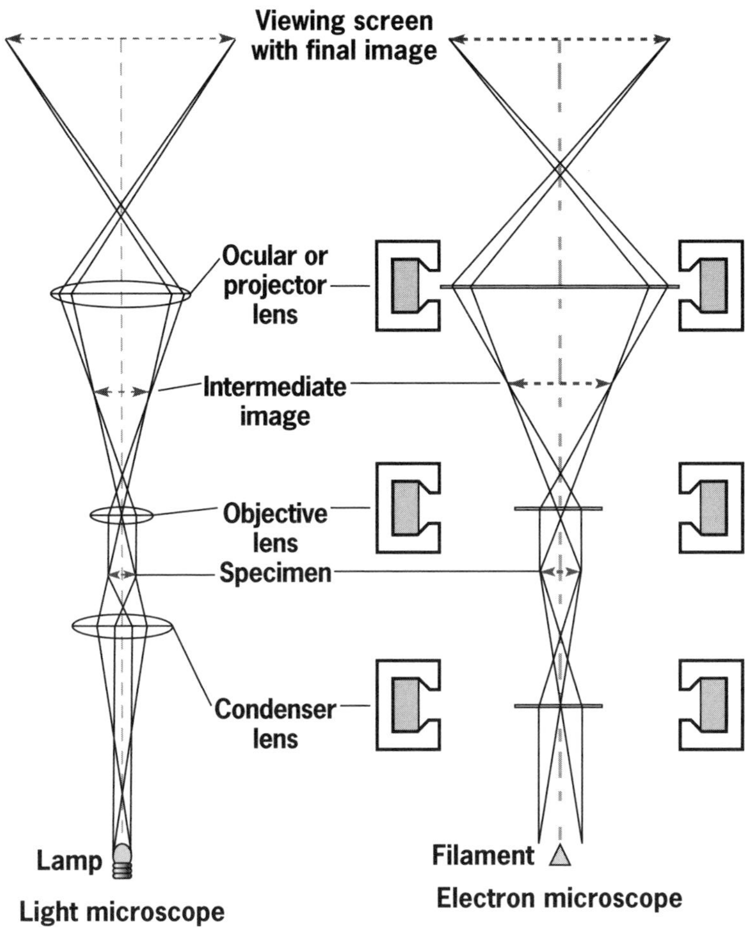

Figure 18.12 A comparison of the lens systems of a light and electron microscope.

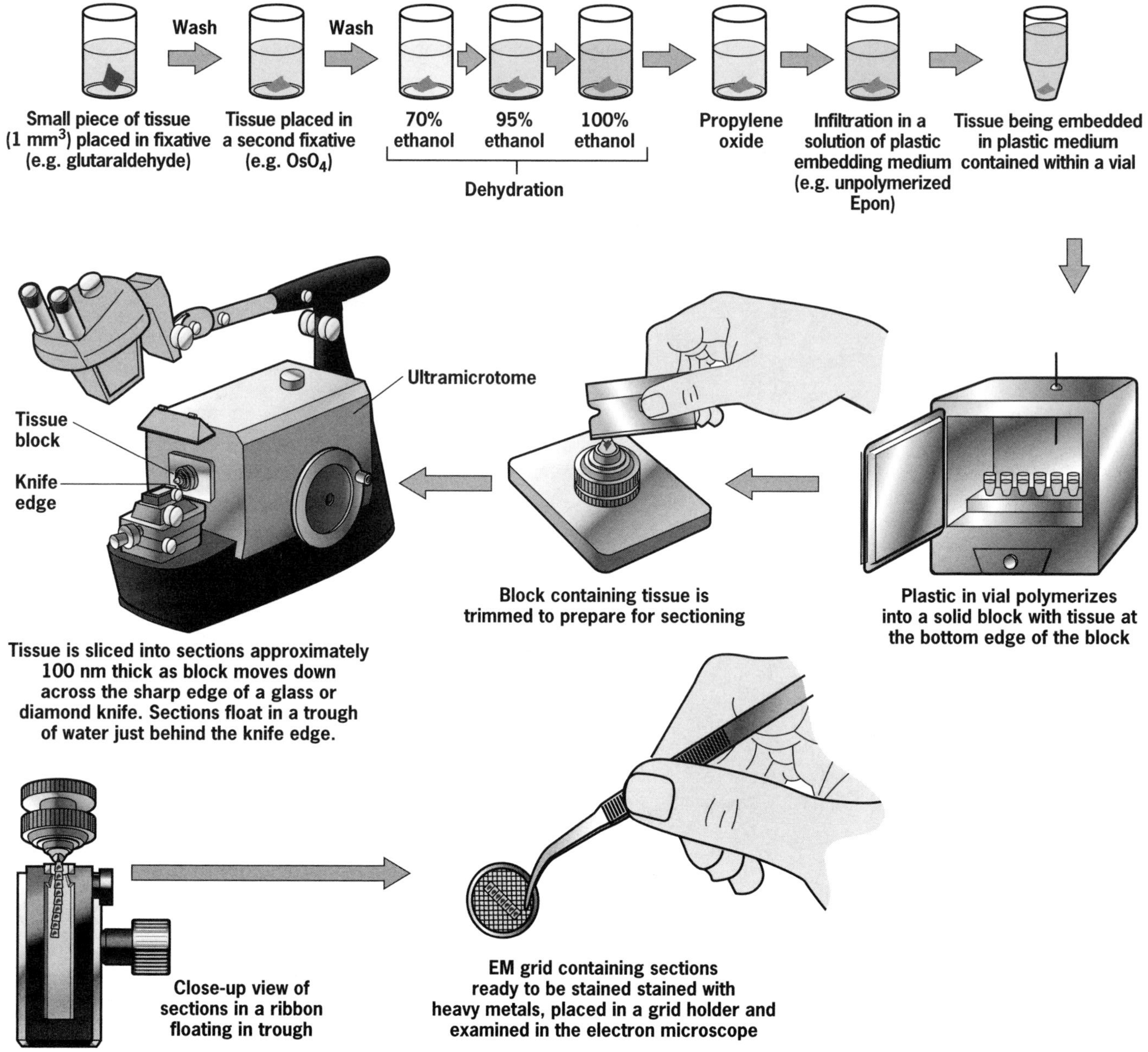

Wash
Wash
Small piece of tissue (1 mm³) placed in fixative (e.g. glutaraldehyde)
Tissue placed in a second fixative (e.g. OsO4)
70% ethanol
95% ethanol
100% ethanol
Dehydration
Propylene oxide
Infiltration in a solution of plastic embedding medium (e.g. unpolymerized Epon)
Tissue being embedded in plastic medium contained within a vial
Ultramicrotome
Tissue block
Knife edge
Block containing tissue is trimmed to prepare for sectioning
Plastic in vial polymerizes into a solid block with tissue at the bottom edge of the block
Tissue is sliced into sections approximately 100 nm thick as block moves down across the sharp edge of a glass or diamond knife. Sections float in a trough of water just behind the knife edge.
Close-up view of sections in a ribbon floating in trough
EM grid containing sections ready to be stained stained with heavy metals, placed in a grid holder and examined in the electron microscope

Figure 18.13 Preparation of a specimen for observation in the electron microscope.

Copyright © 2005 John Wiley & Sons, Inc.

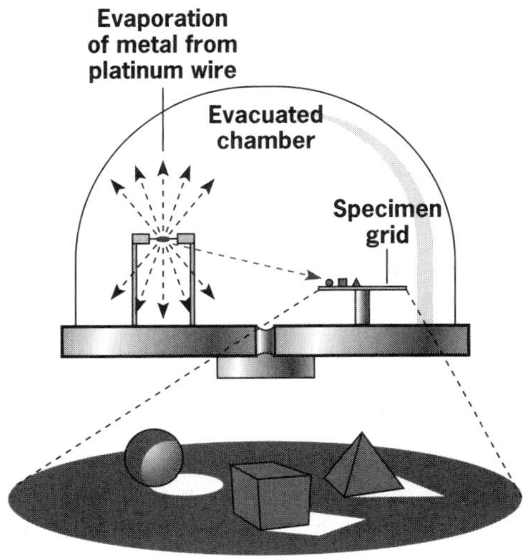

Figure 18.15 The procedure used for shadow casting as a means to provide contrast in the electron microscope.

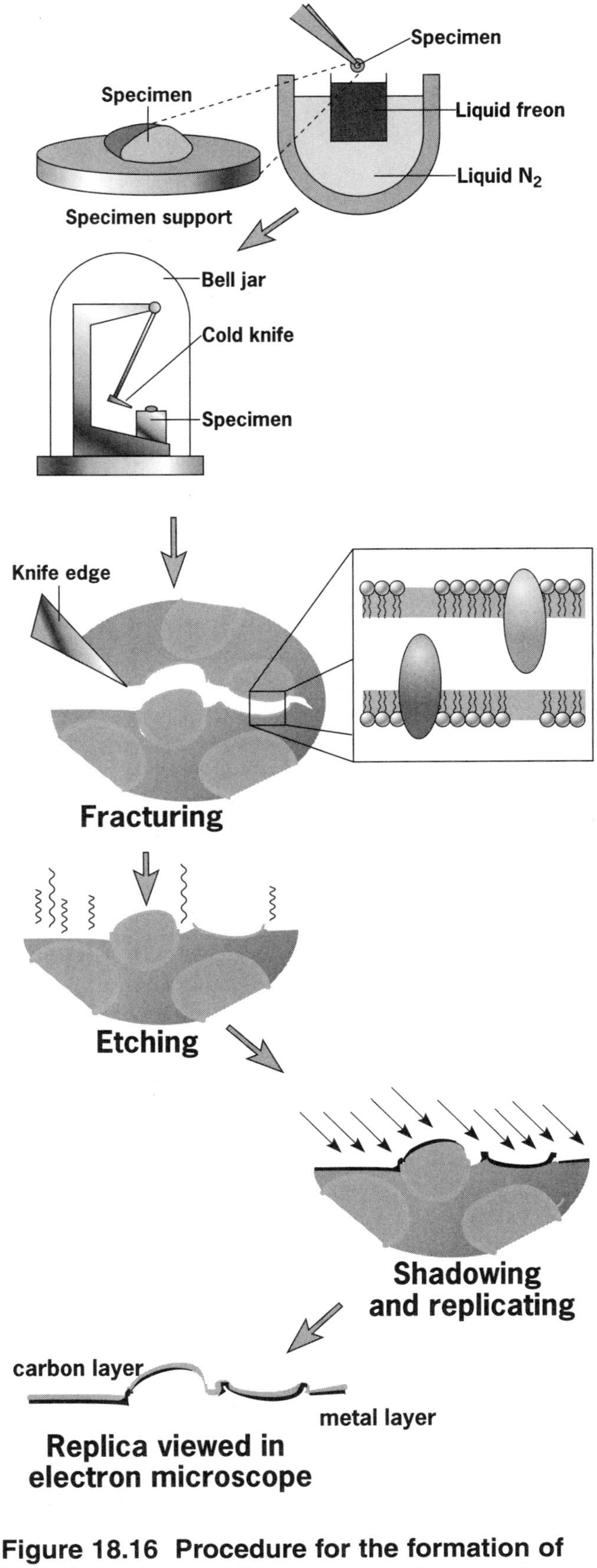

Figure 18.16 Procedure for the formation of freeze-fracture replicas as described in the text.

Copyright © 2005 John Wiley & Sons, Inc.

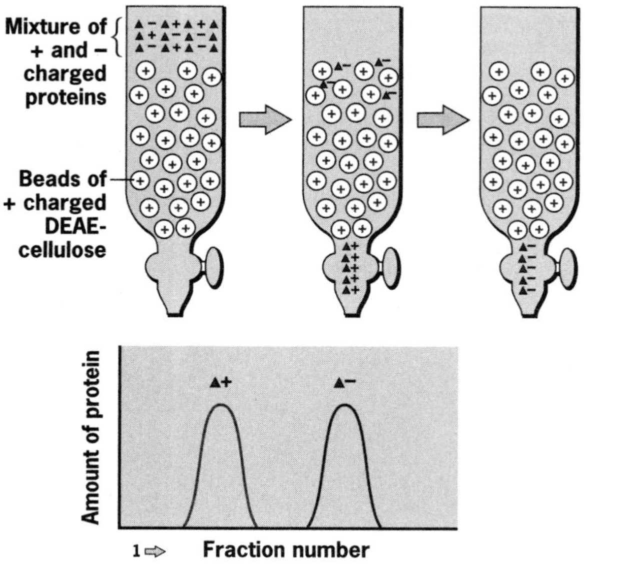

Figure 18.24 Ion-exchange chromatography.

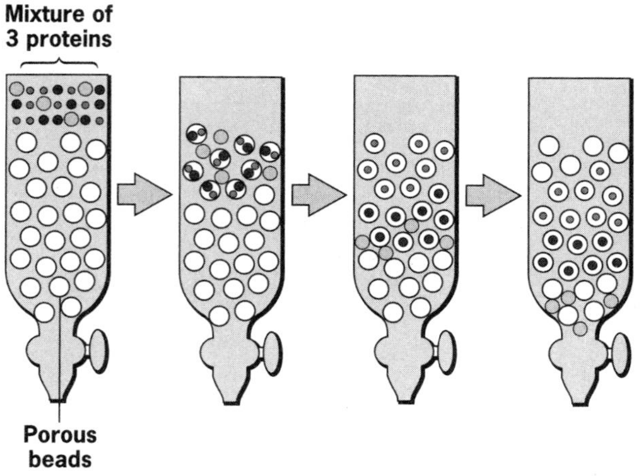

Figure 18.25 Gel filtration chromatography.

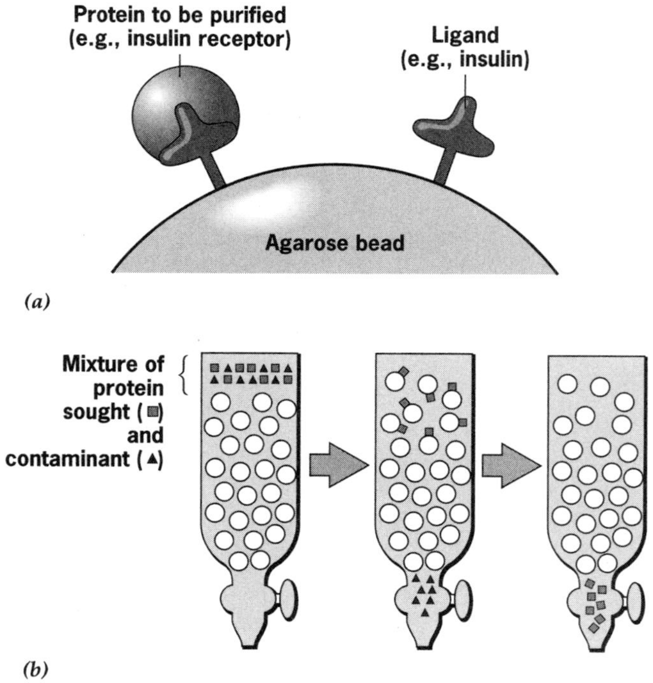

Figure 18.26 Affinity chromatography.

Copyright © 2005 John Wiley & Sons, Inc.

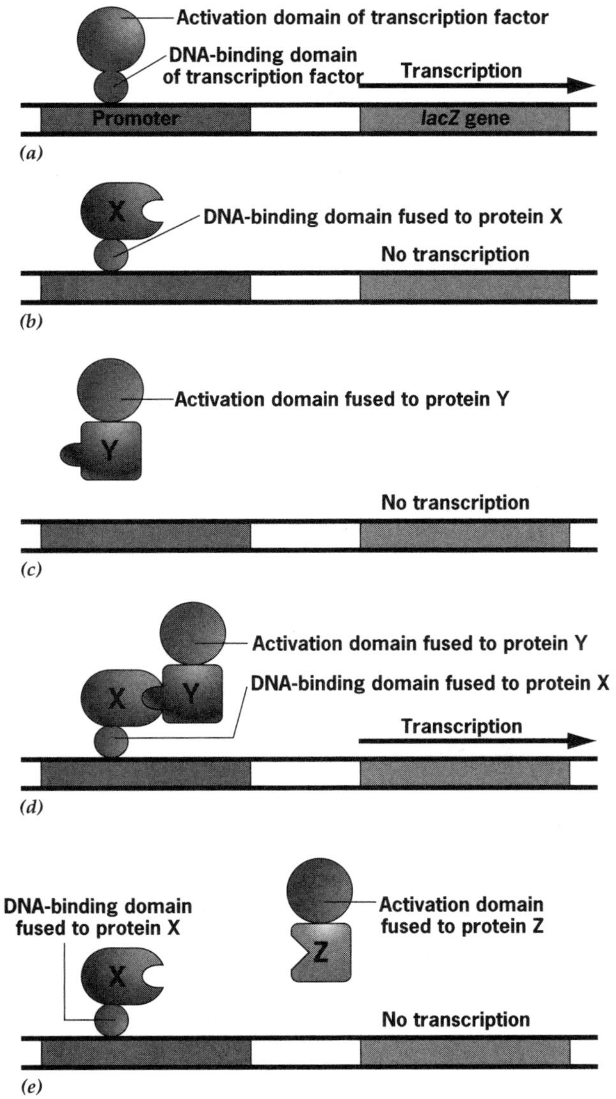

Figure 18.27 Use of the yeast two-hybrid system.

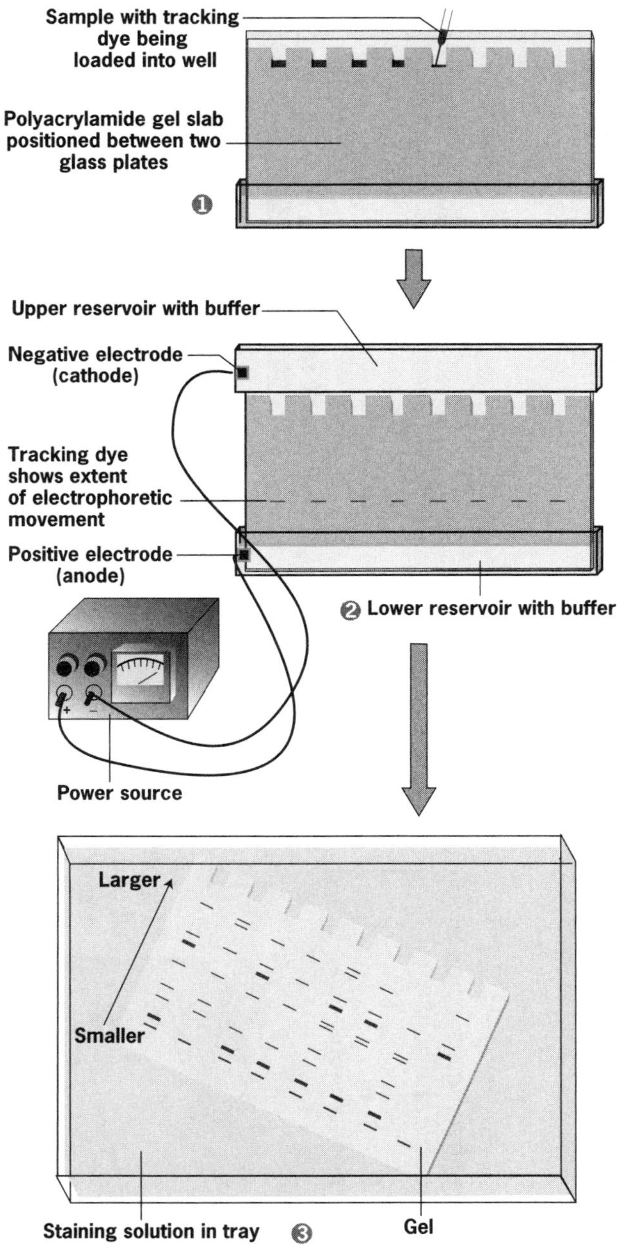

Figure 18.28 Polyacrylamide gel eletrophoresis.

Copyright © 2005 John Wiley & Sons, Inc.

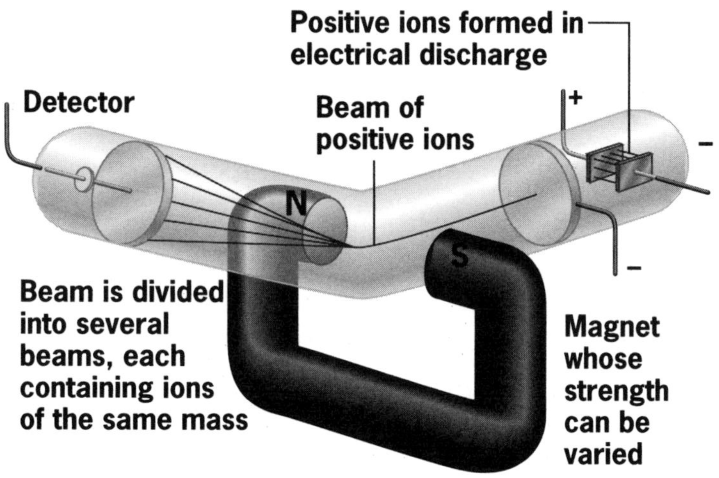

Figure 18.30 Principles of operation of a mass spectrometer.

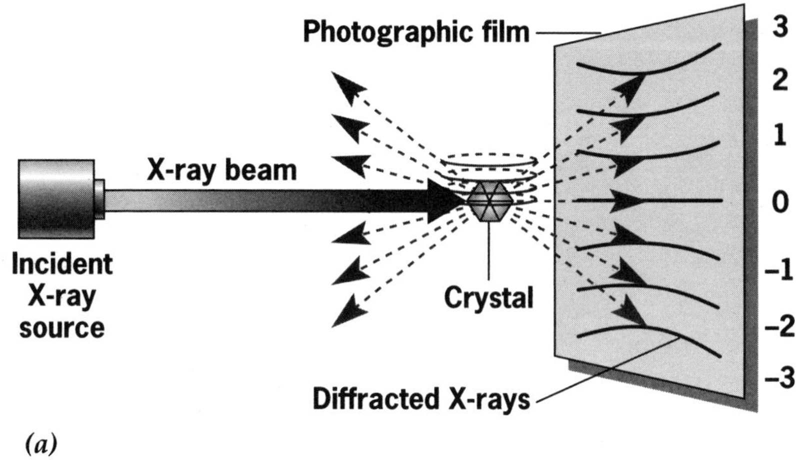

Figure 18.31 X-ray diffraction analysis.

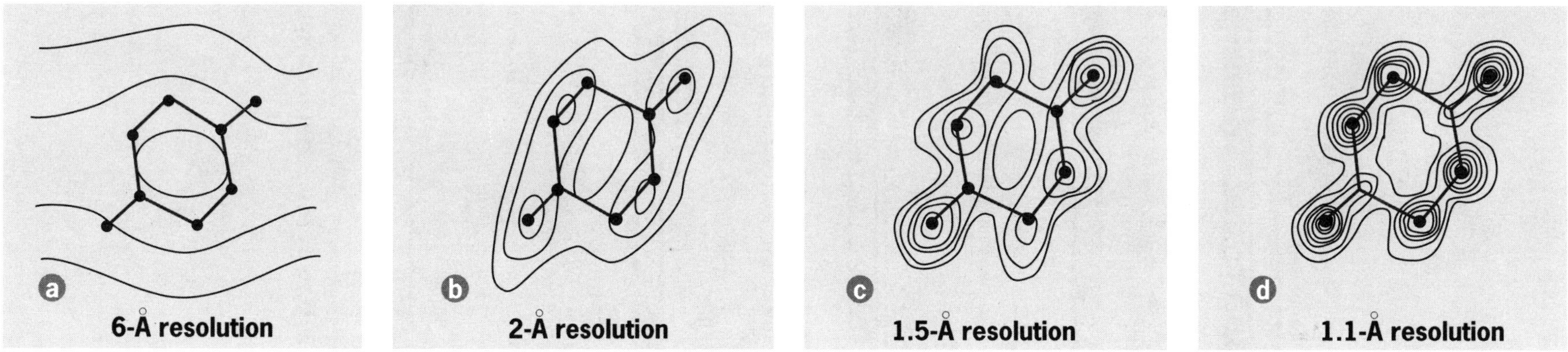

Figure 18.32 Electron density distribution of a small organic molecule (diketopiperazine) calculated at several levels of resolution.

Copyright © 2005 John Wiley & Sons, Inc.

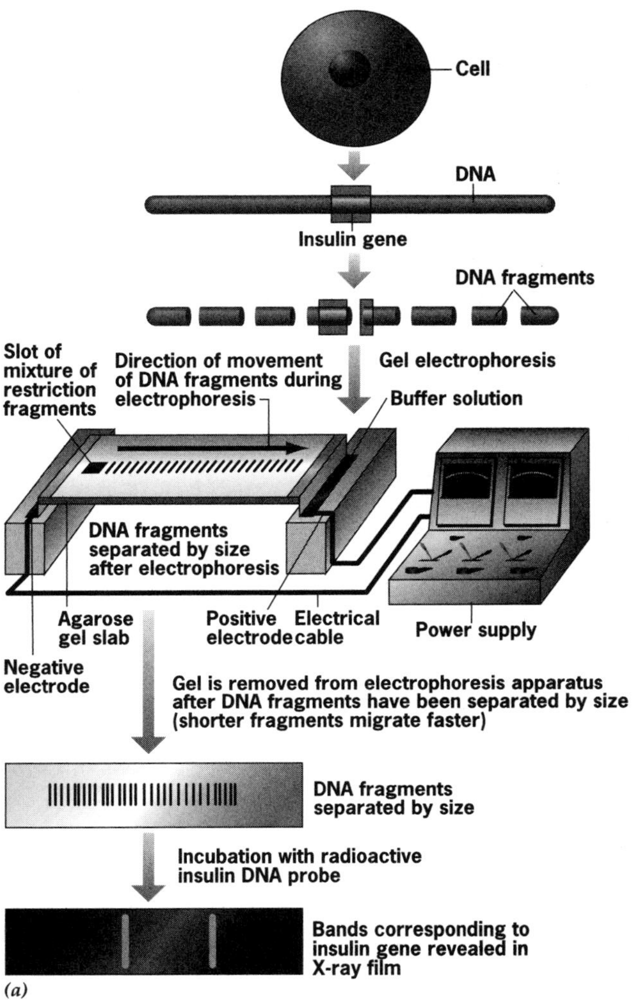

Figure 18.33a Separation of DNA restriction fragments by gel electrophoresis.

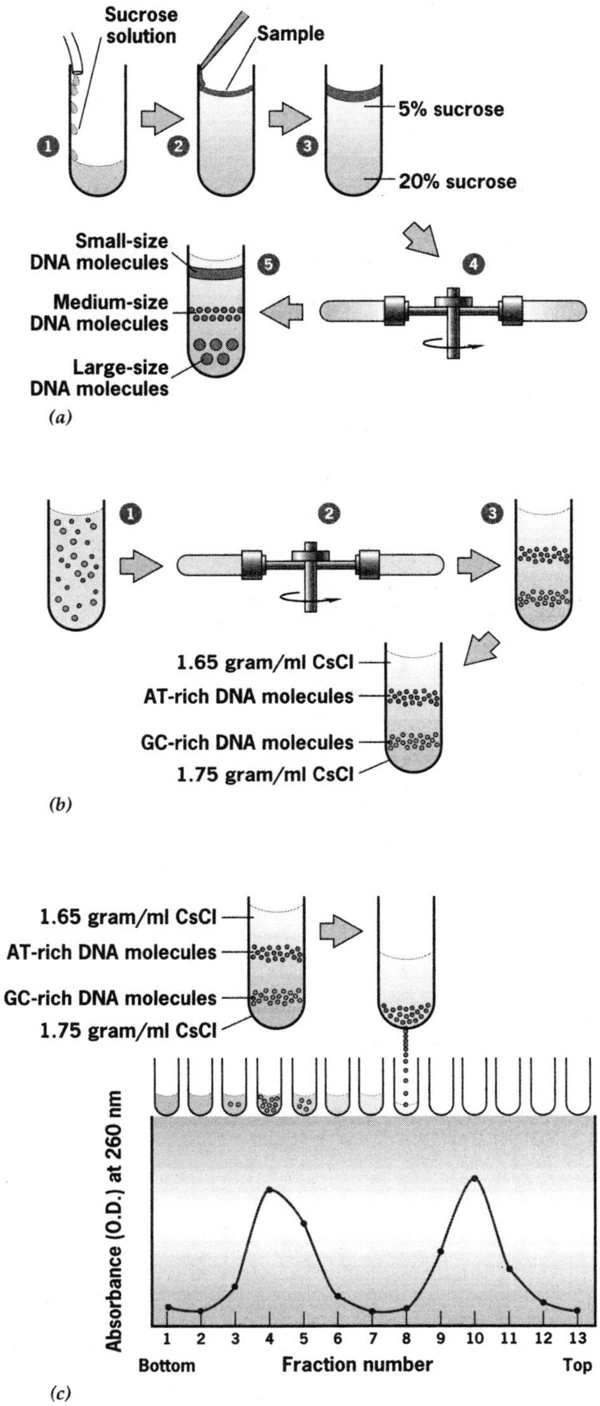

Figure 18.34 Techniques of nucleic acid sedimentation.

Copyright © 2005 John Wiley & Sons, Inc.

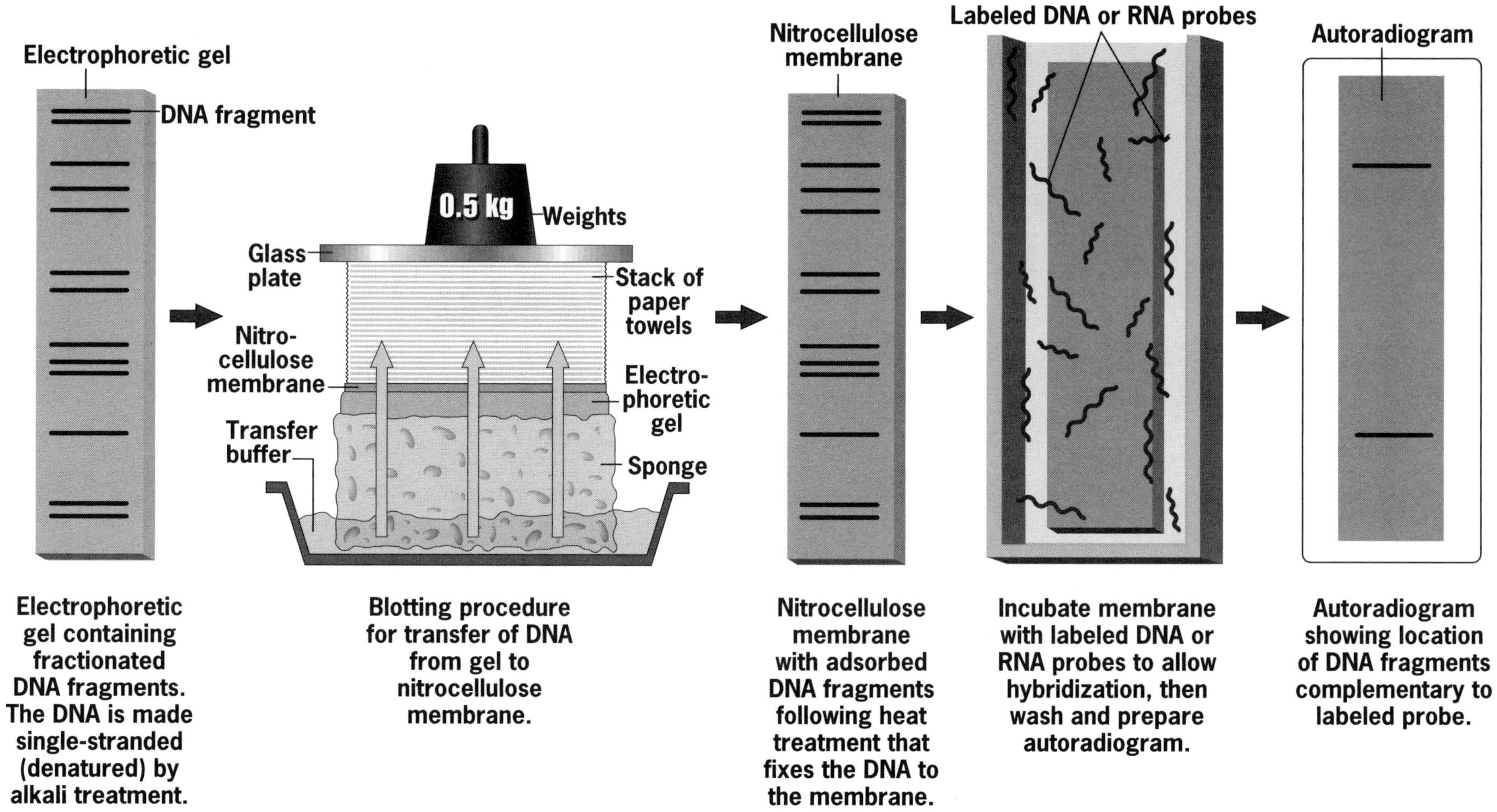

Electrophoretic gel
DNA fragment
0.5 kg
Weights
Glass plate
Stack of paper towels
Nitro-cellulose membrane
Electro-phoretic gel
Transfer buffer
Sponge
Nitrocellulose membrane
Labeled DNA or RNA probes
Autoradiogram
Electrophoretic gel containing fractionated DNA fragments. The DNA is made single-stranded (denatured) by alkali treatment.
Blotting procedure for transfer of DNA from gel to nitrocellulose membrane.
Nitrocellulose membrane with adsorbed DNA fragments following heat treatment that fixes the DNA to the membrane.
Incubate membrane with labeled DNA or RNA probes to allow hybridization, then wash and prepare autoradiogram.
Autoradiogram showing location of DNA fragments complementary to labeled probe.

Figure 18.35 Determining the location of specific DNA fragments in a gel by a Southern blot.

Copyright © 2005 John Wiley & Sons, Inc.

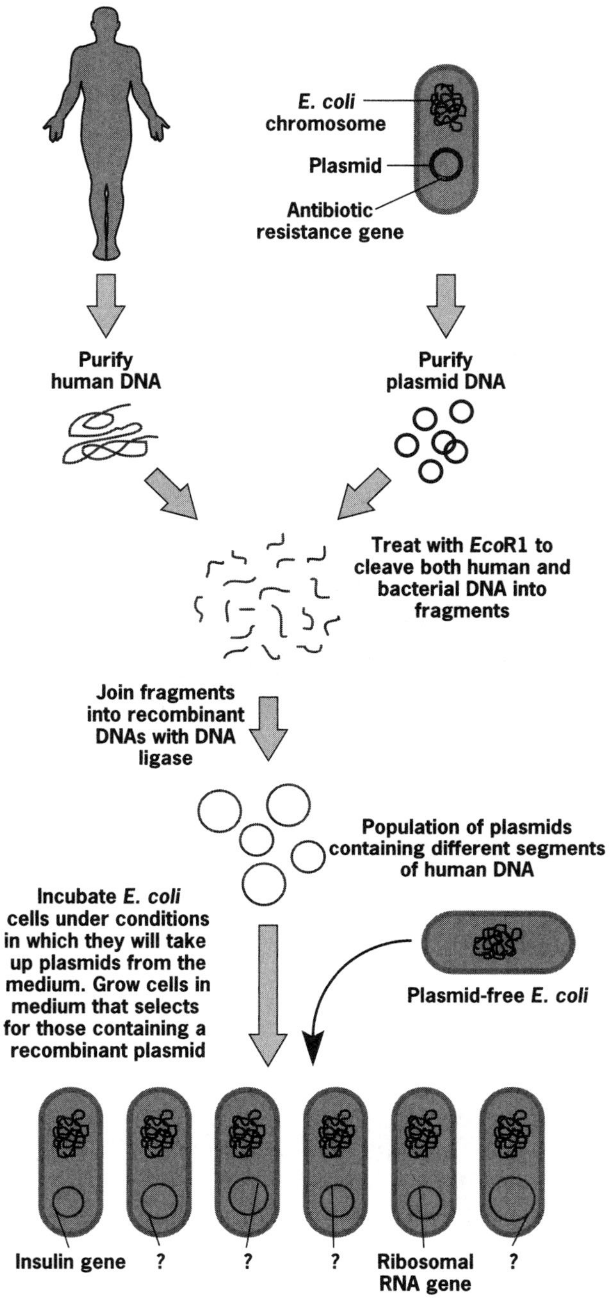

E. coli chromosome
Plasmid
Antibiotic resistance gene
Purify human DNA
Purify plasmid DNA
Treat with EcoR1 to cleave both human and bacterial DNA into fragments
Join fragments into recombinant DNAs with DNA ligase
Population of plasmids containing different segments of human DNA
Incubate E. coli cells under conditions in which they will take up plasmids from the medium. Grow cells in medium that selects for those containing a recombinant plasmid
Plasmid-free E. coli
Insulin gene
?
?
?
Ribosomal RNA gene
?

Figure 18.38 An example of DNA cloning using bacterial plasmids.

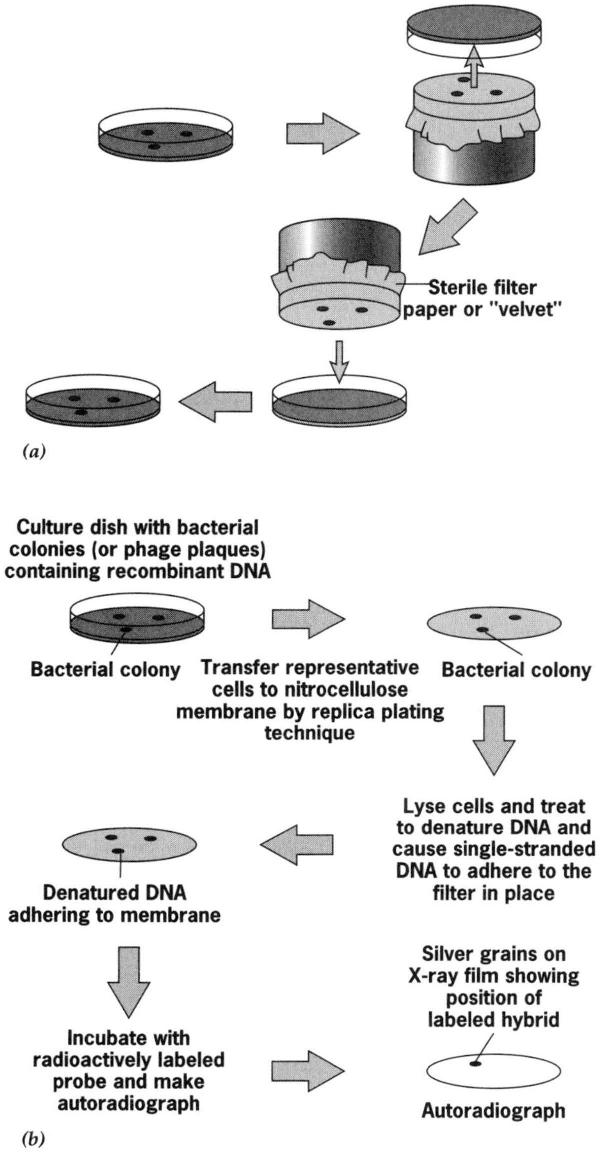

Sterile filter paper or "velvet"
(a)
Culture dish with bacterial colonies (or phage plaques) containing recombinant DNA
Bacterial colony
Transfer representative cells to nitrocellulose membrane by replica plating technique
Bacterial colony
Lyse cells and treat to denature DNA and cause single-stranded DNA to adhere to the filter in place
Denatured DNA adhering to membrane
Silver grains on X-ray film showing position of labeled hybrid
Incubate with radioactively labeled probe and make autoradiograph
Autoradiograph
(b)

Figure 18.39 Locating a bacterial colony containing a desired DNA sequence by replica plating and in situ hybridization.

Copyright © 2005 John Wiley & Sons, Inc.

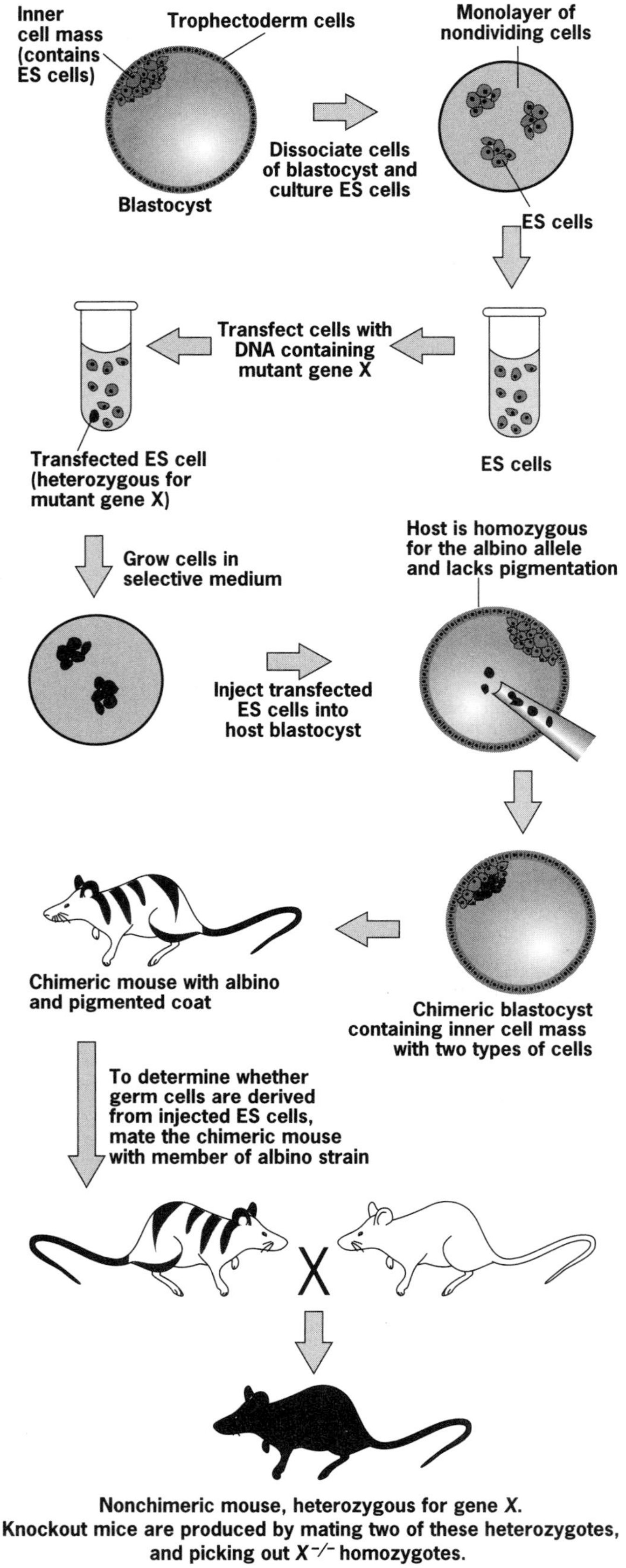

Inner cell mass (contains ES cells)
Trophectoderm cells
Blastocyst
Dissociate cells of blastocyst and culture ES cells
Monolayer of nondividing cells
ES cells
Transfect cells with DNA containing mutant gene X
ES cells
Transfected ES cell (heterozygous for mutant gene X)
Grow cells in selective medium
Host is homozygous for the albino allele and lacks pigmentation
Inject transfected ES cells into host blastocyst
Chimeric mouse with albino and pigmented coat
Chimeric blastocyst containing inner cell mass with two types of cells
To determine whether germ cells are derived from injected ES cells, mate the chimeric mouse with member of albino strain
X
Nonchimeric mouse, heterozygous for gene *X*.
Knockout mice are produced by mating two of these heterozygotes, and picking out $X^{-/-}$ homozygotes.

Figure 18.46 Formation of knockout mice.

Copyright © 2005 John Wiley & Sons, Inc.

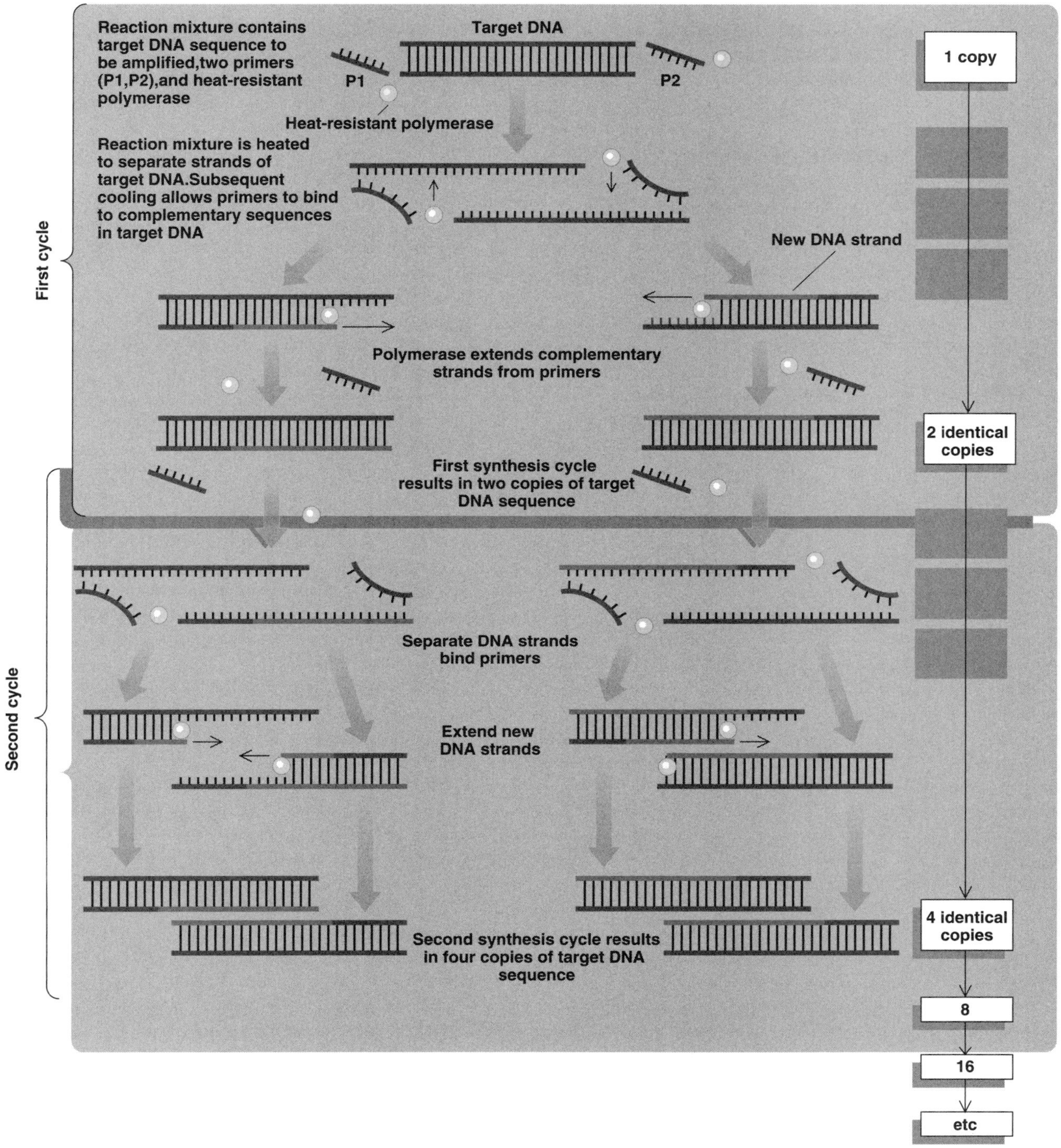

First cycle
Reaction mixture contains target DNA sequence to be amplified,two primers (P1,P2),and heat-resistant polymerase
Target DNA
P1
P2
Heat-resistant polymerase
Reaction mixture is heated to separate strands of target DNA.Subsequent cooling allows primers to bind to complementary sequences in target DNA
New DNA strand
Polymerase extends complementary strands from primers
First synthesis cycle results in two copies of target DNA sequence
Second cycle
Separate DNA strands bind primers
Extend new DNA strands
Second synthesis cycle results in four copies of target DNA sequence
1 copy
2 identical copies
4 identical copies
8
16
etc

Figure 18.48 Polymerase chain reaction (PCR).

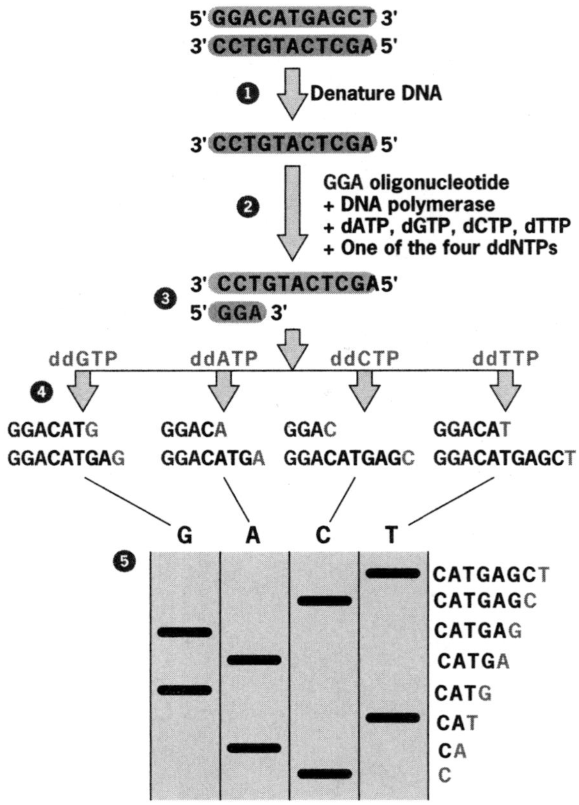

Figure 18.49a DNA sequencing.

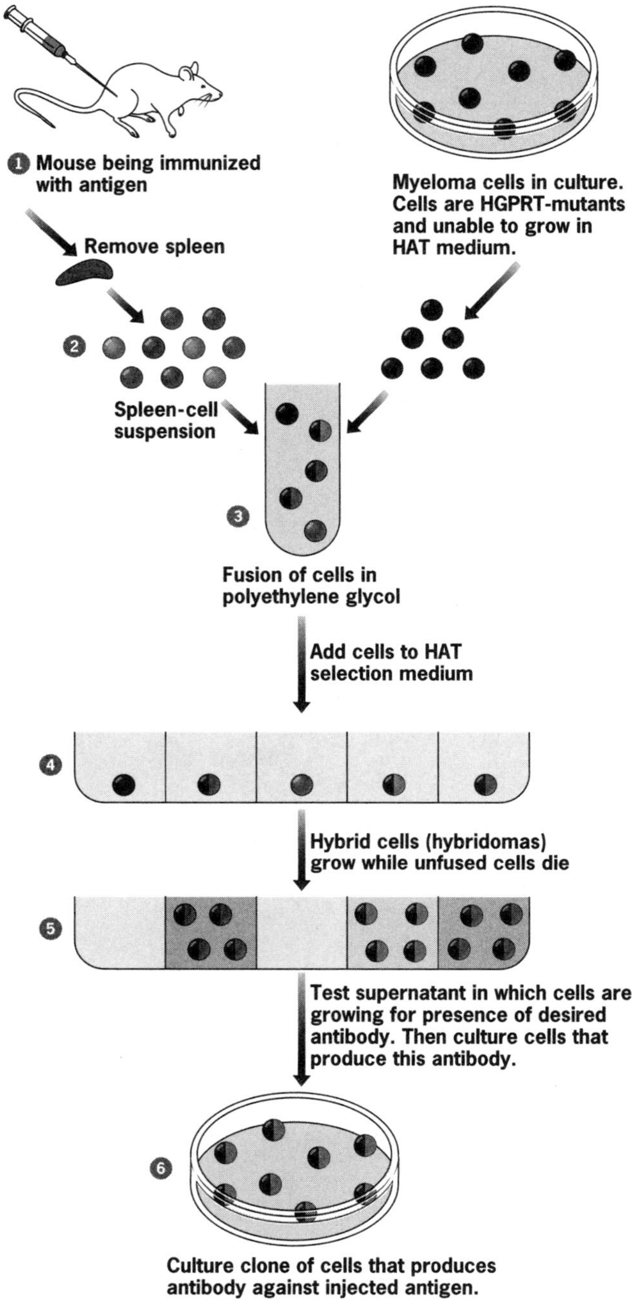

Figure 18.50 Formation of monoclonal antibodies.

Copyright © 2005 John Wiley & Sons, Inc.

✍ Take Note!

✍ Take Note!

✍ Take Note!

✍ Take Note!

✍ Take Note!

✍ Take Note!

✍ **Take Note!**

Your Professor Wants You to *Take Note!*

Are you spending too much time taking notes and missing key concepts during class? Do you get caught up in redrawing figures so that you can study them later? Would you like to become more efficient in your note taking? *Take Note!* can help!

Inside this convenient notebook, you'll find a collection of figures, diagrams, and art that clearly illustrate key concepts featured in the text. Next to each figure, you'll be able to take notes in the space provided during the lecture. Not only will this help you gain a better understanding of the material, but you'll also be better prepared for your exams!

What Makes *Take Note!* So Noteworthy?

- Most pages contain one graphic and all have plenty of space for you to take notes.
- The size of the notebook makes it very easy to use during class.

WILEY

www.wiley.com/college/karp

ISBN 0-471-66909-1

90000

9 780471 669098